D1474510

Technology Strategy for Managers and Entrepreneurs

Technology Strategy for Managers and Entrepreneurs

Scott Shane

A. Malalchi Mixon III Professor of Entrepreneurial Studies
Case Western Reserve University
Weatherhead School of Management

Upper Saddle River, NJ 07458

Library of Congress Cataloging-in-Publication Data

Shane, Scott Andrew
Technology strategy for managers and entrepreneurs / Scott Shane.
p. cm.
Includes bibliographical references and index.
ISBN 0-13-187932-4
1. Technological innovations—Management. 2. Business enterprises—Technological innovations. I. Title.
HD45.S414 2009
658.5'14–dc22

2008003317

Editorial Director: Sally Yagan
Acquisitions Editor: Kim Norbuta
Product Development Manager: Ashley Santora
Editorial Assistant: Elizabeth Davis
Senior Marketing Manager: Jodi Bassett
Marketing Assistant: Ian Gold
Senior Managing Editor: Judy Leale
Associate Managing Editor: Suzanne DeWorken
Project Manager, Production: Karalyn Holland
Senior Operations Specialist: Arnold Vila
Operations Specialist: Carol O'Rourke
Permissions Project Manager: Charles Morris
Cover Design: Bruce Kenselaar
Cover Illustration/Photo: TEK Image/Photo Researchers; DLR Institute of Robotics and Mechatronics; Getty Images, Inc.
Director, Image Resource Center: Melinda Patelli
Manager, Rights and Permissions: Zina Arabia
Manager, Visual Research: Beth Brenzel
Manager, Cover Visual Research and Permissions: Karen Sanatar
Image Permission Coordinator: Angelique Sharps
Photo Researcher: Rachel Lucas
Composition: Integra
Full-Service Project Management: Sharon Anderson, BookMasters, Inc.
Printer/Binder: Edwards Brothers, Inc.

Credits and acknowledgments borrowed from other sources and reproduced, with permission, in this textbook appear on appropriate page within text.

Pearson Education Ltd., London
Pearson Education Singapore, Pte. Ltd.
Pearson Education Canada, Inc.
Pearson Education–Japan
Pearson Education Australia PTY, Limited
Pearson Education North Asia, Ltd., Hong Kong
Pearson Educación de Mexico, S.A. de C.V.
Pearson Education Malaysia, Pte. Ltd.
Pearson Education Upper Saddle River, New Jersey

10 9 8 7 6 5 4 3 2 1
ISBN-13: 978-0-13-187932-4
ISBN-10: 0-13-187932-4

To Lynne, Hannah, and Ryan

Brief Contents

Contents

PREFACE

Technological innovation is one of the most important aspects of business for a student to understand. It is the primary driver of wealth creation for both companies and society at large.

However, innovation is not easy to master. Insights into it come from a wide range of disciplines, including economics, sociology, psychology, strategy, organizational behavior, marketing, and finance. As a result, it is often difficult for students to integrate all of the knowledge we have about innovation into a single, overarching framework.

THE STRUCTURE OF THE BOOK

This book is designed to teach students in business, engineering, and science how to use the strategic management of innovation to enhance firm performance. It is divided into five sections, each reflecting part of an overarching framework that achieves this goal. The first section provides an understanding of how technology evolves over time and the implication of that evolution for companies. The second section describes how companies come up with innovations that meet the needs of their customers. The third section explains how companies capture the value generated from their investment in innovation. The fourth section discusses the development of a technology strategy, while the fifth section discusses its implementation. Although the book provides a theoretical framework to guide students, it also provides practical examples and exercises to ground the effort.

WHY DO WE NEED THIS BOOK?

Other books on the strategic management of technological innovation already exist. So why do we need this one? There are three reasons:

1. **In recent years, academics and practitioners have made rapid progress in understanding the strategic management of innovation, and that information is not reflected in most textbooks.** By incorporating recently learned information, this book ensures that students develop an accurate understanding of the current state of the field.
2. **Many students are interested in technology strategy from the perspective of small, new companies rather than large, established ones.** In fact, the rise of companies like Google and YouTube has made many students more interested in using technology strategy to launch a successful new company than to defend an existing one. Because existing textbooks approach technology strategy only from the point of view of the large incumbent firm, they don't provide students with the kind of information that they need to formulate an innovation strategy for a high-growth technology company.

3. **Neither the instructors teaching courses on technology strategy, nor the students taking them, are happy with existing textbooks.** Almost 40 percent of instructors don't use a textbook to teach these courses because they believe that existing textbooks are inadequate. (In fact, when I surveyed instructors teaching technology strategy courses about the need for a new textbook, many respondents wrote back that "there are no good books on this topic.") Both instructors and students explain that most existing textbooks are dull, weak on student aids, and include few pedagogically useful features. So it should be no surprise that instructors and students both report lower-than-average satisfaction ratings with existing textbooks.

What Are the Guiding Principles Behind the Book?

The writing of this book was governed by three guiding principles: Technology strategy is a unique field; theory and practice both help to explain it; and many disciplines contribute to our understanding of it.

Technology Strategy Is a Unique Field

Recently, many academics have come to view technology strategy as a unique area, different from mainstream strategy, and requiring its own courses and intellectual content. Moreover, the topic is now taught in business schools and schools of engineering and science. I reflect these growing trends by focusing on technology strategy as a distinct subject and by providing an approach that can be understood by students with a wide variety of backgrounds.

Theory and Practice Both Help to Explain It

To understand technology strategy, one needs to learn the theoretical frameworks that underpin it. At the same time, however, it requires application to real examples. The book uses a variety of different tools—from discussion of research to detailed examples to real-world exercises—to balance theory and practice.

Many Disciplines Contribute to Our Understanding of Technology Strategy

Technology strategy is an eclectic field that draws upon well-established disciplines, such as economics, psychology, and sociology, as well as longstanding fields of business, like management, marketing, and finance. Each of these disciplines and fields has something to offer technology strategy, making representation of all of them a guiding principle of this book.

What Makes This Book Unique?

This book has four unique features:

1. **It is designed to help instructors teach.** My own teaching experience tells me that instructors do a better job teaching a class when the textbook that they use is designed to help them. With this idea in mind, I have taken

several steps to make this book "teacher friendly." First, the table of contents was designed from surveys of instructors and a Web search of their syllabi to ensure that the book includes the topics most often covered in technology strategy, management of innovation, and technology entrepreneurship courses. Second, the supplemental materials are designed to enhance the instructor's ability to add value to the class, and include comprehensive suggestions for supplemental readings, cases for discussion, and lesson plans.

2. **The book is written to help students learn.** Students find a textbook more valuable if it is designed to help them learn the material. With this thought in mind, I have taken a number of steps in designing this book. First, it is written in a clear and direct style—one that will communicate with readers rather than bore or irritate them. Second, I have included a number of features designed to help students with their studying. All chapters begin with a Chapter Outline and include brief reviews of Key Points at the end of major sections of the text within chapters. In addition, all important terms are printed in **boldface** and defined in the Key Terms section at the end of each chapter. Third, the book uses a two-color format and a rich illustration package with many charts and graphs specially prepared for the book. Experts on cognition tell us that such an approach facilitates learning. Fourth, all of the exercises, examples, and charts that are based on data shown in the book also come in the form of downloadable Excel spreadsheets. This will minimize the time students need to spend reproducing information and will facilitate the manipulation of data that helps students learn the principles driving the analysis. Finally, each chapter is followed by Discussion Questions, which are designed to stimulate in-class discussion of major points, and several Putting Ideas into Practice exercises, intended to give readers practice with the principles presented.
3. **The book balances theory with practical examples and exercises.** Research helps us to understand the nature of the innovation strategy process, but practical examples and exercises are necessary to keep a book grounded and allow students to relate to it. To balance theory and practice, every chapter of the book draws on the latest research and theoretical frameworks. They also include Getting Down to Business sections that provide detailed examples highlighting the practical implications of key theoretical issues and three Putting Ideas into Practice exercises that help students translate the concepts learned in the chapter into real-world activities. (A list of several Harvard Business School, Ivey Business School, and Darden Business School cases that allow students apply the concepts in the chapter are available to instructors.)
4. **The book examines technology strategy from the perspective of new, small firms as well as large, established ones.** Technology strategy for new, small companies is not the same as technology strategy for large, established firms, with a few zeros dropped off the numbers. For instance, the right way to approach market research, respond to firm evolution, manage new product adoption, and form strategic alliances depends on whether a firm is large and established or small and new. This book identifies the issues for which technology strategy is different for the two types of firms and addresses those differences. Because it examines technology strategy from the perspective of technology entrepreneurs and managers, the book will help to prepare students for both roles.

Supplements

The teaching materials for *Technology Strategy for Managers and Entrepreneurs* include:

- An instructor's manual with suggested cases, additional recommended readings, a class outline, and answers to discussion questions and exercises.
- PowerPoint slides for all lectures, including all figures, charts, graphs, and tables from the book.
- A computerized test bank with questions in different formats (e.g., true/false, multiple choice, essay, etc.).
- Downloadable Excel spreadsheets for all exercises, charts, and tables that require analysis.

Acknowledgments

This book would not have been possible without the help of a lot of people. I would like to thank my students for asking many of the questions that I have tried to answer in this book and for using a draft of the book in class and identifying mistakes that I had made. I would also like to thank all of the scholars and practitioners on whose work I have drawn to provide the framework for this book. While the ideas presented in this book were influenced by many people, several are particularly important: David Audretsch, Clay Christiansen, Richard Foster, Alvin Klevorick, Richard Levin, Geoffrey Moore, Richard Nelson, Everett Rogers, David Teece, and Jim Utterback. The books and articles that these people wrote were extremely valuable in helping me to develop the ideas presented in this book. Lastly, I would like to thank the following people for reading and commenting on earlier drafts of the book: Michael B. Heeley, Colorado School of Mines; Phillip H. Phan, Rensselaer Polytechnic Institute; Aron S. Spencer, New Jersey Institute of Technology; John Lehman, University of Alaska Fairbanks; Shanthi Gopalakrishnan, New Jersey Institute of Technology; Richard T. Hise, Texas A&M University; and Timothy Stearns, California State University–Fresno.

Some Concluding Comments

In writing this book, I have tried to make it as accurate, comprehensive, current, and useful to readers as possible. Was I successful? You tell me. Please let me know your reactions, suggestions, and comments. I would really like your input, and I promise to *listen to it!* Thanks in advance for your help.

Scott A. Shane
Sas46@cwru.edu

Chapter 1

Introduction

Learning Objectives

After reading this chapter, you should be able to:

1. Define *technological innovation*, explain why it is important, and describe how firms use it to achieve their objectives.
2. Describe how technological innovation occurs, and explain the effects that it has on individuals, firms, and society.
3. Define *technology strategy*, and explain why it is important for entrepreneurs and managers to develop technology strategies.
4. Describe the approach to technology strategy taken in the book, and explain why that approach is important.
5. Identify the core areas of technology strategy, and spell out how these different areas of technology strategy affect businesses.
6. Describe how the core areas of technology strategy differ for new and established firms, and explain why these types of firms need to take different approaches to technology strategy.

Technology Strategy: A Vignette[1]

The Xbox 360 game machine, developed by Microsoft to challenge Sony's position as the market leader in the sale of video game consoles, is a prime example of a company's effort to develop an innovative new product as part of its overall technology strategy. It also illustrates many of the topics that are covered in this book.

Video game consoles are a high-technology product, requiring significant investment in research and development to create. Moreover, Microsoft's effort to develop a competitive video game console involves a variety of strategic issues. First, video games involve network effects, which influence the relationship between game and console makers. If Microsoft can sell more Xbox 360 consoles than Sony can sell PlayStation 3 consoles, then video game makers will design the best games for the Xbox. The availability of the best games for the Xbox will increase sales of the consoles, generating a positive returns cycle for the company.

Second, the manufacture of the consoles faces very significant economies of scale. As the volume of production of video game consoles goes up, **costs per unit** (the amount of money that the company needs to spend to create each console) decline. As a result, the largest producers of consoles have significant cost advantages over their competitors. Moreover, initial sales are often made at a loss, as a company ramps up to minimum efficient scale. During that ramp-up process, companies often lose a lot of money. Microsoft, for example, lost $4 billion on the development and production of the Xbox hard drive and microprocessor alone.[2]

Third, success in this business depends on Microsoft's ability to develop the right capabilities to manage a video game business. Traditionally, the company's expertise has been in making computer software. However, video game consoles are pieces of computer hardware. Because the manufacture and sale of computer hardware are very different than the production and sale of computer software, to produce and successfully market the Xbox, Microsoft has to develop very different capabilities than it has needed in its software business.[3]

Fourth, the production of the Xbox requires Microsoft to develop a new supply chain. The Xbox 360 contains 1,700 different parts, which all have to be brought together to make the product. To keep costs down, Microsoft produces the Xbox in China. It uses two factories, each of which serves as a hedge against problems that would shut down the other. Around these two core manufacturers are a host of suppliers of plastic parts for the boxes, capacitors, cooling fans, and other parts whose production needs to be coordinated.[4]

Fifth, the development and production of the Xbox requires Microsoft to manage contractual relationships with other companies. Specifically, it has to coordinate their production of the graphics chip and hard drive used in the Xbox. Its entire effort to sell Xboxes could derail if these producers run into production delays. While diversification protects Microsoft against problems with the production of the hard drive—the console's hard drive is a commodity sourced from a variety of manufacturers—it is vulnerable to problems with the chip, which is custom-made by IBM. The reliance on IBM could prove to be a problem if anything hinders its ability to deliver the key component.[5]

INTRODUCTION

Technological innovation has become an important part of the process by which companies in many industries generate competitive advantage, making it a crucial part of firm strategy. In recent years, many companies have increased their level of technological innovation to produce a greater variety of new products, and to introduce those new products to market faster. In many industries, the share of sales and profits accounted for by products introduced in the past five years has been growing rapidly. In fact, some companies, like 3M, now generate 40 percent of their sales from products that did not exist five years ago.

Companies have also increased their level of technological innovation in response to competition. The reduction of costs and the improvement in quality of products made in lower wage countries, like China and India, have posed a major challenge for firms in developed countries, like the United States and Germany. Many firms from developed countries have responded to this challenge by introducing new products at a faster pace to stay ahead of imitators, and by using technological innovation to reduce their own production costs.[6]

Technological innovation has also increased as more companies that once produced commodity products now seek to differentiate their offerings from those of competitors. The desire of more companies to offer differentiated products has shortened the product life cycle and has increased the importance of investments in new product and process development.[7]

Furthermore, technological innovation has increased as companies have turned intellectual property into a marketable asset. In recent years, the licensing of technology to other companies has become an important revenue stream for many companies, with some, like IBM, adopting the approach that all of its intellectual property is potentially for sale. This marks a major change from only a couple of decades ago when intellectual property was used only as an input into a company's product or service.

In addition, there has been significant growth recently in the formation of high-technology start-ups that use funding by venture capitalists and business angels (individuals who invest their own money in start-up companies, usually by taking an equity stake in them) to introduce high-technology products and compete with established firms. As a result, technological innovation has also been increasing because of entrepreneurs, including those who create spin-off companies, using technology developed at major corporations and universities.

This emphasis on technological innovation as a way to generate or preserve competitive advantage has led to an increased need for managers and entrepreneurs who can develop strategies to successfully manage this activity. While companies can, and do, introduce new products, improve production processes, and target new markets without strategies or plans, companies are better at these activities if they develop, and execute, an effective strategy to undertake them.[8] By combining an understanding of markets and technological evolution with an understanding of firm organization and capabilities in a deliberate and organized manner, managers and entrepreneurs can generate value by developing technology products and services that better meet customer needs, and can become better at capturing that value.

The increased need for managers and entrepreneurs to develop strategies for technological innovation, in turn, has led to an increased demand for business school courses in the management of technological innovation, and for strategic management courses that focus on issues particular to high-tech companies. In short, technology strategy is now an important part of the education of business leaders.

What Is Technological Innovation?

The previous vignette illustrates the importance of understanding technology strategy by highlighting many of the important issues that companies face and that are the subject matter of this book. But before we get into a discussion of what technology strategy is, and how to develop it effectively, we need to lay a little ground work. We have to define *technological innovation* and explain how firms use it to achieve their objectives. After all, strategy is just an approach to achieving a particular goal, making technology strategy nothing more than an approach to using technological innovation to achieve a goal.

So our first step is to define *technological innovation*. That is best done by decomposing the phrase into its two parts, the concepts of *technology* and *innovation*.

Defining Technology

While there has been a tendency for the popular media to use the word *technology* as shorthand for *information technology*, **technology** is much broader than just information technology. It is the application of tools, materials, processes, and techniques to human activity.

Certainly, information technology—the use of zeros and ones in digital form on computers—is an important technology, but there are many other important technologies as well. Biologically based technologies, such as those used to create new drugs or to clean up pollution, are also important. Similarly, mechanically based technologies, such as those that make pumps or valves, matter. New materials, such as those in new ceramic composites, are valuable too.

So when this book discusses technology, we aren't talking just about information technology. Rather, we are talking about a host of technologies, including, but not limited to, new microorganisms, new mechanical devices, new materials, and a variety of other products and processes. So when you see the word *technology* or the phrase *technological innovation* in this book, do not just think of the Internet and computer software, think of processes like nanofabrication (the process of making things less than one micrometer in size) and products like fuel cells, ceramic composites, new drugs, or heart valves.

The use of technology is more prevalent in some industries than in others. Figure 1.1 shows the industries that the U.S. government defines as technology-intensive. Clearly, these industries are the ones in which an understanding of technology strategy is important to entrepreneurs and managers.

FIGURE 1.1 Technology-Intensive Industries

Industry
Aerospace product and parts manufacturing
Agriculture, construction, and mining machinery manufacturing
All other electrical equipment and component manufacturing
Architectural, engineering, and related services
Audio and video equipment manufacturing
Basic chemical manufacturing
Commercial and service industry machinery manufacturing
Communications equipment manufacturing
Computer and office machine repair and maintenance
Computer and peripheral equipment manufacturing
Computer systems design and related services
Data processing services
Educational support services
Electrical equipment manufacturing
Engine, turbine, and power transmission equipment manufacturing
Industrial machinery manufacturing
Management, scientific, and technical consulting services
Manufacturing and reproducing magnetic and optical media
Medical equipment and supplies manufacturing
Motor vehicle body and trailer manufacturing
Motor vehicle manufacturing
Motor vehicle parts manufacturing
Navigational, measuring, electromedical, and control instruments manufacturing
Online information services
Ordnance & accessories manufacturing—ammunition (except small arms) manufacturing
Ordnance & accessories manufacturing—other ordnance and accessories manufacturing
Ordnance & accessories manufacturing—small arms ammunition manufacturing
Ordnance & accessories manufacturing—small arms manufacturing
Other chemical product and preparation manufacturing
Other general purpose machinery manufacturing
Paint, coating, and adhesive manufacturing
Pesticide, fertilizer, and other agricultural chemical manufacturing
Petroleum refineries
Pharmaceutical and medicine manufacturing
Resin, synthetic rubber, and artificial and synthetic fibers and filaments manufacturing
Scientific research and development services
Semiconductor and other electronic component manufacturing
Soap, cleaning compound, and toilet preparation manufacturing
Software publishers

The U.S. government defines these industries as technology intensive because the firms in them devote a large proportion of their revenue to research and development.

Source: Adapted from Science and Engineering Indicators, 2006, http://www.nsf.gov/statistics/seind06.

Defining Innovation

The next important definition is that of **innovation**—the process of using knowledge to solve a problem. Innovation is different from **invention**, which is the discovery of a new idea, because it involves more than just coming up with an idea about how to use knowledge to solve a problem. For example, during the Renaissance, inventors came up with the ideas for parachutes, fountain pens, mechanical calculators, and ball bearings. However, these ideas did not become innovations until much later because they were not technically feasible and could not be implemented at the time that the ideas were discovered.[9]

Defining Technological Innovation

So what is **technological innovation**? Simply put, it is the use of knowledge to apply tools, materials, processes, and techniques to come up with new solutions to problems.[10] Some innovations, like Michael Dell's approach to selling computers—selling personal computers assembled from standard components direct to customers[11]—were not technological innovations because the knowledge used to solve a problem did not involve new technology (tools, materials, processes, and techniques), but, instead, involved new ways of organizing a business. However, other innovations, like genetic engineering, were technological innovations because the knowledge used to solve problems involved new tools, materials, processes, and techniques.

Technological innovation can be planned or accidental. Sometimes the use of technical knowledge to solve problems is purposeful. For instance, many companies invest in research and development with the goal of coming up with an innovative new product or process that will give them an advantage over their competitors.

However, other times technological innovation isn't the result of a deliberate attempt to solve a particular problem. For example, Pfizer was not looking for a solution to the problem of erectile dysfunction when it came up with Viagra. The solution to this problem was merely a side effect discovered in tests of the drug for its intended purpose of treating angina in cardiac patients. Similarly, Alexander Fleming discovered penicillin because a spore of mold contaminated a sample of bacteria that he was using, and inhibited its growth.[12]

This definition of technological innovation has three important implications for understanding technology strategy. First, as Figure 1.2 shows, technological innovation doesn't have to be profitable. Companies can, and do, come up with solutions to problems that they can't make any money exploiting. So a big part of technology strategy is coming up with ways to make money from technological innovation (and another big part of technology strategy is keeping that money, rather than letting competitors take it).

FIGURE 1.2
What Is Technological Innovation?

A technological innovation involves the application of knowledge to solve a problem, and need not work nor result in something of commercial value.

Source: Darby Conley, March 17, 2007; Distributed by UFS, Inc.

Second, there are many different kinds of technological innovations. They can come from any kind of technical knowledge, from knowledge of computer science to knowledge of biology to knowledge of new materials. Moreover, technological innovation can take the form of solutions to a variety of different kinds of problems, from speeding production to introducing new products that meet customer needs to facilitating distribution. The development of a new material for making aircraft lighter or a new way to produce biodiesel are just as much technological innovations as Windows Vista.

Third, there is not a direct, one-to-one, relationship between technological change and new products or processes. Some new technologies may lead to only one, or even no, new products or processes, while others might make possible a very large number of them. Take, for example, the case of radio frequency identification (RFID) technology. This technology has led to the creation of ExxonMobil's speedpass, the E-ZPass highway toll collection system, a system for inventory tracking in libraries, a way to identify the parts and components in computer and automobile manufacturing, a method to record quality problems in disk drive manufacturing, and a way to track products shipped from manufacturers to retailers.[13]

Why Technological Innovation Is Important

Technological innovation is an important source of value creation. The application of knowledge to human activity allows for the more efficient production of existing products and services (that is, production that takes less money or effort) and allows for the creation of products and services that meet needs that were not previously satisfied.[14] For example, technological innovation allows us to provide burn victims with artificial skin, and makes the cost of heating our homes cheaper.

Because of the importance of technological innovation to value creation, technological change has tremendous economic impact. Economists have shown that much of the growth in gross domestic product comes from the use of technology to make more productive use of labor and capital.[15]

Technological innovation is also important because it has a profound effect on the creation of wealth for individual entrepreneurs and corporate shareholders. New technology makes possible the formation and rapid growth of companies like Google and Microsoft that were not around four decades ago, and attributes to the death of leading companies of a previous era, such as Digital Equipment Corporation. In fact, researchers have shown that much of the wealth creation by entrepreneurs is accounted for by the creation of products and services based on technological innovation.[16]

Technological innovation is also important because it has a major impact on our lives. For instance, the way in which we look for information has been dramatically altered by Internet search, how we get around was forever changed by the airplane and the automobile, and life expectancy has been greatly lengthened by medical diagnostics and treatments.

Not all of the impact of technological innovation is good. Research has shown that the rapid and constant technological progression that we have experienced in recent years causes social isolation, increased time pressure, the constant need to multitask, the inability to disconnect from work, and other adverse consequences. As Figure 1.3 illustrates, somewhat humorously, many people have found the loss of privacy and control that technological innovation has generated sometimes outweighs the benefits of the opportunities that technological change has created.

Designing an effective technology strategy entails the creation of new technology products and processes that provide benefit to your firm while minimizing the

FIGURE 1.3
Technological Change Isn't Always Good

Many new communications technologies have an adverse effect on personal privacy.
Source: Berkeleybreathed.com, July 10, 2005, http://cartoonistgroup.com/store/add.php?iid=11027.

adverse effects of technological change on your customers and society at large. Many companies have run into trouble when they have made use of technology to achieve a competitive advantage and that technology has had adverse consequences. For example, the efforts of retailers T.J. Maxx and Marshalls to gain competitive advantage by using information technology to gather customer information that could be used for targeted marketing led to a loss of customer information to identity thieves who could perpetuate fraud against tens of millions of people, alienating large numbers of their customers.[17]

GETTING DOWN TO BUSINESS
Remembering Process Development[18]

Sometimes, **process development**—the anticipation and resolution of problems that arise in the production of a product—is more important than product development as a source of competitive advantage for high-technology firms. The Sensor razor, introduced by Gillette in 1990, is one such example. This product was a new razor that provided a much better shave than alternative razors. The Sensor's advantage in providing a close shave was its use of 23 floating parts, rather than the 5 floating parts of which most other razors are composed.

The idea behind the Sensor razor was not particularly surprising. Most engineers working on developing razors at the companies making them knew that the more floating parts you could put on the razor, the closer a shave it would provide. Although the engineers at many companies knew this about razor design, no other companies developed a razor with more than 5 floating parts. Why?

Increasing the number of floating parts on a razor generated significant production challenges. The mounting of so many parts on the razor bar could not be done by the standard approach of gluing. To develop the Sensor and make it work effectively, Gillette's engineers had to figure out how to mold the plastic blades into the cartridge. And, because the plastic that Gillette used for other razors lost bounce with repeated use, they had to shift to a new resin to make this work.

In addition, the engineers needed to have the blades float on separate springs to get the close shave that would make Sensor better than other razors. This was difficult to do because the blades were as thin as a piece of paper. To make the blades float on the springs, the engineers had to attach the blades to a metal support bar. Then, to keep the blade edges from getting damaged in the process of attaching them to the metal bar, they needed to weld them without using heat. This, in turn, required them to create a non-heat-producing laser spot welder that could make very fast intricate welds with very low tolerances.

All of this process development made it possible for the company to develop an innovative product that met customer needs better than the alternatives offered by competitors. Moreover, Gillette's process development provided the company with a better manufacturing process than its competitors, thus creating an additional competitive advantage. In short, Gillette's process development capabilities drove the development of an important new product, the Sensor razor.

Key Points

- Technological innovation is the use of technical knowledge to come up with solutions to problems.
- Technological innovation is important to entrepreneurs and managers because it provides a mechanism to create and preserve competitive advantage.

Why Technology Strategy?

While the previous section of this chapter discussed the importance of technological innovation to companies and outlined some basic forms that technological innovation can take, it did not say anything about why companies need to develop a strategy to manage innovation. However, many observers believe that companies can, and should, develop technology strategies. A **technology strategy** is the approach that a firm takes to obtaining and using technology to achieve a new competitive advantage, or to defend an existing technology-oriented competitive advantage against erosion.

Technology strategy is different from overall business strategy in several important ways:

1. Technology strategy has to deal much more with issues of uncertainty than general business strategy because technological change is highly uncertain. A very small portion of new technology ideas result in new products or processes.
2. Technology strategy involves the use of intellectual property management to capture financial returns to a much greater degree than general business strategy.
3. Technology strategy involves the creation of new products and services that are sometimes new to the world, which demand different mechanisms for assessing market needs and designing products than is the case with general business strategy.
4. Technological change occurs in ways that influence the design of effective strategies, and create business dynamics that are different from those that exist when new technology is not important.

5. Organizations faced with a great deal of technical change need to be structured, manage human resources, and design business models in ways that are different from organizations that are not faced with technical change.
6. Making decisions about technology projects requires the use of different decision-making tools than is the case with nontechnology projects.
7. Technological change opens up opportunities for new, high growth businesses in ways not possible in other settings.
8. High-technology businesses face many strategic issues, like standards and increasing returns (you will learn more about this in later chapters), that are much rarer with low technology businesses.

In short, technology strategy differs from general business strategies in enough ways to make a book (and a course) focused on it useful to students.

Purpose of the Book

This book will provide you with a set of analytical tools that will help you to become a successful technology strategist. Using these tools will require you to understand a variety of things, including the sources of opportunity for technological innovation and how to identify them; the assessment of customer needs; the product development process; technology evolution; new product adoption and diffusion; increasing returns, network effects, and technical standards; trade secrets, trademarks, copyrights, patents, and other types of intellectual property protection; firm capabilities and competitive advantage; management of the innovation process; and organizational form and structure. Each of the following chapters explains some of these ideas and presents specific tools and techniques to help you to manage them.

Approach to Technology Strategy

Before we turn to a discussion of the topics of subsequent chapters, four observations about the approach to technology strategy taken in this book are important to make. First, this book is inherently interdisciplinary. Technological innovation is a general management problem whose solution requires an understanding of psychology, economics, strategy, finance, organization behavior, and marketing. Therefore, the concepts that are discussed in the following chapters are drawn from a wide variety of fields, including economics, psychology, sociology, strategic management, finance, and operations management.

Second, this book discusses technology strategy in all types of businesses in all kinds of industries, from aerospace to environmental consulting to motor vehicles. This means that the issues that are discussed, and the examples that are given, are not limited to those industries commonly thought of as "high tech," like computer software. As a result, the focus of this book is on issues that transcend industries, like the management of intellectual property and the use of incentives to motivate people to come up with and implement new ideas.

Third, the book discusses technology strategy in service businesses as well as manufacturing businesses. Technology strategy for services is different from technology strategy for products because services are intangible and cannot be examined before customers purchase them and are inseparable—they are used in the same place that they are produced.[19] As a result, innovation in services involves greater joint activity between companies and their customers, has goals that are more difficult to measure, is more difficult to standardize, and involves outputs that cannot be maintained in inventory.[20]

This book's coverage of service businesses is important because technological innovation is not something that only manufacturing firms undertake. In fact, research and development (R&D) spending in services now accounts for 30 percent of total R&D expenditures. Moreover, services are a growing part of the economies of most developed countries. In the United States, for instance, services now account for 68 percent of the gross domestic product. Furthermore, in some industries, like biotechnology and software, technological innovation in services is central to competitive advantage.

Fourth, this book examines technology strategy in the context in which it occurs. Technological innovation does not occur in a vacuum, but, instead, is affected by the environment in which it occurs. Therefore, this book considers the effects of the industrial, geographic, political, regulatory, cultural, competitive, and economic context when discussing different aspects of technology strategy. For example, in Chapter 8, when the book discusses the role of patents, it explains how differences in the legal environments in developing and developed countries affect decisions about how to manage intellectual property.

Technology Strategy in Start-ups and Large, Established Firms

While technological innovations are often developed and exploited by large, established corporations, sometimes people start new companies to create and exploit these innovations, making the creation of new companies a very important mechanism through which technological innovations get exploited. For example, in 2003, researchers estimated that just those start-up companies funded by venture capitalists since 1970 employed 10 million people, or 9.4 percent of the private sector labor force in the United States, and generated $1.8 trillion in sales, or 9.6 percent of business sales in this country.[21] In 2000, the 2,180 publicly owned companies that received venture-capital backing between 1972 and 2000 comprised 20 percent of all public companies in the United States, 11 percent of their sales, 13 percent of their profits, 6 percent of their employees, and one-third of their market value, a figure in excess of $2.7 *trillion* dollars.[22]

However, most textbooks approach the topic of technology strategy exclusively from the perspective of the large, established firm. This perspective is limited for those students who want to work in young companies or even start one of those companies some day. It is also limited for those students who want to work for large companies because those companies often need to compete with, or partner with, small, start-up companies.

This book takes a different perspective. It looks at technology strategy issues from the perspective of both the small, new firm and the large, established one. For many aspects of technology strategy, this only means clarifying that the strategy is the same for all firms, whether they are large and established or small and new. For example, the demand for technical standards to permit different companies to link their components together is true regardless of whether an industry is composed primarily of large, established firms, or small, new ones.

However, for other aspects of technology strategy, this means discussing two approaches to the same issue. For instance, the way in which large, established companies with deep pockets protect their intellectual property is not the same as that used by cash-poor small, young businesses.

Where aspects of technology strategy are different for large, established firms and small, new companies, the book specifically identifies and discusses those differences. For example, as Figure 1.4 shows, R&D intensity (the amount of money spent

FIGURE 1.4
R&D Intensity by Firm Size for the Period 1999–2003

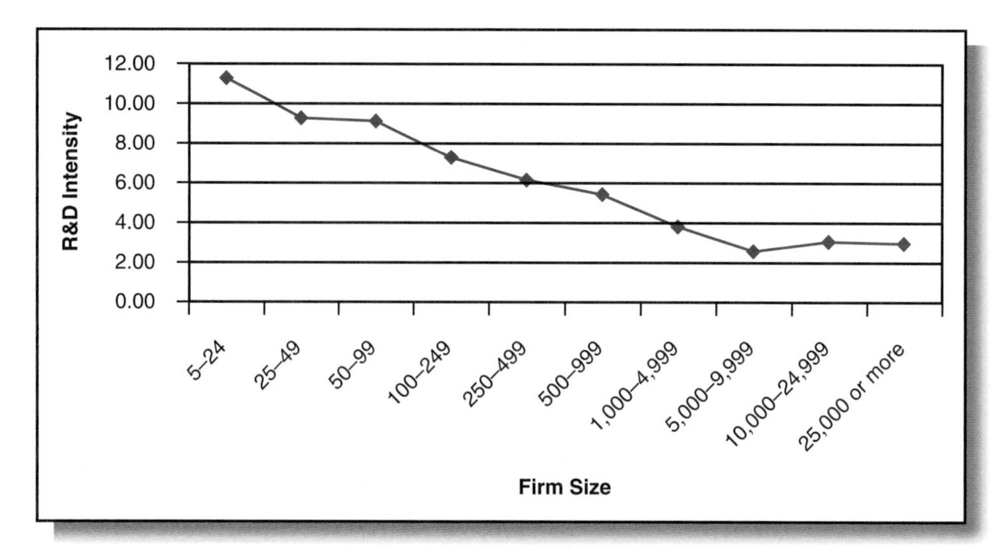

Technology strategy is different in large and small firms for many reasons, including the tendency of small firms to spend more of their revenue on R&D than large firms.

Source: Compiled from data in Science and Engineering Indicators, 2006, http://www.nsf.gov/statistics/seind06/c4/c4s3.htm.

on R&D as a percentage of sales) is much higher for small firms than for large ones. This means that the approach to generating competitive advantage in new, small firms depends much more on the use of legal barriers to imitation and products with attributes that competitors cannot deliver than is the case for established, large firms.

A Preview of the Chapters

The book is divided into four sections. The first section discusses the way in which technologies change and how those changes affect the sources of opportunity for technological innovation. The second section focuses on identifying innovation projects that meet customer needs and result in the development of new products and services. The third section examines the way in which companies capture the returns to investments in innovation. The fourth section examines the process of implementing technology strategy, exploring the management of human resources, organizational structure, and decision-making processes.

Understanding Technological Change

An effective technology strategy involves an understanding of the evolution of technology, the adoption of new technology over time, and the role of technological opportunities in innovation. Chapter 2 provides a framework for understanding the evolution of technology, and presents several useful tools for managing it, including Foster's technology S-curve, Abernathy and Utterback's model of technology evolution, and modifications to it to incorporate the reverse product cycle, architectural innovation, disruptive innovation, and competence-enhancing and competence-destroying innovation. Chapter 3 discusses technology adoption and diffusion, and examines the distribution of adopters of new technology products and services; the transition from the initial adopters of your product to the majority of the market; and the dynamics of market growth. Chapter 4 examines the role of technological opportunities in innovation, examining the sources of opportunities, the entities that engage in technology development, the reasons for investing in research and development, and the different forms that innovation can take.

Coming Up with Innovations

An effective technology strategy depends on the effective use of tools to make decisions about innovation projects, to assess customer needs, and to develop new products. Chapter 5 explains how to make decisions about innovation projects, examining approaches that you can take to manage uncertainty, the tools that you can use to evaluate innovation projects, and the ways you can manage an innovation project portfolio. Chapter 6 looks at the tools and techniques that you need to identify customer needs and the features that will satisfy them for both technology push and market pull innovations. It focuses on why and how companies identify real needs for new products and services, the characteristics that satisfy key stakeholders, profitable solutions to meet customer needs, and effective prices for new products and services. Chapter 7 discusses the role of product development in technology strategy, explaining why rapid product development is important and how to reduce cycle time. It describes key tools to identify the right features for your products, and to manage the product development process.

Benefiting from Innovation

An effective technology strategy requires firms to **appropriate**, or capture (rather than allowing them to go to their competitors), the financial returns to their investments in R&D through effective decision making about the legal protection of intellectual property, effective use of nonlegal mechanisms to deter imitation, the use of a company's capabilities to create competitive advantage, and an understanding of the role of technical standards and increasing and decreasing returns. Chapter 8 discusses patents, focusing on what is patentable, the use of patents, and the decision whether or not to patent inventions. Chapter 9 discusses how companies use trade secrets, copyrights, trademarks, and domain names as part of their technology strategies, as well as the issues raised by differences in intellectual property protection across countries. Chapter 10 explains the nonlegal ways that companies appropriate the returns to investment in innovation, including controlling resources, being a first mover, developing a brand name reputation, moving up the learning curve, exploiting complementary assets, and taking advantage of economies of scale, and brings together the intellectual property issues discussed in Chapters 7 and 8 to present a model of when a company should focus on the control of complementary assets and when it should focus on being innovative. Chapter 11 discusses the ways that companies create and sustain competitive advantage in high-technology industries, explaining how to use industry analysis to identify favorable positions in the value chain in attractive industries, and describing how companies exploit their core competencies to innovate repeatedly in their lines of business. Chapter 12 discusses the creation and exploitation of technical standards, focusing on strategy toward standards-setting battles and the choice between open and closed technical standards. Chapter 13 explains how increasing returns businesses differ from decreasing returns businesses, discusses the sources of increasing returns, and identifies specific strategic actions that enhance performance in industries based on increasing returns.

Formulating Technology Strategy

Developing an effective technology strategy also includes making decisions about how to organize a company for innovation. Chapter 14 explores the macro structural issues, describing the choice between contractual and vertically integrated modes of doing business; identifying several common contractual forms, including licensing, joint ventures, strategic alliances, outsourcing, and contract manufacturing; discussing the advantages and disadvantages of different modes of doing business; and explaining how to set up effective collaborative organizational arrangements. Chapter 15 examines the management of human resources for effective innovation, discussing the role of corporate culture in enhancing technological innovation; effective human resource management techniques for R&D personnel; ways to enhance employee creativity; and the different types of product development teams and the ways to manage them. Chapter 16 discusses the relationship between organizational structure and innovation, explaining how to design organizations to make them more innovative; comparing the innovation process in large and small firms; explaining why people start companies to exploit innovations that they identified during their prior employment; and discussing the advantages and disadvantages of corporate venturing, and the different ways in which companies engage in it.

Key Points

- Technology strategy is the approach that firms take to obtaining and using technology to achieve a new competitive advantage or to defend an existing competitive advantage against erosion.
- Technology strategy helps companies in many industries to perform better.
- Technology strategy involves a variety of issues, which can be categorized as those related to understanding technical change, meeting the needs of customers, capturing the value created from innovation, and implementing a technology strategy.

Discussion Questions

1. What is technology? What are some examples of different types of new technology? What do they have in common? How are they different?
2. What is technological innovation? Why is it important to firms?
3. What is technology strategy? Why do firms in high-technology industries need to develop technology strategies?
4. What are some key aspects of technology strategy that entrepreneurs and managers need to learn? Why are these aspects central to technology strategy?

Key Terms

Appropriate: To capture the returns from investment in innovation and keep them from going to competitors.

Cost per Unit: The amount of money that a company needs to spend to create one of a product.

Innovation: The process of using knowledge to solve a problem.

Invention: The discovery of a new technical idea.

Process Development: The anticipation and resolution of problems that arise in the production of a product.

Technological Innovation: The use application of tools, materials, processes, and techniques to come up with new solutions to problems.

Technology: The application of tools, materials, processes, and techniques.

Technology Strategy: The approach of firms to obtaining and using technology to achieve a new competitive advantage or to defend an existing competitive advantage against imitation.

NOTES

1. Adapted from Guth, R. 2005. Getting Xbox 360 to market. *Wall Street Journal*, November 18: B1, B5.
2. Ibid.
3. Ibid.
4. Ibid.
5. Ibid.
6. Schilling, M. 2005. *Strategic Management of Technological Innovation*. New York: McGraw-Hill Irwin.
7. Allen, K. 2003. *Bringing New Technology to Market*. Upper Saddle River, NJ: Prentice Hall.
8. Clark, K., and T. Fujimoto. 1991. *Product Development Performance*. Boston: Harvard Business School Press.
9. Mokyr, J. 1990. *The Lever of Riches*. New York: Oxford University Press.
10. Afuah, A. 2003. *Innovation Management*. New York: Oxford University Press.
11. http://en.wikipedia.org/wiki/Dell#Origins_and_evolution
12. Gilbert, S. 2006. The accidental innovator. *Working Knowledge*, July 5, http://hbswk.hbs.edu/item/5441.html.
13. Srivastava, B. 2004. Radio frequency ID technology: The next revolution in SCM. *Business Horizons*, 47(6): 60–68.
14. Mokyr, J. 1990. *The Lever of Riches*. New York: Oxford University Press.
15. Schilling, M. 2005. *Strategic Management of Technological Innovation*. New York: McGraw-Hill Irwin.
16. Eckhardt, J. 2003. *When the Weak Acquire Wealth: An Examination of the Distribution of High Growth Startups in the U.S. Economy*, Ph.D. Dissertation, University of Maryland.
17. Sidel, R. 2007. Giant retailer reveals customer data breach. *Wall Street Journal*, January 18: D1, D6.
18. Adapted from Hammonds, K. 1990. How a $4 razor ends up costing $300 million. *Business Week*, 3143: 62–63.
19. Mohr, J., S. Sengupta, and S. Slater. 2005. *Marketing of High Technology Products and Innovations* (2nd edition), Upper Saddle River, NJ: Prentice Hall.
20. Ettlie, J. 2000. *Managing Technological Innovation*. New York: John Wiley.
21. Venture Impact 2004, http://www.nvca.org/pdf/VentureImpact2004.pdf.
22. Gompers, P., and J. Lerner. 2001. *The Money of Invention: How Venture Capital Creates New Wealth*. Boston: Harvard Business School Publishing.

Chapter 2

Technology Evolution

Learning Objectives

After reading this chapter, you should be able to:

1. Explain why technology tends to develop in an evolutionary manner.
2. Graph a technology S-curve and describe what it measures.
3. Define a shift in the S-curve, and explain who shifts the S-curve.
4. Spell out the pros and cons of using technology S-curves as a management tool.
5. Describe the Abernathy-Utterback model of technology evolution and explain how it is useful to technology strategy.
6. Define a dominant design and interpret its effect on technology development and industry competition.

7. Define radical and incremental technological change and explain how the two types of innovation affect technology strategy.
8. Summarize the modifications that researchers have made to the Abernathy-Utterback model, and explain why they were needed, and how they are helpful.
9. Describe architectural innovation and explain how it influences technology strategy.
10. Identify the types of innovation that new and established firms are each better suited to develop, and explain why this is the case.

Technology Evolution: A Vignette[1]

The history of photography provides a good example of technology evolution. In 1839, the first photographic technique, the daguerreotype, was developed in France. This technology produced images on copper plates that had been sensitized and coated with silver. In the 1850s, a new technology for photographic images was developed. With it, a photographer would use silver nitrate to condition a glass plate coated with a substance called collodion just before taking a photograph. By exposing the glass to light, a negative image would be developed and then fixed on photosensitive paper through exposure to sunlight.

This technology, called "wet plate," was followed by the development of "dry plate" technology in the 1870s. With dry plate technology, photosensitized glass plates were covered with a dry gelatin emulsion. Because dry plates could be produced in factories, they could be made on a larger scale and at a lower cost than wet plates. Therefore, dry plate technology was an important commercial advance.

Celluloid roll film was the next technology to be developed. In 1889, George Eastman introduced a photosensitive celluloid film that was rolled into the back of a camera. The development of roll film led to a dramatic growth in the market because it allowed ordinary people to take photographs. On the back of this technology, Kodak grew from a tiny start-up into a multinational corporation.

Celluloid roll film technology remained the dominant technology in cameras for approximately 100 years. But in the 1990s, Apple and Sony both introduced cameras that used digital images that could be viewed without any processing or developing.

This example describes the evolution of a single technology, and points out some of the important aspects of technology evolution that will be discussed in this chapter—the incremental nature of most technological innovation and the tendency of radical technological changes to be introduced by firms outside the industry.

FIGURE 2.1
Technological Changes That Kodak Has Faced

While exaggerating to make the point humorously, the figure shows that Kodak has faced waves of technological change.

Source: www.cartoonstock.com.

INTRODUCTION

New technologies tend to evolve in ways that allow you to use tools and techniques to forecast their evolution. By forecasting technology evolution, you can develop a strategy that considers the emergence of technologies, changes in the dominant strategies and structures of other firms, and shifts in the nature of competition between firms.[2]

This chapter provides a framework for understanding the evolution of technology, and presents several useful tools for managing it. The first section describes the evolutionary pattern of technological development. The second section defines radical and incremental technological change. The third section discusses Foster's technology S-curve. The fourth section describes the Abernathy-Utterback model of technology evolution. The fifth section discusses the modifications that different scholars have made to the basic Abernathy-Utterback model to incorporate the reverse product cycle, architectural innovation, disruptive innovation, and competence-enhancing and competence-destroying innovation.

EVOLUTIONARY PATTERNS OF DEVELOPMENT

Many researchers view technological advance as a process that begins with basic scientific discoveries and ends with commercial products that are adopted by a wide range of customers. When new technology is first invented, it emerges in a primitive form in which its usefulness is unclear, often even to the inventors of the technology.

Over time, as researchers work on the technology, it evolves and develops, with its performance improving to the point at which someone identifies a commercial use for it, and introduces a product based on the technology to the market. The companies developing products based on the technology then refine it, and introduce additional generations of products that are based on improved versions of the technology. As the technology gets better, it meets the needs of a wider range of potential customers, and products based on the technology are adopted by more and more users.

This technological evolution depends on the process through which scientific advance occurs. New advances are made as researchers seek answers to current technical problems, building on prior knowledge that has accumulated.[3] This process leads scientists and engineers to work within particular frameworks, or **paradigms**, which influence how technical problem solving occurs. These frameworks affect both the choice of problems that scientists and engineers work on and the ways that they go about answering them. For example, the invention of streptomycin in the 1940s led to the development of a paradigm of developing pharmaceuticals through synthesis and subsequent testing of large numbers of organic molecules.[4] That is, after this paradigm was developed, scientists in all pharmaceutical firms focused on synthesis and testing of organic molecules to develop new drugs until a new paradigm emerged.

The evolution of technology is not entirely driven by scientific advance. Social, economic, and political forces affect the path that technological advance takes, each in somewhat different ways. The economic environment influences the nature of incentives created for technological advance, thus affecting the evolutionary path by increasing the rewards for development to go in certain directions and not in others. Political forces have an effect because they influence the rules of the society in which the technological advance takes place. As a result, they lead the technology to develop in ways consistent with society's rules and not in ways inconsistent with it. Social forces matter because technological development takes place in a human environment in which the fears, motivations, and attitudes of people come into play. Consequently, technology evolution tends to move toward things that people are supportive and accepting of and away from things that they are afraid of or intolerant of.[5]

The tendency for scientists and engineers to work within technological paradigms leads to the creation of **technology trajectories**, which researchers define as paths of improvement of a technology on some performance dimension.[6] Take, for example, a key technological trajectory in the computer industry—the trend toward smaller and more powerful microprocessors. In 1965, Gordon Moore, the founder of Intel, made a statement that has become known as Moore's Law. He argued that the number of transistors that could be packed onto an integrated circuit at the minimum cost for the component will double every two years. As Figure 2.2 shows, over time, engineers have developed better ways of packing more processors onto each microchip, and the chips have become more powerful along the trajectory laid out in Moore's law.

Although technology paradigms have a valuable focusing effect on research, they also have an important downside. They tend to limit the alternatives that researchers are willing to consider when problem solving, and, consequently, often keep researchers from identifying fundamentally different, and better, alternatives.[7] For example, early on in his career, Robert Langer, an MIT chemical engineering professor and one of America's most prolific inventors, developed a new type of plastic that could be used to deliver a large molecule drug. When Langer presented this finding to the scientific advisory board of a leading pharmaceutical

FIGURE 2.2
The Technological Trajectory in Microprocessors

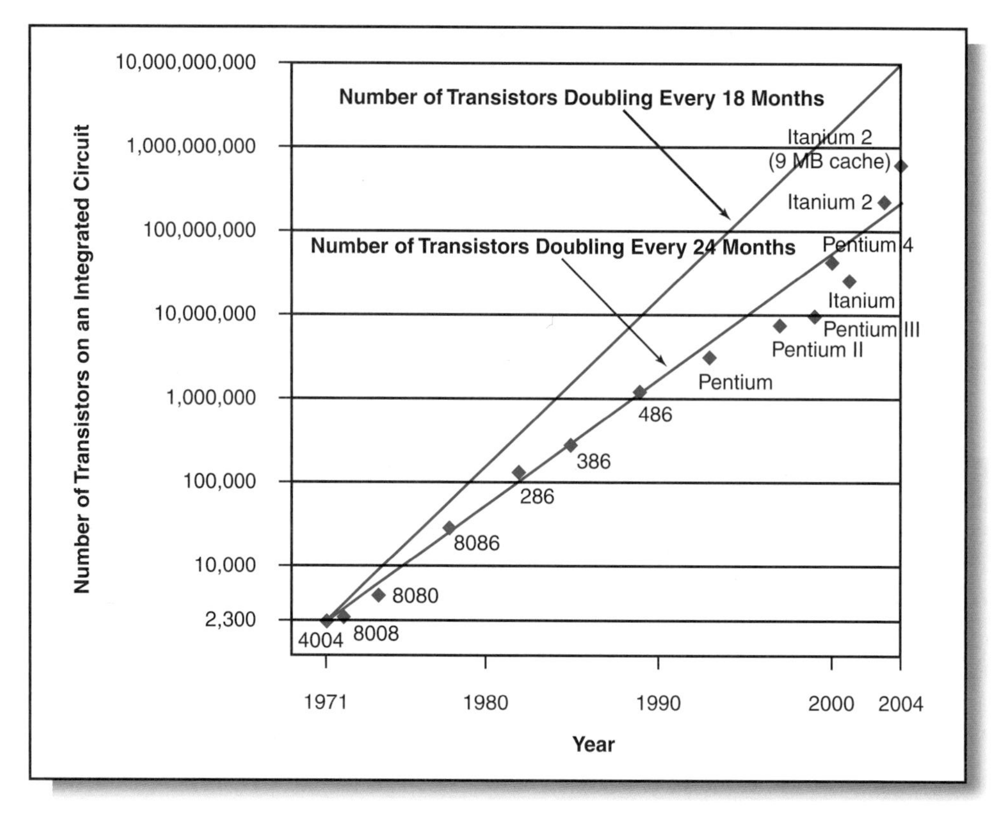

Moore's Law is perhaps the most famous example of a technological trajectory; it indicates that the number of transistors on an integrated circuit has doubled every two years since the 1970s.

Source: http://en.wikipedia.org/wiki/Image:Moore_Law_diagram_%282004%29.png.

company, two Nobel-prize-winning members of that board responded that Langer's discovery was impossible. Similarly, when Langer presented the same finding to a conference of polymer chemists, they responded that what he had just presented was contrary to the established literature, and was just plain wrong. However, Langer went on to garner many patents and found several companies based on this discovery, which subsequently was considered a paradigm-shifting breakthrough in drug delivery.[8]

As the previous paragraph alludes, at discrete points in time, opportunities appear to fundamentally change the technological paradigms within which scientists and engineers operate. For example, the opening vignette in this chapter describes the paradigm shift to digital technology from chemically based film in the 1990s. The identification of these paradigm shifts is very important because they fundamentally alter the ways in which products and services are developed, and so have profound effects on technology strategy.

Unfortunately, identifying paradigm shifts is much more difficult than Figure 2.3 suggests. Innovators often don't know in advance if they have come up with paradigm shifting technologies or if the reason that the new technologies are poorly received is just that they are bad ideas. Take, for example, the Segway Human Transporter, a device that famed venture capitalist John Doerr said would be more

FIGURE 2.3
Paradigm Shifts

Technology entrepreneurs and managers wish that the identification of paradigm shifts were this easy; in reality, they are very difficult to spot in advance.

Source: New Yorker Magazine. 1975. *The New York Album of Drawings, 1925–1975.* New York: Penguin Books.

important than the development of the Internet. While the founder of this device, Dean Kamen, thought it would shift the paradigm in how human beings get around, the Segway transporter turned out to be inferior to foot power, electric vehicles, scooters, and mopeds for most of the users for which it was intended. As a result, despite the near $100 million invested in the technology, only a couple thousand of the devices were ever developed, and the device did not change the transportation paradigm.[9]

Radical and Incremental Technological Change

Most technological innovation is **incremental**, and involves small improvements to existing technologies. A good example of an incremental innovation is Intel's 486 microprocessor. This microprocessor was faster than previous generations of Intel's products, but the changes that made it faster represented relatively small improvements to its basic technology.

Other times, technological innovation is **radical**, and involves a fundamentally new way of solving a problem. A good example of a radical technology is the digital camera, which is based on fundamentally different technical principles from traditional chemical film technology.

While research has shown that only a small percentage of all new technologies are radical,[10] they often have enormous impact (see Figure 2.4) because they are high risk–high return developments. Moreover, their creation and implementation requires fundamentally different strategies and organizational structures from those used for more incremental products.[11] For example, as we will see in Chapter 6, the techniques for gathering market research data and the tools for making decisions about projects that work well for incremental innovations work poorly for radical innovations.[12]

FIGURE 2.4
Radical Technologies That Have Changed Markets

Aspirin	Internet	Personal Digital Assistant
Automobile	Jet Airplane	Superconductor
Beta Blockers	Medical Diagnostic Imaging	Television
Computer Operating Systems	Mobile Telephone	Transistor
Digital Music	Microprocessor	Video Cassette Recorder
Fiber Optics	Organic Fibers	Xerography
Genetic Engineering	Personal Computer	

Some new technologies have dramatically changed markets because they were based on fundamentally different ways of solving problems.

Sources: Adapted and compiled from Rothaermel, F. 2000. Technological discontinuities and the nature of competition. *Technology Analysis and Strategic Management*, 12(2): 149–160; Brown, S., and Eisenhardt, K. 1997. The art of continuous change: Linking complexity theory and time-paced evolution in relentlessly shifting organizations. *Administrative Science Quarterly*, 42(1): 1–34; Christiansen, C. 1998. Valurec's venture into metal injection molding. Harvard Business School Teaching Note, Number 698–002; Miller, W. 2006. Innovation rules! *Research Technology Management*, 49(2): 8–14; Markides, C., and Geroski, P. 2005. *Fast Second: How Smart Companies Bypass Radical Innovation to Enter and Dominate New Markets.* San Francisco: Jossey-Bass.

TECHNOLOGY S-CURVES

Technology S-curves are graphical representations of the development of a new technology. They compare some measure of performance (for example, speed, cost, or capacity) with some measure of effort (for instance, R&D dollars, or person-hours devoted to research).

These graphs are called S-curves because the relationship between effort and performance is typically S-shaped. Initially, performance improvements per unit of effort are small because there are many things that you need to learn before you can improve the performance of new technologies significantly. Moreover, when you first begin to develop a new technology, you probably don't understand the key drivers of performance. Consequently, your efforts to improve the technology involve much trial and error, with many dead ends, before you figure out how to generate performance improvements. As a result, initially you get very little performance improvement for each unit of effort.

But once you identify the key drivers of performance, rapid improvement tends to follow and you see a big performance improvement for each unit of effort.[13] For instance, the first airplanes required a significant investment in research on wing and engine design to achieve a few additional feet of flight. However, once the basic design of the airplane was identified, the performance of aircraft increased dramatically.

While these improvements to the technology can continue at a rapid pace for a while, at some point the technology reaches its physical limits, and diminishing returns begin to set in, leading the curve to take on the upper portion of the S-shape. That is, you start to need a lot of effort again to get each unit increase in effort. For instance, conventional semiconductors have been improved over the past three decades by doubling the number of transistors per square inch of silicon every couple of years. However, the dimensions of semiconductors have now shrunk to the point that the photographic techniques that semiconductor manufacturers employ to produce them can no longer be used to make them smaller, causing the law of diminishing returns to kick in. (Figure 2.5 shows the technology S-curve for Intel's semiconductors.)

FIGURE 2.5
S-Curves in Microprocessors

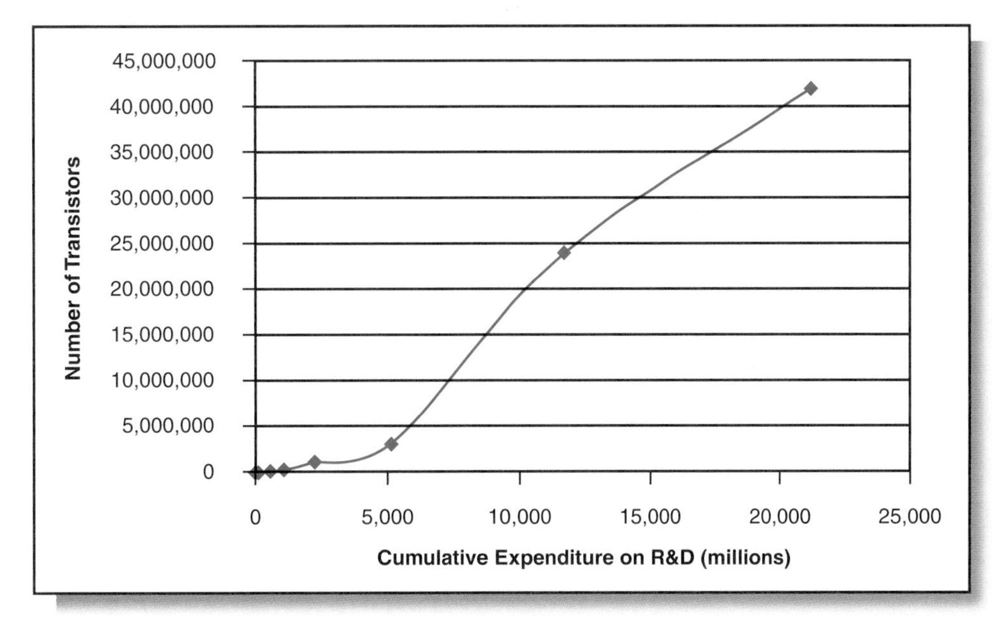

Intel's development of different generations of microprocessors shows the traditional S-shaped curve of technology development; investment in the development of the technology initially leads to a small improvement in performance—in this case the number of transistors on a chip—followed by rapid acceleration in performance and, ultimately, diminishing returns.

Source: Adapted from data contained in Schilling, M. 2005. *Strategic Management of Technological Innovation.* New York: McGraw Hill; www.icknowledge.com/trends/uproc.html.

You need to understand how technology advances along an S-curve to formulate an effective technology strategy because this pattern of technical advance affects who innovates successfully in an industry. Technological improvements along an S-curve tend to be incremental, building on prior developments, and taking place within an existing paradigm. Consequently, established firms, with experience operating in the industry, tend to be the ones that make these types of technological advances. First, they have existing technical, market, and organizational capabilities, which they can use to make improvements to the underlying technology with which they are working.[14] Second, they have an existing customer base that provides them with information about market needs, which they can use to introduce new versions of their products and services that appeal to customers.[15] Third, they have access to internal cash flow to invest in the improvement of the technology without having to raise money from external investors. Because external investors have less information about the development of new products than the companies developing those products, they demand a premium for financing new product development, relative to the cost of using internal cash flow, putting new companies at a financial cost disadvantage in developing new products and services.

Shifting S-Curves

When an existing technology reaches the point of diminishing returns, it becomes difficult to achieve very marked performance improvement by making incremental advances to it. At this time, a new technology is often developed to challenge the existing technology.[16]

The new technology typically differs from the old one in certain important ways. When it is first introduced, the new technology is usually inferior to existing technology on key dimensions, such as quality, performance, or reliability.[17] For example, the quality of printed pictures of the first digital cameras (which cost about $1,000) was inferior to that of disposable 35 millimeter cameras (which cost about $10).

However, because the new technology is a radical change from the existing technology (it is based on fundamentally different technical principles), it has greater potential for performance improvement than the existing technology. For instance, in the 1990s, the rate of improvement of digital camera technology was greater than that of chemical film technology. As a result, the performance of digital film technology reached a point where it surpassed the chemical film alternative.

The introduction of new technologies can be represented graphically by a shift in the S-curve. As Figure 2.6 shows, traditional telecommunications switching technology has reached the point of diminishing returns, leading an alternative called Voice-over-Internet protocol (VOIP) to emerge. VOIP is less expensive than traditional telephone technology because the voice messages are divided into packets of data, which are sent over the most efficient route to the receiver and reassembled at the other end, rather than being sent directly over dedicated lines.[18]

As Figure 2.6 indicates, VOIP began with a lower level of performance than traditional telecommunications technology but then improved at a higher rate than traditional telecommunications technology to ultimately surpass it on the performance dimensions that customers care about. Once a new technology improves to the point where it surpasses the old technology on the dimensions that customers value, customers shift to the new technology in large numbers. The new technology then takes off, and the old technology goes into decline.[19]

FIGURE 2.6
Shifting S-Curve in Telecommunications

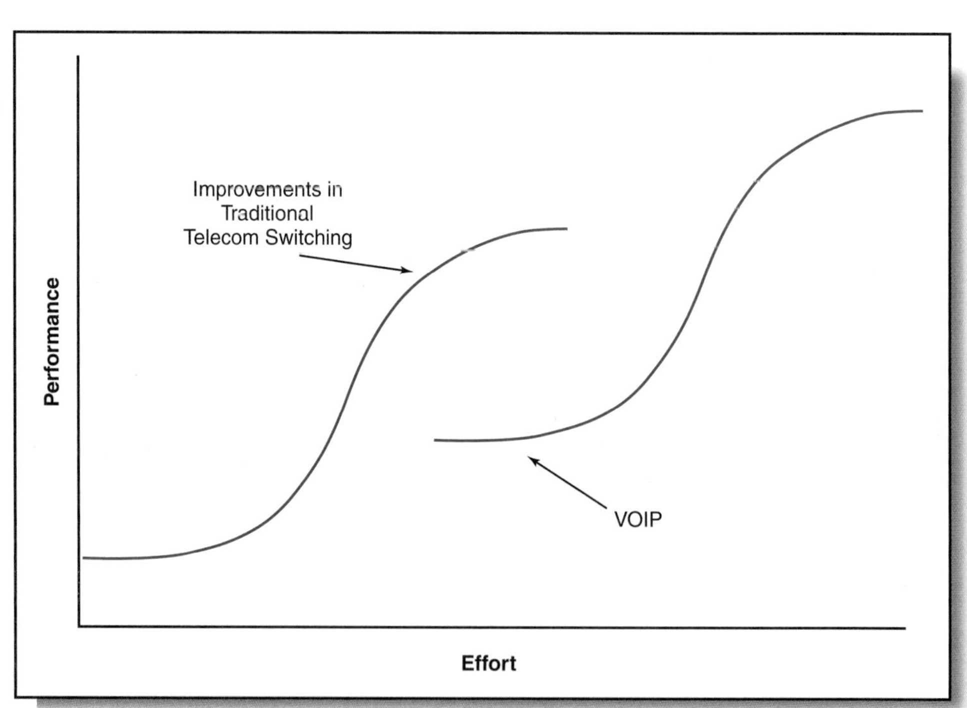

This figure shows a shifting S-curve in telecommunications; initially the performance of VOIP was inferior to the technology that it replaced, but its rate of improvement has been higher, allowing it to overtake the performance of the traditional telecom switching technology.

Because technologies eventually reach a point of diminishing returns, companies using a particular technology often investigate alternative technologies that can provide greater performance improvement in the future. For instance, decreasing returns in microchip density have led Hewlett-Packard to use nanotechnology to develop tiny junctions of titanium and platinum wires a few atoms wide.[20] The goal of this effort is to fit more transistors on the chip, in this case increasing the number of transistors eight times, while reducing energy use.[21] Because these approaches to making semiconductors are fundamentally different from current approaches to producing them, they are more likely to permit higher transistor density than incremental improvements to existing technology.

Who Shifts the S-Curve?

Even though incumbent firms typically investigate new technologies when the core technology with which they are working reaches the point of diminishing returns, new entrants tend to be the ones to introduce products and services based on new technologies. Even when they have developed the new technologies, incumbent firms tend to persist with investments in their current technologies,[22] and rarely shift to newer technologies until the new technologies challenge their ability to serve their core customers with their core products.[23] For example, AT&T was slow to shift to VOIP even though its researchers invented the voice compression technology that made Internet telephone calls possible;[24] and Eastman Kodak was slow to shift to digital cameras even though researchers at that company invented the digital camera in 1976.[25]

So why do new entrants introduce the new technologies that shift the technology S-curve? Basically, there are six reasons:

1. **Incumbents have no incentive to introduce the new technology:** The performance of new technology is typically inferior to the performance of existing technology when it is first introduced, giving incumbent firms the option of changing to a technology that will produce less income (sometimes no income) or making an incremental change to their existing technology.[26] For instance, the major semiconductor firms are not switching from dynamic random access memory (DRAM) chips to magnetic random access memory chips (MRAM) even though the latter permit data to be stored longer. Why? Because the first generation of MRAM chips cost $25 each, and store four megabits of data, while DRAM chips cost $5 each, and store 512 megabits of data. Thus, the storage capacity and price of MRAM chips is not yet competitive with DRAM chips.[27]
2. **Incumbents have investments in existing technology:** Incumbents have to write-off their investments in the old technology to adopt the new technology, and the size of these write-offs can be very large when new technologies require fundamentally different assets to be produced. For example, many of the incumbent telephone companies did not want to launch VOIP phone service because they had large investments in copper phone lines that Internet phones would make obsolete.[28]
3. **Products based on the new technology cannibalize incumbents' sales:**[29] (**Cannibalization** occurs when a firm sells a new product that takes sales from one or more of its existing products.[30]) If a product based on a new technology completely cannibalizes existing sales, then introducing a product based on a new technology generates additional costs, but no extra revenue, making it financially unattractive.[31] For example, the makers of copper cable for

TABLE 2.1 The Failure to Foresee the Value of New Technologies
Technological evolution is hard to forecast; even very knowledgeable people fail to see the value of new technologies in their fields.

Technology	Quote	Person
Phonograph	The phonograph . . . is not of any commercial value.	Thomas Alva Edison, inventor of the phonograph, 1880
Airplane	Heavier-than-air flying machines are impossible.	Lord Kelvin, British mathematician, physicist, and President of the British Royal Society, 1895
Computer	I think there is a world market for about five computers.	Thomas J. Watson, Chairman of IBM, 1943
Personal Computer	There is no reason for any individual to have a computer in their home.	Ken Olson, President of Digital Equipment Corporation, 1977
Telephone	The "telephone" has too many shortcomings to be seriously considered as a means of communication. The device is inherently of no value to us.	Western Union internal memo, 1876
Television	Television won't be able to hold on to any market it captures after the first six months. People will soon get tired of staring at a plywood box every night.	Darryl Zanuck, Head of 20th Century Fox Films, 1946
Movies with Sound	Who the hell wants to hear actors talk?	Harry M. Warner, 1927

Sources: Adapted from Schoemaker, P., and V. Mavaddat. 2000. Scenario planning for disruptive technologies. In G. Day and P. Shoemaker (eds.) *Wharton on Emerging Technologies*. New York: John Wiley; and Mohr, J., S. Sengupta, and S. Slater. 2005. *Marketing of High Technology Products and Innovations* (2nd edition). Upper Saddle River, NJ: Prentice Hall.

telecommunications resisted the shift to fiber optic cable because the latter was a substitute for the former and selling fiber optic cable would completely cannibalize their sales to telecommunications companies.

4. **Managers at incumbent firms do not see the new technology as a threat (see Table 2.1):** To maintain focus on their core activities, established firms create routines that lead managers to filter out information about alternatives to their core technologies, which leads them to view new technologies negatively.[32] For example, Epson did not shift to the production of ink-jet printers after Hewlett-Packard introduced them because its management believed that ink-jet printers were not a cost-effective alternative for most consumers, and that ink-jet printers would not challenge the dot-matrix printer business.[33]
5. **Incumbent firms can improve the performance of their old technologies:**[34] For instance, Jim Utterback, a professor of technology management at MIT, has shown that the producers of mechanical typewriters, sailing ships, mechanical watches, and the telegraph all improved those products significantly in response to the introduction of the new radical technologies that ultimately replaced them.[35]
6. **Incumbent firms face organizational obstacles to changing their core technologies:** When a new technology comes along, organizations often have to change their structures to fit the new technology. This causes fighting within the organization because some managers have to lose political power and influence.[36] These managers resist the changes, even if they are necessary for the organization to survive. For example, Polaroid failed in its efforts to exploit the shift to digital camera technology because its managers resisted making the people in charge of electronic digital signal processing, software, and storage technologies more central to the organization than the people in charge of optics and film technology.

Using S-Curves as a Management Tool

Richard Foster, a management consultant who has written extensively about technology S-curves, suggests that you can use them as a tool to predict when your company should invest in the development of a new technology. For example, Hewlett-Packard used the technology S-curve to determine that it should invest in the development of a line of photo printers back in 1995, when traditional film cameras still were dominant, and digital cameras were very expensive.[37] By plotting data on the relationship between the amount of investment that has been made in a technology against data on key performance indicators, you can determine where a current technology lies on the S-curve, and whether it has reached the point of diminishing returns.[38] If plots of the technology S-curve reveal that your company's core technology has reached the point of diminishing returns, then you should prepare for a technological shift by investing in the development of alternative technologies.

However, there are several important limitations to technology S-curves as a tool to help formulate your firm's technology strategy. First, plots of technology S-curves will tell you that your company needs to change from the old technology to the new technology, but not when your company should make that change. Few technologies have known performance limitations, making it difficult to determine the amount of improvement to the old technology that is possible once the new one has been introduced. Take, for example, the internal combustion engine and the fuel cell. It is not possible to predict when companies should switch from making vehicles with internal combustion engines to ones with fuel cell engines because we do not know when the performance of the fuel cell will exceed the performance of the internal combustion engine. The inability of S-curves to predict *when* you should shift from an old technology to a new one is an important limitation because shifting too early will saddle your company with a new technology that is inferior to the old one, while switching too late will allow other firms to move up the learning curve and position their product or service as the dominant design or technical standard in the industry.

Moreover, in many cases, several things have to happen before a new technology can "take off," and it is hard to know when that confluence of events will occur. Take, for example, digital camera technology. The replacement of traditional film cameras with digital cameras did not take off until personal computer manufacturers had developed computers with the processing and storage capability for basic imaging, and digital cameras could be designed to use computer storage and processing.

Second, your decision to switch to a new technology shouldn't just depend on the old technology having reached the point of diminishing returns. It should also depend on your identification of a need for a product based on the new technology, your ability to get customers to switch to products based on the new technology, your company's capability to produce products based on the new technology, and the expected rate of diffusion of products based on the new technology.[39] Existing technologies will often reach a point of diminishing returns long before you can offer a viable alternative based on a new technology. For example, the shift to the technology of e-books has been very slow even though improvements to paper books long ago reached the point of diminishing marginal returns.

It can be particularly difficult to get industrial customers to switch to products based on a new technology because product improvements based on existing technology are much easier to integrate into their product development and production activities than are new products based on fundamentally new technology. If your customers need to change their infrastructure or production and product development processes to adopt products based on a new technology, they are going to resist that change. As a result, you will risk losing your existing customers by making the shift to a new technology instead of improving your old technology.[40]

Third, switching to the new technology may not make sense because the new technology is initially inferior to existing alternatives and, therefore, cannot be used to serve the mainstream of the market. To use new technology successfully, you first have to focus on segments of the market that may be less profitable, while investing in the further development of the new technology to get it to the point where it is competitive with existing technologies in mainstream markets.[41] For example, the companies that exploited VOIP telephone service first had to target niche markets and improve the technology before they could offer good enough call quality to serve the mainstream of the telecommunications market. The need to focus on lower margin niche markets for an indeterminate amount of time might make the shift to the new technology too costly for your company.

Fourth, as the manager of an incumbent firm, you might be able to deal with decreasing marginal returns to your technology's performance without immediately changing to a new technology. You can improve your product architecture, enhance components,[42] identify new applications for your products, or even fight the adoption of new technology with legal action or public relations efforts, to maintain, or even increase, your sales. As a result, you might be better off waiting to shift your company to the new technology until after it offers better performance than the existing technology. Your company's control over assets in manufacturing, marketing, and distribution, and your brand name, could allow you to catch up to new firms that were the first to adopt the new technology. For instance, AT&T is betting on the value of its brand name with consumers to allow it to catch up to Packet8 and Voice Pulse, VOIP start-ups that have seen rapid growth in their number of customers in the last few years.[43]

Key Points

- New technologies evolve in an evolutionary manner, along technological trajectories, constrained by the paradigms within which scientists and engineers work.
- Incremental change involves small changes to a technology within an existing paradigm; radical change involves large changes to a technology and a break with the existing paradigm.
- The development of new technologies tends to follow an S-shaped pattern: In the initial period, large amounts of effort yield small performance improvements; in the intermediate period, small amounts of effort yield large performance improvements; and in the final period, large amounts of effort again yield small performance improvements.
- New entrants typically shift the S-curve because the new technologies start with lesser performance than existing alternatives, require existing assets to be written off, cannibalize the sales of existing products, and are often dismissed by managers of incumbent firms, who face significant organizational pressures to maintain existing technology and who can improve their existing technology.
- Incumbent firms can use technology S-curves as a management tool to predict when to invest in a radical new technology; however, the use of S-curves for this purpose faces important limitations, including the inability to identify when to switch technologies, the failure to incorporate all of the factors that matter to the decision to switch, the need for adopters of the new technology to focus on niche markets before tackling the mainstream of the market, and the existence of alternative ways to respond to the introduction of new technology.

The Abernathy-Utterback Model

The late Bill Abernathy, a professor at Harvard Business School, and Jim Utterback, a professor at the Sloan School of Management at MIT, proposed a model of technology evolution that expands upon the basic ideas of the technology S-curve. According to Abernathy and Utterback, technology evolves through periods of incremental innovation, interrupted by periods of radical innovation. The development of a radical innovation leads to a **fluid phase** in an industry, during which time many firms enter and compete on the basis of different product designs. Eventually, the firms in the industry converge on a dominant design, which results in the **specific phase**, during which time only incremental innovation occurs. After a while, the cycle repeats with the development of a new radical innovation, which introduces a new fluid phase.

The Nature of Innovation and Competition

This evolutionary cycle that professors Abernathy and Utterback described affects the nature of innovation and firm competition. To see how the evolutionary cycle affects the nature of innovation, you need to understand the difference between product and process innovation. When technological innovation involves the creation of new goods and services sold to customers, **product innovation** is the term used to describe it. The DVD player is an example of a product innovation because technical knowledge was used to create something new that could be sold to customers.

When technological innovation involves problem solving that improves the method of creating or delivering a product or service, it is called **process innovation**. For example, Boeing engaged in process innovation when its engineers figured out how to make the wings of the new Dreamliner aircraft out of a composite material in place of aluminum to make the plane lighter and more fuel efficient. The innovation in the materials used to make the wing was a process innovation because it did not lead to a new product or service purchased by customers, but, instead, led to a better way of making an existing product or service.

The initial phase of an industry is called the fluid phase. This phase is a period of high uncertainty. During this phase, markets are relatively small, customers do not have clearly defined preferences, and the diffusion of the industry's core technology is quite limited.

Companies that enter the industry during the fluid phase engage primarily in product innovation, competing on the basis of the novelty of their product designs. For instance, in the beginning of the twentieth century, when the automobile industry had not yet converged on a dominant design of a four-wheeled vehicle with an enclosed body and a steering wheel, some cars were sold with three wheels instead of four, with open bodies instead of closed bodies, and with joysticks instead of steering wheels.

During the fluid phase, production is relatively inefficient. Initial entrants are often small and use a variety of different production processes. They engage in very little process innovation and tend to use generic inputs and production equipment.[44]

The fluid phase of technology evolution comes to an end when the industry converges on a **dominant design**, or a common way that all companies producing a product will design that product.[45] The internal combustion engine is a good examples of a dominant design. For decades, all automakers have produced vehicles using this design; the original steam and electric vehicle alternatives to this design are now long gone. (The fact that car companies are now considering making electric vehicles again, approximately one hundred years after the industry abandoned that design alternative, is evidence of the cycle of radical and incremental innovation that Abernathy and Utterback identified.)

The rise of the dominant design leads to the specific phase of industry evolution. During this phase, competition shifts away from design uniqueness to production cost, as firms focus on achieving production efficiencies with the common design.[46] The standardization that comes from convergence on a dominant design allows for the introduction of interchangeable parts, and an agreed-upon format for components,[47] both of which permit investment in specialized long-lived assets, such as manufacturing equipment,[48] and efficient volume production. As a result, production shifts to large firms that can exploit economies of scale, allowing the cost of production and prices to fall.[49]

Ironically, the diffusion of new products tends to take off during the specific phase, when customer preferences stabilize and product variety is reduced. Most customers will not adopt new products until there are clear features and metrics on which those products can be evaluated, and these features and metrics tend not to be identified until after a dominant design emerges.

When firms innovate during the specific phase, they tend to focus on process innovation, leading the rate of product innovation to decline dramatically during this phase of industry evolution.[50] (Figure 2.7 shows how the timing of product and process innovation changes over the industry life cycle.)

The personal computer industry provides a good example of how the dominant type of innovation changes as an industry shifts from the fluid to the specific phase. In the early days of the personal computer business, many different companies competed on the basis of different product designs, with companies

FIGURE 2.7
The Timing of Product and Process Innovation

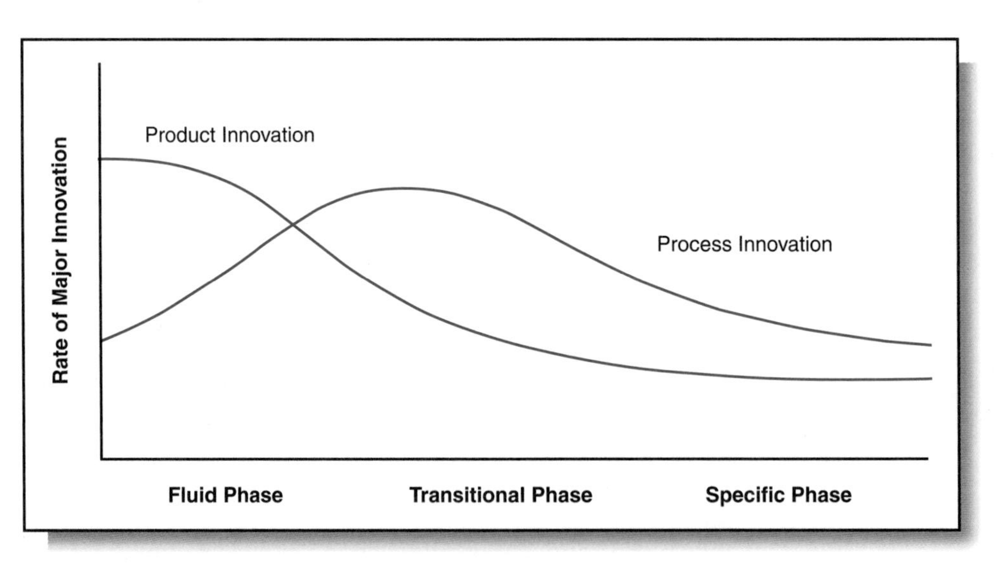

The Abernathy-Utterback model holds that the level of product innovation tends to decrease over time, while the level of process innovation tends to peak in the transitional phase of industry evolution.

Source: Utterback, J. 1994. *The Dynamics of Innovation*. Boston: Harvard Business School Press.

offering a variety of disk formats, microprocessors, and operating systems.[51] Then a dominant design emerged and personal computers became standardized. As a result, companies in the industry reduced spending on product innovation and focused their attention on efforts to reduce manufacturing, distribution, and marketing costs.

New Firm Performance

The Abernathy-Utterback model provides insight into when new and established firms are most successful in technology-intensive industries. In general, the fluid phase of an industry is the most favorable to new firms, while the specific phase is most favorable to incumbent firms. Several factors account for this pattern.

1. Before the establishment of a dominant design, new firms can operate without adopting the same product design as more experienced firms. However, once a dominant design has been adopted, new firms must adhere to the standard product design in the industry. Because established firms have greater experience working with this design than new firms do, new firms are disadvantaged when all firms have to use the same product designs.[52]
2. Before the establishment of a dominant design, firms operate on a small scale to minimize technical uncertainty.[53] Moreover, they tend to have nonhierarchical organization structures because these structures facilitate product design and development. However, once a dominant design has been established industry-wide, the production process becomes standardized, and competition shifts to production efficiency and economies of scale.[54] To support this need for efficiency, organization structures become more hierarchical and bureaucratic, which favor large, established businesses.
3. Before a dominant design emerges, learning curves are weak, allowing new firms to enter without operating at a severe competitive disadvantage. However, after a dominant design emerges, learning curves become more pronounced. Because efficient manufacturing, effective selling, and careful response to customer complaints all involve learning by doing, established firms, which have more operating experience than new firms, perform better.

Number of Firms in the Industry

The number of firms in an industry also tends to change over the different phases of industry evolution. In almost all cases, during the fluid phase, there is a high rate of new firm entry and a low rate of new firm exit, leading to a large increase in the number of firms in the industry.[55] For example, during the fluid phase of the automobile industry (from 1895 to 1923), the number of automobile manufacturers increased from 0 to 75.

Once the industry converges on a dominant design, a shakeout ensues as the industry consolidates around the small number of firms that are able to develop products that fit the dominant design.[56] As Table 2.2 shows, the magnitude of the shakeout is quite dramatic in most industries, with the drop in the number of firms from the peak level averaging 52 percent, and going as high as 87 percent.[57]

TABLE 2.2 The Severity of the Shakeout

This table shows the results of a study measuring the number of firms that exited several industries and the rate of decline to a stable number of firms; it shows that all industries experience a shakeout of firms, but the severity of the shakeout varies substantially across industries.

Product	Peak Number of Firms	Decline to Stable Level	Number of Firms Remaining
DDT	38	87%	5
Streptomycin	13	85%	2
Penicillin	30	80%	6
Automobile tires	275	77%	63
Radio transmitters	76	72%	7
Saccharin	39	72%	11
Electric blankets	17	65%	6
Freezers	61	62%	23
Phonograph records	49	61%	19
Windshield wipers	51	59%	21
Electric shavers	32	56%	14
Photocopy machines	43	53%	20
Adding machines	55	51%	27
Fluorescent lamps	34	41%	20
Piezo crystals	45	38%	28
Outboard motors	21	38%	15
Polariscopes	16	38%	16
Cryogenic tanks	84	35%	55
Jet engines	29	31%	20
Cathode ray tubes	39	28%	28
Zippers	49	18%	40
Shampoo	114	4%	109
Average		**52%**	

Source: Adapted from Klepper, S., and E. Graddy. 1990. The evolution of new industries and the determinants of market structure. *Rand Journal of Economics*, 21(1): 32.

Limitations to the Model

While the Abernathy-Utterback model is quite useful, it does not predict the patterns of industry evolution equally well in all manufacturing industries. It appears to work best in assembled products in which customer tastes are homogenous, and worst in nonassembled products, like rayon or glass, or in products that involve nonassembled components, like integrated circuits.[58] With these technologies, firms engage in less product innovation and more process innovation, altering how innovation and competition change as industries transition from the fluid phase to the specific phase of industry evolution. Perhaps more importantly, the model is not very effective in predicting patterns of industry evolution in service businesses, a point discussed in more detail next.

Key Points

- According to Abernathy and Utterback, technology evolves through periods of incremental change, interrupted by radical change, which affects both the nature of innovation and firm competition.

- During the fluid phase, markets are uncertain, customer preferences are ill-defined, diffusion of the technology is limited, firms are small, and most innovation takes the form of new products.
- The fluid phase ends with convergence on a dominant design, which ushers in the specific phase, during which time firms shift to process innovation, produce standardized products, and strive for production efficiencies based on scale economies and interchangeable parts.
- New firms perform best in the fluid phase of industry evolution because they are better than established firms at product innovation and worse than established firms at efficient production based on scale economies.
- During the specific phase, a shakeout typically occurs, with approximately half of the firms exiting the industry; those firms least able to fit their operations to the dominant design tend to exit.
- The Abernathy-Utterback model holds best in assembled manufacturing in which consumers have homogenous tastes; it holds less well in nonassembled manufacturing and manufactured products based on nonassembled components; and it does not hold in services.

GETTING DOWN TO BUSINESS

The History of Electric Vehicles[59]

While most people think that the use of electricity to power hybrid vehicles is a novel idea, electric-powered vehicles actually were once the most common type of automobiles in the United States. In fact, at the turn of the twentieth century, most experts predicted that the electric engine would become the dominant design in automobiles, largely because of their greater range and power. Of course, that didn't happen. So why did the internal combustion engine become the dominant design for automobiles, and the electric engine fall by the wayside for close to 100 years?

Technical and social factors, combined with increasing returns to adoption, were largely responsible for the convergence on internal combustion as the dominant design in automobile engines. Even though electric vehicles had the technical superiority of range and power, they had an important weakness. They weren't very good for touring, and the main purpose for automobiles at the turn of the twentieth century was fun. (Horse-powered wagons still did most of the actual work). The problem that electric vehicles faced for touring was that there was no way to recharge batteries. So if you got lost or were slow getting back home, you could run out of battery power and have no way to refuel. In contrast, with a gasoline-powered car, you could just pack some extra gas in a tank and go off with little worry that you would run out of fuel.

The initial advantage of gasoline-powered cars for touring was reinforced by increasing returns to adoption. Because more people selected gasoline powered vehicles, the fueling stations that developed as a complement to automobiles focused on gasoline, not electricity, and the mechanics that repaired vehicles for others concentrated their skills on the more popular type of vehicle. These two trends reinforced the initial tendency of people to choose gasoline-powered vehicles, and the internal combustion engine became the dominant design. Once this happened, most of the producers of electric vehicles closed down.

Once the dominant design emerged, the automobile industry changed dramatically. The companies founded to provide gasoline-powered vehicles did much better than the ones founded to offer electric-powered vehicles because it was hard for the makers of electric vehicles to transition to another core technology. As a result, many of the electric vehicle makers exited the industry, and the gasoline-powered vehicle makers grew into large, hierarchical organizations that mass produced cars and took advantage of economies of scale. New firms stopped entering the industry in large numbers, and the few large companies in the industry competed amongst themselves, largely on the basis of process innovation, efficiency, and marketing. And for many decades, no manufacturer offered a car powered by electricity.

Modifications to the Abernathy-Utterback Model

Although the Abernathy-Utterback model has proven to be quite useful in explaining the evolution of technology, researchers have identified four important modifications that you need to understand: Barras's reverse product cycle theory; Tushman's model of competence-destroying and competence-enhancing innovation; Christensen's model of value networks and disruptive innovation; and Henderson and Clark's model of architectural innovation.

Reverse Product Cycle Theory

After Abernathy and Utterback introduced their model, researchers wondered if it applied to service businesses as well as to the manufacturing businesses on which it had been developed and tested. When Dr. Richard Barras, a London-based researcher, looked at this question, he concluded that the answer was no. A different model operated in service industries, which he called the "reverse product cycle."

According to Barras, service industries typically adopt new technologies that are first developed in a goods industry.[60] For instance, the insurance industry adopted computer technology that was developed by the manufacturers of mainframe computers.

This adoption of technology from a goods industry leads to the first stage of the reverse product cycle. The adopted technology is employed to make existing services more efficient, thereby reducing costs.[61] These initial innovations are typically incremental improvements.[62] For example, the insurance industry first used mainframe computer technology to create computerized records of insurance policies.[63]

In the second stage of the reverse product cycle, the new technology is used to make the service more effective. The innovations that are introduced at this stage tend to focus on changing the processes used to serve customers, thereby enhancing quality.[64] For example, in the 1980s, insurance companies began to use information technology to offer better claims service, with greater hours and more alternatives.[65]

In the third stage of the reverse product cycle, the technology is used to create new services, making changes at this stage radical.[66] For example, in the 1990s, insurance companies used network computing to offer online quotation services to their customers.[67]

Comparing the reverse product cycle model with the Abernathy-Utterback model, you can see that the timing of radical and incremental change is different in manufacturing and service businesses: Manufacturing industries start with radical innovation and move to incremental innovation, while service industries start with incremental innovation and move to radical innovation.[68] As a result, the timing of advantage for new and established firms is different in manufacturing and services. In manufacturing, new firms are most advantaged when new technology is first developed and a dominant design has not yet emerged. By contrast, in services, established firms initially benefit the most from the new technology because it reduces the cost of delivering existing services. It is only later on, when the technology permits the creation of new services, and barriers to entry have been reduced by the decline in costs, that advantage shifts to new companies.[69]

Competence-Enhancing and Competence-Destroying Innovation

In the years after Abernathy and Utterback published their model, researchers noticed an important puzzle that the model could not explain. In some cases, incumbent firms had little trouble transitioning to new radical technologies. For instance, General Electric was able to shift from making medical diagnostic devices based on X-ray technology to those based on CAT scans to those based on MRIs, even though all three technologies are fundamentally different.[70]

This puzzle led Professor Michael Tushman of Harvard Business School and his students to modify the Abernathy-Utterback model to explain why incumbent firms were sometimes able to transition to radical new technologies and other times were not. Professor Tushman and his colleagues explained that radical new technology does not always undermine the capabilities of incumbent firms because sometimes it can be competence-enhancing. A radical technology is **competence-enhancing** if it makes use of existing knowledge, skills, abilities, structure, design, production processes, and plant and equipment; whereas, it is **competence-destroying** if it undermines the usefulness of these things. For example, the shift from vacuum tubes to integrated circuits in computers destroyed existing competencies because it rendered obsolete the expertise that firms had developed in vacuum tubes.

As Tushman explained, established firms are able to transition to a radical technology when that technology is competence-enhancing but fail to do so when it is competence-destroying. When radical technological change is competence-enhancing, incumbent firms invest in its development because they have an incentive to do so. Moreover, they have the capabilities to develop that technology successfully. For example, incumbent firms were able to transition to the use of turbofans in jet engines because this technology was competence-enhancing and drew heavily on the skills that jet engine manufacturers already had.[71]

However, incumbent firms cannot transition easily to competence-destroying radical technological change, often leaving that type of technology to firms outside the industry, or, in many cases, to start-up companies. Even when incumbent firms invent competence-destroying technologies, they rarely introduce them. Competence-destroying radical technological change tends to be financially costly for incumbent firms, giving them little incentive to invest in their development.[72] Moreover, their existing capabilities do not help them make use of those technologies, and, in many cases, actually hinder their ability to develop them by requiring incumbent firms to unlearn their old ways of doing things.[73] For instance, *many* incumbent firms in the printing industry were unable to transition to digital printing because this technology was competence-destroying and required the firms to develop skills in database systems, which they did not have, to manage the printing process.

Architectural Innovation

Rebecca Henderson, a faculty member at the Sloan School at MIT, and Kim Clark, a faculty member at the Harvard Business School, offered a different modification to the Abernathy-Utterback model from that offered by Tushman and his students. Henderson and Clark noticed that incumbent firms often failed to manage the transition from one technology to another, even when those technologies were not radical. Xerox, for example, had severe problems developing small copiers based on the underlying plain-paper copier technology that it had pioneered.

To explain why incumbent firms were tripped up by what seemed to be rather incremental technological changes, Henderson and Clark developed a more fine-grained taxonomy of innovation than the one on which the Abernathy-Utterback model is based. In this taxonomy, two additional types of innovation not mentioned by Abernathy and Utterback are present: **modular innovation** and **architectural innovation**. A modular innovation is one that changes the components from which the innovation is created, but not the linkages between those components. For example, the digital telephone is a modular innovation because it and the analog telephone have fundamentally different components, but the same linkages between them.

An architectural innovation is one that changes the linkages between the components, but leaves the components themselves intact.[74] For example, the portable DVD player is an architectural innovation. The design of its components is the same as with the desktop DVD player, but the architecture—how a built-in screen is linked to the rest of the device, how the device is powered, and so on—is different.[75] The addition of modular and architectural innovation to the Abernathy and Utterback taxonomy leads to the two-by-two matrix shown in Figure 2.8.

Henderson and Clark explain that incumbent firms often fail in the face of architectural innovation for three reasons. First, incumbent firms often lack the right external linkages to gather information about a new technology architecture emerging in an industry. Because organizations adopt external ties that are appropriate for their current activities, their external linkages facilitate the exchange of information with suppliers and customers about their current product architecture and discourage the exchange of information about other possible architectures.[76]

Second, incumbent firms often lack the capacity to recognize the value of information about architectural innovation that is presented to them. As we will see in

FIGURE 2.8
Henderson and Clark's Taxonomy of Innovations

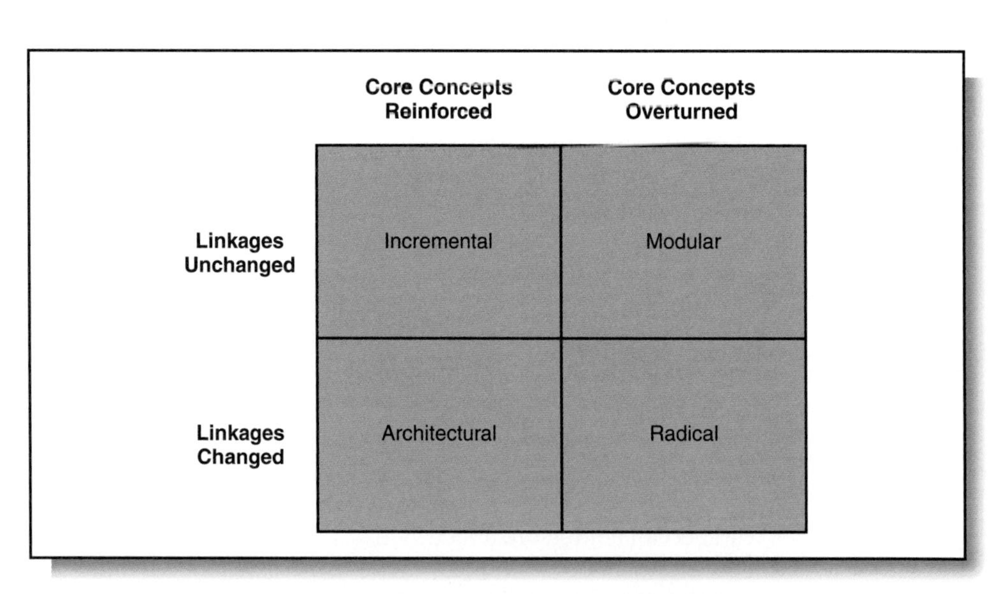

Henderson and Clark developed a taxonomy of innovation based on the degree of change to core concepts and the linkages between them; they found that established firms have many problems responding to architectural innovation.

Source: Adapted from Henderson, R., and K. Clark. 1990. Architectural innovation: the reconfiguration of existing product technologies and the failure of existing firms. *Administrative Science Quarterly*, 35(1): 9–30.

greater detail in Chapter 4, companies need to have existing, related knowledge to recognize the value of information about architectural innovation. However, once an industry has converged on a dominant design, firms accept the product architecture as given, and focus their attention on improving components or technological processes. As a result, they stop making investments in alternative product architectures, which keeps them from developing the background necessary to absorb external information about new architectures.[77]

Third, even when incumbent firms recognize the value of architectural innovation, they often have difficulty making use of it because adopting an architectural innovation typically requires a company to restructure. Organizations typically align their structures to their product architecture (see Figure 2.9). For instance, automobile manufacturers typically develop a steering gear organization composed of a steering column group, a rack-and-pinion group, a power steering group, and a tie rod group because this parallels the architecture of a steering column.[78] Thus, to change its product architecture, a company must also change its supporting organizational structure, changing who reports to whom and who talks to whom inside the organization. This, of course, is a difficult and costly activity.

Disruptive Technology and Value Networks

Clayton Christensen, a Harvard Business School professor, was puzzled by a different problem with the Abernathy-Utterback model than that which puzzled Tushman, and Henderson and Clark. Christensen noticed that the incumbent firms that were unable to adopt the radical new technologies were often the very firms that invented them. This pattern suggested that a lack of technological capability could not be the explanation of the firms' failure to transition to the new technology.

Christensen believed that the source of the problem was the willingness of a company's customers to adopt products and services based on the new technology, and

FIGURE 2.9 Organizational Structure Mirrors Product Architecture

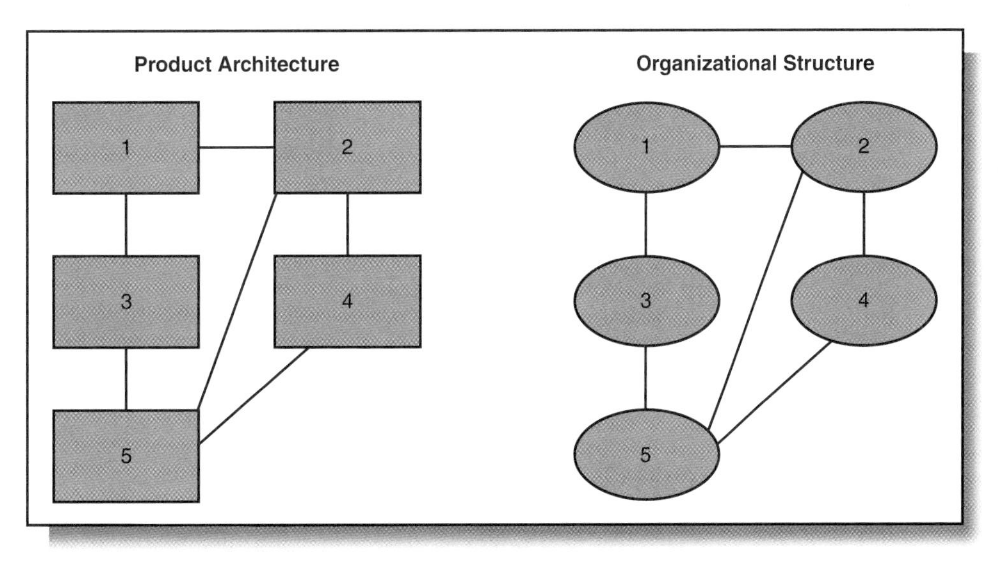

One reason why architectural innovation is difficult for incumbent firms is that organizational structure evolves to mirror product architecture, requiring the organizational structure to change for the product architecture to change.

Source: Adapted from Christensen, C. 1997. *The Innovator's Dilemma*. Boston: Harvard Business School Press.

not on the company's technical capabilities. He found that engineers at incumbent firms often developed radical, or what he called disruptive, technologies. When the companies produced new products based on the new technologies and showed them to their existing customers, the customers rejected them because they were typically inferior to existing technology on some dimension that was important to the customers.[79] For example, Teradyne's semiconductor testing-equipment company was unable to adopt complementary metal-oxide semiconductor (CMOS) chips and Windows-based software in their testing equipment because their primary customers were worried about the reliability of CMOS chips and software compatibility.[80]

As a result, the primary customers of incumbent firms would tell the marketing personnel to make incremental improvements products based on their old technology.[81] The companies would then focus on these incremental improvements to ensure that they satisfied their mainstream customers on whom their sales and profits depended.

Often the employees that developed the new technologies did not agree with the decision to focus on incremental improvements, and quit to start new companies to exploit the new technology.[82] These employees-turned-entrepreneurs had no more success serving the mainstream customers of their former employers than their former employers themselves, and had to turn to new markets to find customers interested in buying their products. The new companies often found these customers in segments of the market that had not been served well previously.[83]

For example, take the case of a radical new technology used in memory chips, MRAM, which combines magnetized material and silicon to allow data to be stored magnetically, rather than electronically. Magnetic storage allows data to be stored indefinitely, which is not possible with DRAM chips, and also allows data to be accessed more quickly. Because MRAM chips are much more expensive than DRAM chips, the producers of cell phones and personal computers, which are big users of DRAM chips, are not interested in the product. However, Freescale Semiconductor, the company selling MRAM chips, has found a market in the makers of home security devices, networking devices, and server systems, which are not adequately served by DRAM chips because they need permanent storage and rapid data access.[84]

In Christensen's model, the new firms provide products for these niche market segments, which are often the least attractive segments of the market, because they have no choice; their products lack the features necessary to satisfy the mainstream of the market. Incumbent firms often cede the niche markets to the new firms, and focus on the mainstream of the market, because the niches are not worth serving if the incumbent firms have to compete to serve them.[85]

Over time, the new firms invest in the development of the new technology. Because the potential for improvement of the new technology is higher than the potential for improvement of the old technology (see the S-curves in Figure 2.6 again if you do not realize why), the new firms are eventually able to produce products that are better than those of the incumbent firms on the dimensions that matter to mainstream customers. The new entrants then target the mainstream of the market, which is more attractive than the segments that the new entrants had initially chosen to serve. Because the new entrants now can provide mainstream customers with the product attributes that they want at a lower price, the new entrants now can take customers away from the incumbent firms.

Take, for example, the history of computer workstations. Initially, workstations built around the microprocessor did not have enough computing power to meet the needs of minicomputer users. As a result, companies like Sun Microsystems and Silicon Graphics went after different segments of the computing market—ones that were not terribly attractive. Because minicomputer makers like Wang and Digital

Equipment Corporation found those segments to be unattractive, they accommodated the workstation companies by ceding them the segments. Because the rate of increase in the computing power of the microprocessor was faster than that of the minicomputer, the workstation manufacturers were able to expand from their initial market niche and take the minicomputer market away from minicomputer manufacturers. The end result was the demise of companies such as Wang and Digital Equipment Corporation.[86]

A similar pattern can be observed in minimills, which used a technology for making steel from scrap to compete with traditional integrated mills. As Figure 2.10 shows, the minimills entered with lower quality steel than the integrated steel mills. At first the steel's surface quality was only good enough to take customers for rebar, a type of steel used to reinforce concrete. However, over time the minimills improved the surface quality of their steel to the point where it was better than that of the integrated steel mills and took the steel mills' primary automotive and appliance customers.[87]

Christensen's research provides important implications for your technology strategy, whether you are managing a new entrant or an incumbent firm. If your firm is a new entrant, his research shows that it will perform better if you target a new or underserved segment of the market with a disruptive innovation, rather than targeting an already satisfied segment of the market with that innovation. If your technology allows your customers to accomplish things that they cannot accomplish with existing products, you can get a foothold in the market and get the chance to improve your core technology to the point at which you can target segments that previously preferred the older technology.[88]

If your firm is an incumbent, Christensen's research shows that it will perform better if your core technology conforms to the features of the dominant design because deviation from the dominant design will make it harder for you to satisfy

FIGURE 2.10
Upmarket Migration in Steel

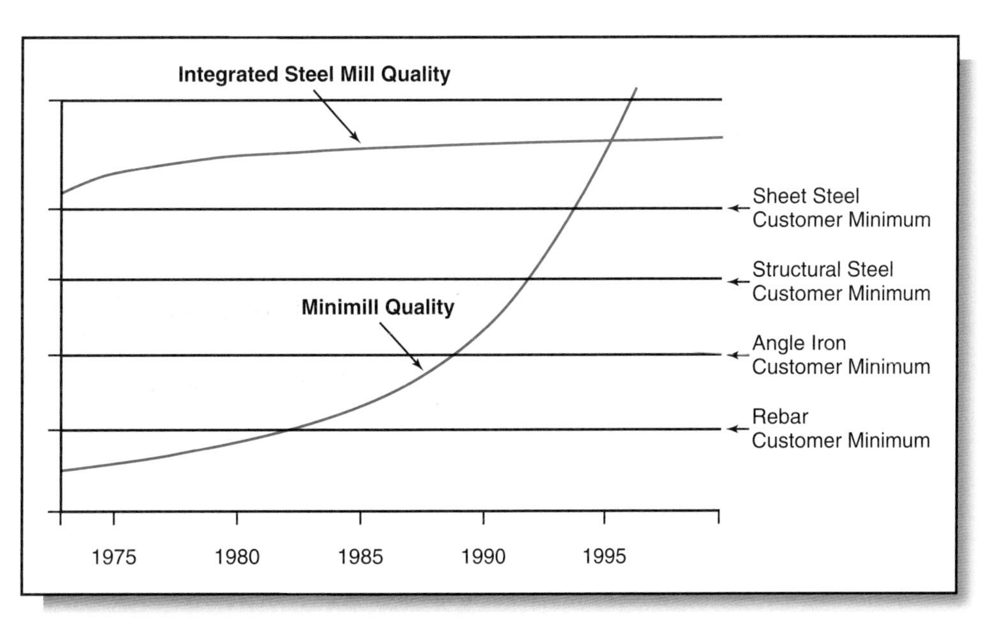

The minimills entered segments of the steel industry that had low demands for quality and expanded to segments with higher demands for quality over time by improving the quality of their steel.

Source: Adapted from Christensen, C. 1997. Continuous Casting Investments at USX Corporation, Harvard Business School Teaching Note, 5-697-066, April 24.

customers once that design has emerged.[89] Moreover, it indicates the perils of listening too closely to what your primary customers say.[90] Disruptive innovations typically appeal to segments of the market that are underserved, and not to the mainstream of the market.[91] So, if you serve the mainstream of the market, listening to your customers will lead you to ignore disruptive technology that you should adopt.

Furthermore, as the manager of an incumbent firm, you can avoid the outcomes that befell the integrated steel mills and the minicomputer manufacturers by developing a new company to exploit the disruptive technology, rather than ignoring it, or trying to develop the technology within the confines of your existing organization. Creating a new business allows you to avoid the budget-driven, planning-oriented, approaches of large, established businesses that often hinder the development of disruptive technologies.[92] Establishing a new business also allows you to exploit the disruptive technology on a small scale, so you can tinker with it, and get it right before you need to serve large numbers of customers, or spend large amounts of money.[93] And pursuing the new technology through an independent company allows you to focus on new customers rather than on existing ones,[94] which facilitates the identification of previously underserved market segments.[95]

For example, when Teradyne wanted to introduce a new semiconductor-testing equipment product using CMOS chips and Windows-based software, it established an independent company to pursue customers that it had not served previously. When Teradyne's mainstream customers rejected the new product, the company still had a set of customers to whom it could sell the new product and off of whom it could drive improvements to the technology.[96]

Key Points

- The reverse product cycle theory holds that innovation is different for services than for manufacturing; for services, process innovation precedes product innovation.
- Radical new technology can be competence-enhancing or competence-destroying; incumbent firms may invent competence-destroying technologies, but they rarely introduce them because they lack the capabilities to do so.
- Incumbent firms transition to competence-enhancing radical technology without much difficulty.
- Architectural innovations are innovations that change how components are linked together but leave the components themselves intact.
- Incumbent firms often fail in the face of architectural innovation because they lack the right external ties to learn about the new architecture and the absorptive capacity to recognize its value, and because they are unwilling or unable to restructure to make use of the new architecture.
- Incumbent firms sometimes have trouble exploiting a radical technology, not because they have insufficient technical capability, but because their customer base does not want them to transition to the new technology.
- Mainstream customers often restrict incumbent firms to making incremental improvements to existing technology, allowing new firms to exploit disruptive technology in small, undesirable market niches.
- New firms often enter undesirable niches because they have no choice, but improve their technology so that they can move upmarket to target the more desirable mainstream of the market.
- Incumbent firms need to establish or invest in new organizations if they want to exploit new disruptive technologies successfully.

Discussion Questions

1. What forces explain the timing of major technological transitions? Can you predict and plan for such transitions? If so, how and when should you try?
2. How should you determine what to map on the vertical axis of an S-curve?
3. When is a technology S-curve analysis helpful? How should it be used? What are the limitations of the technology S-curve as predictive tool? Is there something better? If so, what is it?
4. How are the approaches to managing innovation the same and different for new and incumbent firms? Are these patterns the same in all industries? If not, how do they vary across industries?
5. Are some types of innovation better and worse for entrepreneurs founding new firms? If so, why? If not, why not?
6. Why do incumbent firms often fail in the face of radical technological innovation? Why do these firms fail in the face of architectural innovation? Why don't incumbent firms always fail in the face of radical technological innovation?
7. What are the similarities and the differences in how Abernathy and Utterback, Barras, Henderson and Clark, Tushman and his students, and Christensen view technological change?
8. What is the right strategy for new and established firms in response to a disruptive technological change? Why?

Key Terms

Architectural Innovation: An innovation that changes the way that the components of a system link together.

Cannibalization: The sales of a new product that come at the expense of sales of a firm's existing products.

Competence-Destroying: A change that undermines the usefulness of existing knowledge, skills, abilities, structure, design, production processes, and plant and equipment.

Competence-Enhancing: A change that increases the usefulness of existing knowledge, skills, abilities, structure, design, production processes, and plant and equipment.

Dominant Design: A common way that all companies producing a given product will design their version.

Fluid Phase: The period of technology evolution before a dominant design emerges.

Incremental Innovation: An innovation that makes a small improvement on an existing knowledge base.

Modular Innovation: An innovation that changes the components of a system but not the way that components are linked together.

Paradigm: The framework within which technical problem solving occurs.

Process Innovation: An innovation that enhances the way a good or service is made.

Product Innovation: An innovation that leads to the creation of new goods and services sold to customers.

Radical Innovation: An innovation that draws on a fundamentally new knowledge base.

Specific Phase: The period of industry evolution after a dominant design emerges.

Technology S-Curve: A graphical representation of the development of a new technology, showing the relationship between some measure of performance and some measure of effort.

Technology Trajectory: The path of improvement of a technology on some performance dimension.

Putting Ideas into Practice

1. **Technology S-Curves** Select a technology with which you are familiar. Identify measures of performance of the technology and of the cumulative amount of effort to develop the technology over time (e.g., R&D expense, person-hours devoted to the technology, etc.). Putting the performance measure on the vertical axis and the effort measure on the horizontal axis, plot the relevant S-curve for that technology. Then identify the point on the S-curve at which the technology is currently. Once you are done graphing the technology S-curve, use the figure you created to answer the following questions: Is the technology/industry subject to "natural" technological limits? Why or why not? Has it experienced disruptions? Is it likely to do so soon? How will customer characteristics evolve as companies move along the technology S-curve? During a disruption, how will the customers of the old and new technology differ?

2. **Dominant Designs** For an industry/technology with which you are familiar, identify a dominant design. Then examine the industry/technology before and after the dominant design emerged. Document the effect of the dominant design on (1) changes in the number of firms in an industry, (2) the relative investment by firms in product and process innovation, (3) the nature of competition between firms, (4) the average profit margin of firms, and (5) the typical organizational structure.
3. **Developing Fuel Cell Vehicles** General Motors has built a fuel cell vehicle, which it plans to introduce commercially in 2011. (Fuel cell engines run on hydrogen. The only exhaust they produce is water. The facilities used to make these engines look like semiconductor factories. The key component is a platinum-carbon liquid used for the membranes that has to be mixed in a dust-free clean room.[97]) Identify the advantages and disadvantages that General Motors has at developing this technological innovation. Then take the perspective of an entrepreneur who has founded a new auto company; identify the advantages and disadvantages that the start-up has at developing the same innovation. Explain how the advantages and disadvantages of the entrepreneur and General Motors are the same and different.

NOTES

1. Adapted from Utterback, J. 1995. Developing technologies: The Eastman Kodak story. *McKinsey Quarterly*, 1: 131–144.
2. Narayanan, V. 2001. *Managing Technology and Innovation for Competitive Advantage*. Upper Saddle River, NJ: Prentice Hall.
3. Dosi, G. 1988. Sources, procedures, and microeconomic effects of innovation. *Journal of Economic Literature*, 26: 1120–1171.
4. Markides, C., and P. Geroski. 2005. *Fast Second: How Smart Companies Bypass Radical Innovation to Enter and Dominate New Markets*. San Francisco: Jossey-Bass.
5. Mokyr, J. 1990. *The Lever of Riches*. New York: Oxford University Press.
6. Schilling, M. 2005. *Strategic Management of Technological Innovation*. New York: McGraw-Hill.
7. Dosi, Sources, procedures, and microeconomic effects of innovation.
8. Cooke, R. 2001. *Dr. Folkman's War: Angiogenesis and the Struggle to Defeat Cancer*. New York: Random House.
9. Day, D. 2004. Segway human transporter: More than a cool invention? *Richard Ivey School of Business Case*, Number 905M45.
10. 1998. *Marketing News*, March 30.
11. Ettlie, J. 2000. *Managing Technological Innovation*. New York: John Wiley.
12. Miller, W. 2006. Innovation rules! *Research Technology Management*, 49(2): 8–14; 1998. *Marketing News*, March 30.
13. Chandy, R., and G. Tellis. 2000. The incumbent's curse? Incumbency, size, and radical product innovation. *Journal of Marketing*, 64: 1–17.
14. Ibid.
15. Bower, J., and C. Christensen. 1996. Disruptive technologies: Catching the wave. *Journal of Product Innovation Management*, 13(1): 75–76.
16. Foster, R. 1986. *Innovation: The Attacker's Advantage*. New York: Summit Books.
17. Ibid.
18. Brown, K., and A. Latour. 2004. AT&T will offer Internet phone calls in selected markets. *Wall Street Journal*, March 30: B1, B2.
19. Chandy and Tellis, The incumbent's curse?
20. Clark, D. 2005. H-P team claims a milestone toward successor to the transistor. *Wall Street Journal*, February 1: B5.
21. Clark, D. 2007. H-P touts advance in chip technology. *Wall Street Journal*, January 16: B5.
22. Brown, R. 1992. Managing the "S" curves of innovation. *Journal of Consumer Marketing*, 9(1): 61–72.
23. Mitchell, W. 1989. Whether and when? Probability and timing of incumbents' entry into emerging industrial subfields. *Administrative Science Quarterly*, 34(2): 208–230.
24. Rhoads, C. 2005. AT&T inventions fueled tech boom and its own fall. *Wall Street Journal*, February 2: A1–A12.
25. Zimmerman, R., and J. Bandler. 2004. Kodak sues Sony in patent dispute. *Wall Street Journal*, March 10: B4.
26. Allen, K. 2003. *Bringing New Technology to Market*. Upper Saddle River, NJ: Prentice Hall.
27. Clark, D. 2006. Freescale to sell memory chips using breakthrough technology. *Wall Street Journal*, July 10: B8.
28. Belson, K. 2005. Cable's new pitch: Reach out and touch someone. *New York Times*, May 8: 5.
29. Afuah, A. 2003. *Innovation Management*. New York: Oxford University Press.
30. Conner, K. 1988. Strategies for product cannibalism. *Strategic Management Journal*, 9: 9–26.
31. Chandy, R., and G. Tellis. 1998. Organizing for radical product innovation: The overlooked role of willingness to cannibalize. *Journal of Marketing Research*, 35: 474–487.

32. Chandy and Tellis, The incumbent's curse?
33. McGrath, R., I. MacMillan, and M. Tushman. 1992. The role of executive team actions in shaping dominant designs: Towards the strategic shaping of technological progress. *Strategic Management Journal*, 13: 137–161.
34. Foster, *Innovation: The Attacker's Advantage.*
35. Utterback, J. 1994. *Mastering the Dynamics of Innovation.* Cambridge: Harvard Business School Press.
36. Allen, *Bringing New Technology to Market.*
37. Tam, P. 2004. As cameras go digital, a race to shape habits of consumers. *Wall Street Journal*, November 19: A1, A10.
38. Foster, *Innovation: The Attacker's Advantage.*
39. Christensen, C. 1999. *Innovation and the General Manager*. New York: McGraw-Hill.
40. Christensen, C. 1998. Valurec's venture into metal injection molding. *Harvard Business School Teaching Note*, Number 698–002.
41. Day, G., and P. Schoemaker. 2000. Avoiding the pitfalls of emerging technologies. In G. Day and P. Schoemaker (eds.), *Wharton on Managing Emerging Technologies*. New York: John Wiley.
42. Christensen, *Innovation and the General Manager.*
43. Brown, K., and A. Latour. 2004. AT&T will offer Internet phone calls in selected markets. *Wall Street Journal*, March 30: B1, B2.
44. Afuah, *Innovation Management.*
45. Utterback, *Mastering the Dynamics of Innovation.*
46. Afuah, *Innovation Management.*
47. Ibid.
48. Teece, D. 1986. Profiting from technological innovation: Implications for integration, collaboration, licensing and public policy. *Research Policy*, 15: 285–305.
49. Klepper, S., and E. Graddy. 1990. The evolution of new industries and the determinants of market structure. *RAND Journal of Economics*, 21(1): 27–44.
50. Gort, M., and S. Klepper. 1982. Time paths to the diffusion of product innovations. *Economic Journal*, 92(367): 630–653.
51. Freiberger, P., and M. Swaine. 1984. *Fire in the Valley: The Making of the Personal Computer*. Berkeley, CA: Osborne/McGraw-Hill.
52. Utterback, *Mastering the Dynamics of Innovation.*
53. Mueller, D., and D. Tilton. 1969. Research and development costs as a barrier to entry. *The Canadian Journal of Economics*, 2(4): 570–579.
54. Utterback, *Mastering the Dynamics of Innovation.*
55. Gort, M., and S. Klepper. 1982. Time paths to the diffusion of product innovations. *Economic Journal*, 92(367): 630–653.
56. Ibid.
57. Klepper, S., and E. Graddy. 1990. The evolution of new industries and the determinants of market structure. *Rand Journal of Economics*, 21(1): 32.
58. Teece, D. 1986. Profiting from technological innovation: Implications for integration, collaboration, licensing and public policy. *Research Policy*, 15: 285–305.
59. Adapted from Kirsch, D. 2000. *Electric Vehicles and the Burden of History.* New Brunswick, NJ: Rutgers University Press.
60. Damanpour, F., and S. Gopalakrishnan. 2001. The dynamics of adoption of product and process innovations in organizations. *Journal of Management Studies*, 38(1): 46–65.
61. Barras, R. 1986. Towards a theory of innovation in services. *Research Policy*, 15: 161–173.
62. Damanpour and Gopalakrishnan, The dynamics of adoption of product and process innovations in organizations.
63. Barras, Towards a theory of innovation in services.
64. Damanpour and Gopalakrishnan, The dynamics of adoption of product and process innovations in organizations.
65. Barras, Towards a theory of innovation in services.
66. Damanpour and Gopalakrishnan, The dynamics of adoption of product and process innovations in organizations.
67. Barras, Towards a theory of innovation in services.
68. Ibid.
69. Barras, R. 1990. Interactive innovation in financial and business services: The vanguard of the service revolution. *Research Policy*, 19: 215–237.
70. Afuah, *Innovation Management.*
71. Schilling, *Strategic Management of Technological Innovation.*
72. Lynn, G., J. Morone, and A. Paulson. 1996. Marketing and discontinuous innovation: The probe and learn process. *California Management Review*, 38(3): 8–37.
73. Afuah, *Innovation Management.*
74. Tushman, M., and W. Smith. 2002. Organizational technology. In J. Baum (ed.), *The Blackwell Companion to Organizations*. Cambridge, UK: Blackwell: 286–414.
75. Henderson, R., and K. Clark. 1990. Architectural innovation: The reconfiguration of existing product technologies and the failure of established firms. *Administrative Science Quarterly*, 35: 9–30.
76. Ibid.
77. Ibid.
78. Christensen, C. 1997. *The Innovator's Dilemma.* Boston: Harvard Business School Press.
79. Ibid.
80. Bower, J. 2005. Teradyne: The Aurora project. *Harvard Business School Case*, Number 9–397–114.

81. Christensen, *Innovation and the General Manager*.
82. Christensen, *The Innovator's Dilemma*.
83. Gilbert, C. 2003. The disruption opportunity. *Sloan Management Review*, Summer: 27–32.
84. Clark, Freescale to sell memory chips using breakthrough technology.
85. Christensen, *The Innovator's Dilemma*.
86. Christensen, C. 1999. Hewlett Packard's Merced Division. *Harvard Business School Case*, Number 9–699–011.
87. Christensen, *Innovation and the General Manager*.
88. Christensen, C., F. Suarez, and J. Utterback. 1998. Strategies for survival in fast-changing industries. *Management Science*, 44(12): S207–S220.
89. Ibid.
90. Paap, J., and R. Katz. 2004. Anticipating disruptive innovation. *Research Technology Management*, 47(5): 13–22.
91. Christensen, C., M. Johnson, and D. Rigby. 2002. How to identify and build disruptive new businesses. *Sloan Management Review*, Spring: 22–31.
92. Ibid.
93. Gilbert, The disruption opportunity.
94. Chandy and Tellis, Organizing for radical product innovation: The overlooked role of willingness to cannibalize.
95. Christensen, *The Innovator's Dilemma*.
96. Bower, J. 2005. Teradyne: The Aurora project. *Harvard Business School* Case, Number 9–397–114.
97. Boudette, N. 2006. GM hopes engine of the future sells cars now. *Wall Street Journal*, November 29: B1, B2.

Chapter 3

Technology Adoption and Diffusion

Learning Objectives

After reading this chapter, you should be able to:

1. Identify the different groups of adopters of new technology products and services, and understand the key factors influencing their willingness to adopt.
2. Explain why adopters of new technology products and services tend to be normally distributed, and why the proportion of the market adopting a new technology product or service at a point in time is typically S-shaped.
3. Define *crossing the chasm*, and explain why it is important for firms.
4. Figure out how to cross the chasm.
5. Explain why forecasting demand is difficult, but important.
6. Define *technology diffusion*, describe the typical diffusion pattern, and identify the factors that influence technology diffusion.

7. Understand how information and product diffusion models predict the rate and functional form of diffusion.
8. Use the Bass model to estimate the rate of diffusion of a new technology product or service.
9. Use the Delphi Technique to estimate the rate of diffusion of a new technology product or service.
10. Explain why complementary technology has a profound effect on technology diffusion.
11. Define *technology substitution*, and explain how substitution affects the rate of adoption of new technology products and services.
12. Explain why it is important to estimate how long it will take for technology substitution to occur, and how technology substitution can affect the strategies of new entrants and incumbent firms.

Technology Adoption and Diffusion: A Vignette

The concept of technology diffusion is well illustrated by the adoption of digital cameras. As Figure 3.1 shows, the sales of consumer digital cameras have grown dramatically since Apple Computer brought out the first one, the QuickTake 100, in 1994. The initial diffusion of the technology was slow from 1994 until 1996, when digital camera sales increased from zero to 400,000 units. The pace increased a little from 1996 to 1998, as sales grew to 1.1 million units per year. In 1999, the pace of diffusion accelerated, with sales almost doubling to 2.0 million units. Sales more than doubled in 2000, reaching 5.0 million units, and then almost doubled again over the next two years, to reach 9.3 million units by 2002. As a result, by 2003, digital camera sales in the United States had surpassed chemical film camera sales, with 12.5 million digital cameras being sold versus 12.1 million traditional film cameras.[1] By 2006, sales of digital cameras had grown to 30 million per year.

Digital cameras also illustrate the concept of technology substitution. As sales of digital cameras have grown, sales of 35 millimeter traditional film cameras have fallen, dropping below 8 million in 2004, a decline of 20 percent in just two years.[2] Even more dramatic has been the decline in sales of one of the most popular chemical film cameras, Kodak's Advanced Photo System, which dropped from 10.5 million units annually in 2000 to less than 2 million in 2003.[3]

The substitution of digital camera technology for traditional camera technology also has led to a decline in the sales of photographic film. Sales of chemical film peaked at a high of 1.05 billion rolls in 2000 and have declined since. In 2005 alone, the major producer of photographic film in the United States, Kodak, faced a decline in traditional film sales of more than 25 percent.[4]

FIGURE 3.1
U.S. Digital
Camera Sales

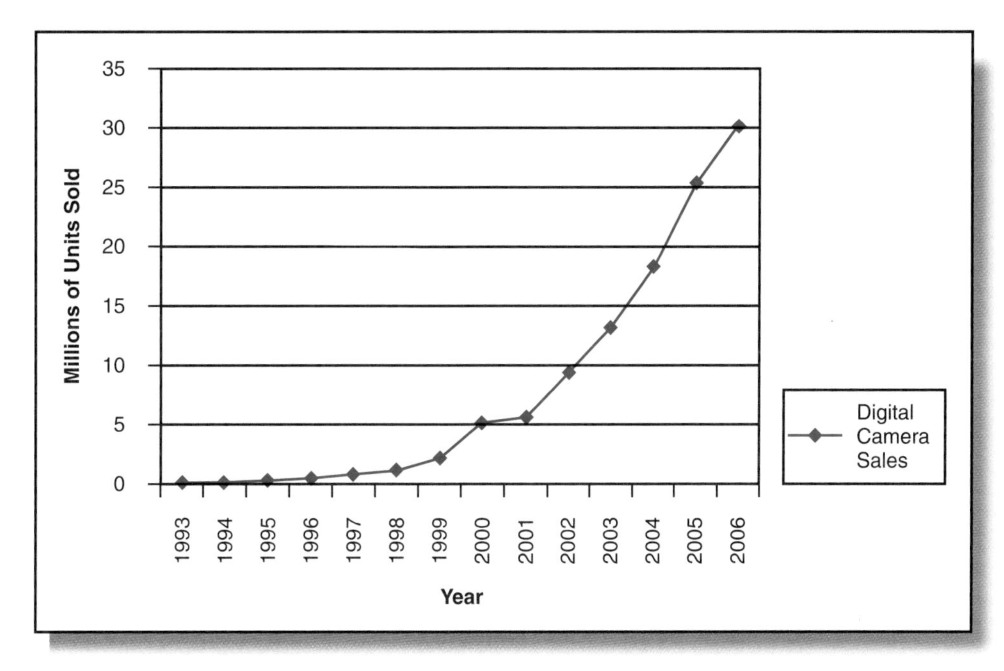

Digital camera shipments follow the typical accelerated pattern of technology diffusion over time.
Source: Adapted from http://www.infotechtrends.com and http://www.dataquest.com/press_gartner/quickstats/cameras.html.

Introduction

To formulate an effective technology strategy, you need to understand the patterns of technology diffusion and customer adoption. You need to figure out the likely distribution of adopters of your products and services to target your customer base effectively. You also need to follow particular strategies to transition from innovators to the majority of the market. Finally, you need to accurately forecast the growth of demand for your products and services.

This chapter discusses technology adoption and diffusion. The first section describes the distribution of adopters of new technology products and services, and explains how that distribution affects technology strategy. The second section explains how to transition from the initial adopters of your product to the majority of the market. The third section discusses the dynamics of market growth, focusing on the factors that influence technology diffusion and substitution, and the tools that you can use to estimate the size of the market for, and the pace of diffusion of, your new products.

Distribution of Adopters

While you might be excited when your business sells a new product or service to its first customers, meeting the needs of just a few customers is usually not enough for your company to succeed. Rather, you probably need to achieve broad **adoption** (the decision of customers to purchase a new product or service) of your new products or services by the mainstream of the market. If you don't sell to the mainstream, you probably won't be able to achieve sufficient sales volume to take advantage of economies of scale in production and distribution, making your cost structure uncompetitive. And you will likely get sidetracked into making custom products for your initial customers, which will keep you from developing standardized versions of your product. Moreover, your failure to transition to the mainstream market will be particularly problematic if your business is new and is financed with venture capital; the cost of private capital virtually necessitates the reduction in costs and increase in margins that come from selling to the mainstream of the market.

So, how can you achieve widespread adoption of your new product or service? Basically, you need to understand why and when different groups of customers adopt new products, and you need to make sure that your products possess the characteristics necessary to gain acceptance from each segment of the market. The first step in this process lies in understanding patterns of customer adoption of new technology products and services.

These patterns are grounded in some basic mathematics. Although there is some variation in the distribution of adopters of new technology products and services,[5] the single most common pattern is a normal distribution, as shown in Figure 3.2. Why? Because most patterns of human behavior are **normally distributed**, with a small portion of people doing things early, a small portion doing things late, and most people doing things in the middle.[6]

FIGURE 3.2
The Normal Distribution of Adopters

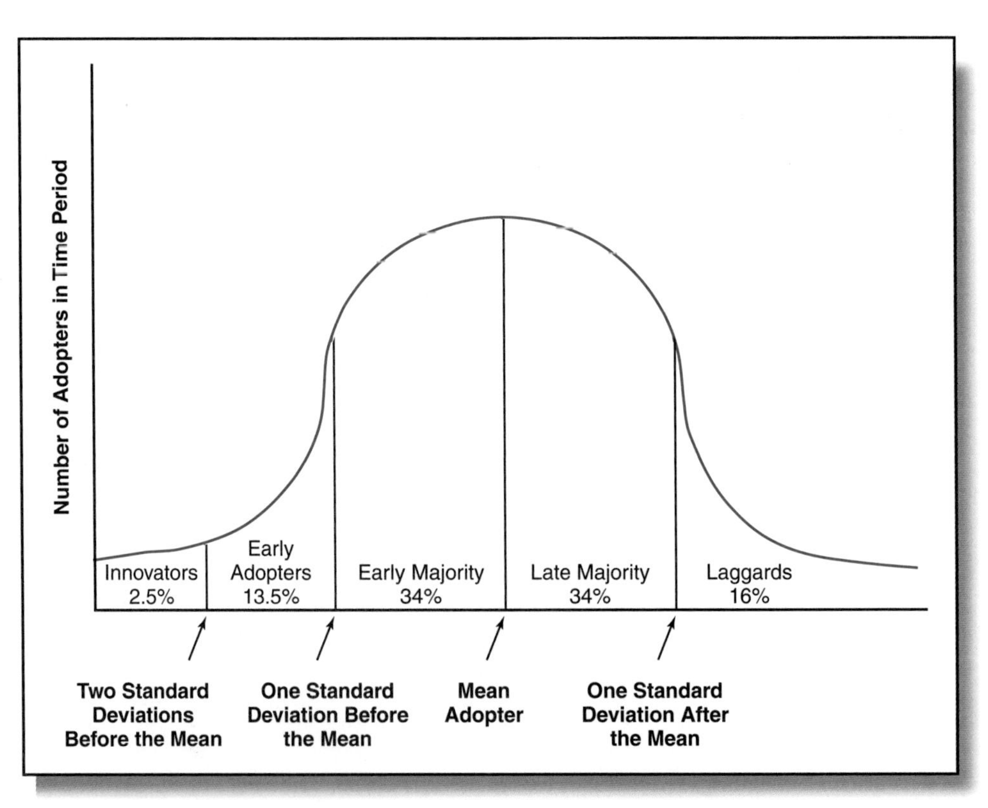

The adopters of most new technology products are normally distributed; there are few innovators and early adopters, but also few laggards, with most people adopting in between.

Source: Adapted from: Rogers, E. 1983. *Diffusion of Innovations*. New York: Free Press.

For example, the distribution of adopters of CT scanners was normally distributed. Just less than 4 percent of hospitals saw the need for this technology right after it was introduced and became innovators. Another 14 percent were early adopters, purchasing on the heels of the innovators, while just over two-thirds were in either the early or late majority, and just under one-fifth were laggards who adopted the technology after all of the other hospitals.[7]

Groups of Adopters

The different groups shown in the different sections of the adoption bell curve in Figure 3.2 adopt new technology products at different times, and for different reasons. When a new technology product is first introduced, a small number of customers, called **innovators**, adopt it immediately. This segment of the market is very eager for new technology, "needing" new devices as soon as they are made available. They are intrigued by technology and like to explore it.[8]

While most potential customers require a lot of information about the potential value of a new product or service before adopting it, this portion of the market is savvy enough to evaluate new technology products and services with little information, using intuition to make decisions about a new product's value. They often pursue products before their formal marketing, sometimes even buying prototypes and beta versions. Take DVD players as an example. You probably know someone who bought one when they were first introduced because that person just had to have the technology to play with.

Innovators are not very price sensitive, and are willing to pay a lot of money for new technology products and services.[9] For instance, some people in California have adopted electric cars even though these cars cost much more to lease than cars with internal combustion engines, and there are few places to recharge the vehicle batteries.[10]

A second group of customers, called the **early adopters**, follow the innovators. This group of customers is larger than the initial group of innovators. Why? Because more people are willing to purchase new products and services as information about them becomes known, and uncertainty about their value declines. And the adoption of new technology products and services by the innovators generates information about the value of the new products to potential early adopters, which helps them to make the adoption decision. The early adopter segment of the market does not have as much knowledge of the potential value of new products as the innovator segment, but it is technically savvy, and responds to marketing with a great deal of technical content.

The early adopters are an important group because the effort to sell to them often provides evidence that can be used to sell products to the majority of the market. Take, for example, the experience of a medical device company named Imalux that has developed a new technology called optical coherence tomography, which can be used to capture high-resolution images of tissue that cannot be seen with alternative technologies. Imalux is targeting urologists because they already use fiber optic devices to see inside the body. The company hopes to use information from the opinion leaders in that segment to generate evidence of effectiveness, which can be used to formulate the value proposition for the majority of the market.[11]

A third segment of the market is the **early majority**, which tends to adopt a new technology product slightly before the average for the market. This segment sees value in technology but is driven by practical considerations.[12] It views the early adopters as an important reference group and looks at their adoption decisions as indicators of the value of the new technology.[13] For instance, the people who waited until cell phones had shrunk to handheld sizes and were reasonably priced were the early majority of the cell phone market.

To convince the early majority to adopt your product, you need to change your product's features and the way that you sell it. The early majority wants to see a whole product solution and a value proposition before buying.[14] So you will probably need to add features to your product and revamp your marketing message to sell to this group. Your new advertising message should not emphasize technical content, but should focus on the completeness and ease of use of your product.

You also should try to achieve a dominant position in the market. The early majority prefers to purchase from the market leader because the leader's products tend to be the most reliable, and the most likely to provide support in the aftermarket.[15]

A fourth segment is the **late majority**, and is roughly the same size as the early majority. This segment is often more skeptical about the value of technology than the early majority, and tends to adopt it only after strong evidence shows that the value of adoption is greater than the cost of adoption, and that other customers have adopted the technology successfully.[16] In general, adoption by the late majority is driven by considerations other than the value of the technology itself.[17] For instance, the late majority adopted CDs in place of cassettes and records when it became difficult to buy music in the older forms and the availability of music became a factor in the type of devices used to play music.

The late majority is price sensitive and has strong needs for service and support. To sell to this segment successfully, not only must you offer a whole product, but also your whole product must not be expensive, and must be reliable and easy to use. The most important advertising message for this segment is reliability and low cost.[18]

A final market segment, called the **laggards**, follows the late majority. This segment will not adopt a new technology product until it is well established, preferring to avoid adoption of these products as long as possible, and often adopting only when the alternatives to adoption are no longer available.[19] This group is smaller than the majority of the market, and is generally equal in size to the innovators and early adopters combined.[20]

I bet you know a laggard in some technology area. Perhaps your parents have never caught on to CDs and MP3 players and still have cassettes in their basement. In the business of music-playing devices, those people still playing cassettes on their stereos would be considered laggards.

S-Curves of Adopters

Understanding that the typical pattern of adoption of new products and services is normally distributed is important for you, as a technology strategist, for several reasons. First, it points out that different groups of customers adopt new products at different points in time for different reasons. Innovators typically purchase new products because they need to explore the uses of new technology, while laggards usually adopt them because they have no alternative. The motivations of the different market segments indicate that you need to design variations of your new product to meet the needs of these different segments.[21]

Second, this distribution provides important information about the right promotional strategy for your new product. For the innovator segment, advertising may not be necessary, and for the early adopters and the early majority, promotion based on references from satisfied customers might be very effective. However, for the late majority of the market, you need to shift to more informational and price-based advertising.[22]

Third, the normal distribution of adopters indicates the appropriate pricing strategy for your new product. For the innovators and early adopters, who are not very price-sensitive, a high price might be the way to go. For the majority of the market, you might want to reduce your price to spur adoption. For the laggards, you will need to cut your price because this group is quite price-sensitive.

Take, for example, the change in online shopping between the late 1990s and today. In the late 1990s, Internet shoppers tended to be young, financially well-off males who were technology savvy and price insensitive. They were quite willing to purchase products from independent online retailers like Amazon.com. Today, as many more people make online purchases, the typical customer is female, less well-off, and less technology savvy. For these customers, lower prices and brand name reputations attract them, which has allowed companies like the Gap and Wal-Mart to catch up to the online retail pioneers.[23]

Fourth, understanding the distribution of adopters provides you with information that will help you to estimate demand growth over time. As Figure 3.3 indicates, the normal distribution of adopters translates into an S-shaped pattern of demand growth over time because the bell curve shows the rate of adoption while the S-shaped curve shows the cumulative adoption.[24] (For those of you used to thinking in calculus terms, the bell curve is the derivative of the S-curve, and the S-curve is the integral of the bell curve.)

FIGURE 3.3
The Adoption S-Curve

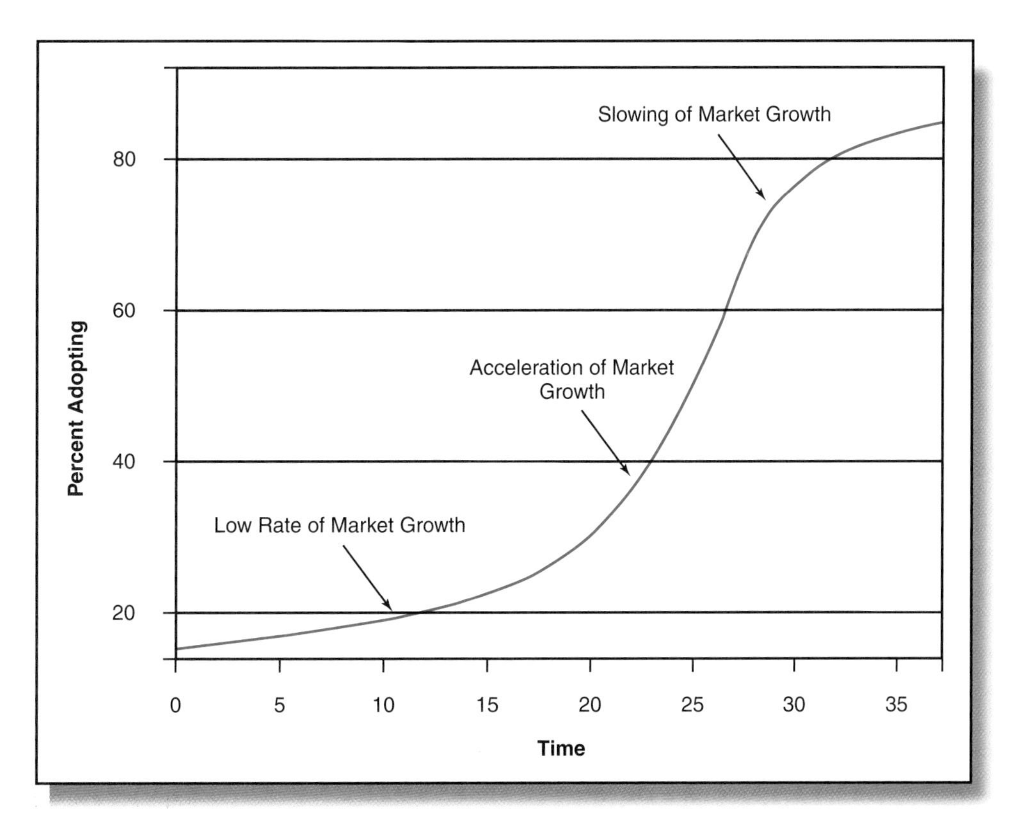

The percent of the market that has adopted a new technology product at a given point in time tends to follow an S-shaped pattern, displaying slow growth initially, then accelerating, and then decelerating as the market becomes saturated.

The S-shaped pattern indicates that there is an acceleration point at which a market **takes off**. Consequently, you need to be prepared for this acceleration by ensuring that you have adequate supply of raw materials and labor, and sufficient manufacturing and distribution infrastructure, to meet demand. Moreover, even when adoption patterns do not follow a normal distribution, you can use deviations from that baseline to forecast the growth in demand for your new technology product over time.

Fifth, the normal distribution of adopters tells you something about the financial attractiveness of a market at different points in time. In the first stage of the product life cycle, when customers are innovators and early adopters, revenues are low and costs are high, making profit margins relatively small. In the second stage of the life cycle, when you transition to serving the majority of the market, revenues rise faster than costs, and profit margins increase.[25] In the third stage of the life cycle, when customers are laggards, revenues fall and costs increase, leading profit margins to decline.[26]

Key Points

- The most common pattern of adoption of new products and services is a normal distribution.
- Adopters can be divided into five segments: innovators, early adopters, early majority, late majority, and laggards; each group adopts products at different points in time, and for different reasons.
- The normal distribution of adopters leads to an S-shaped pattern of cumulative adoption over time.
- Markets for new technology products "take off" when they transition from early adopters to the early majority of the market.
- The normal distribution of adopters is important to technology strategy because it highlights the different motivations of each group of adopters; illustrates the right promotional strategy to reach each of them; provides information about how to price products; helps to estimate demand growth; and provides information about the financial attractiveness of the market over time.
- The adoption S-shaped curve is different from the technology S-curve described in Chapter 2. The adoption S-curve measures the percent of the market adopting a new product as a function of time. The technology S-curve described in Chapter 2 measures the level of performance of a technology as a function of the amount of effort put into its development.

Crossing the Chasm

The discussion of the distribution of adopters in the previous section indicates the importance of transitioning to the majority of the market, or what many people call the "mass market." For most products, markets are too small for you to be successful by just selling to innovators and early adopters. To achieve an adequate return on your investment in product development, you need to sell to the mass market. If you don't, you will end up engaged in costly customization for the innovator and early adopter segments. Moreover, the faster you move to the mass market, the sooner you can sell in larger volume, take advantage of scale economies, and make greater profits. So how do you transition to the mass market?

Identifying the Take-Off Stage

The first step in reaching the mass market is identifying that a new product has reached the take-off stage. You can figure this out in a couple of different ways. First, you can look at the past demand for the product. Is the rate of change in demand accelerating? If it is, that signals that a product is reaching the take-off stage. Second, what do new customers for the product look like and how do they differ from existing customers? When the market is about to take off, the customer base for a product begins to shift away from the innovators and early adopters. If new customers are focused on the value proposition from purchasing your product or service or care a lot about the ancillary attributes of your product package, then your product or service is probably reaching the take-off stage.

You don't have to be completely reactive about your product reaching the take-off stage; there are things you can do to push your product to take off. First, any actions that you take—developing a marketing message based on the value proposition, getting early adopter testimonials, etc.—that make the mainstream market more interested in the product will increase the odds that the product will take off. Second, changes that you make to improve your product and lower its cost will make the product more attractive to a wider range of the market, and more likely to take off.

How to Cross the Chasm

Marketing consultant Geoffrey Moore termed this transition **crossing the chasm** (see Figure 3.4) because of the differences in the adoption decisions of early adopters and the majority of the market.[27] As the previous section of the chapter indicated, to sell your new product to the majority of the market, you need to show how it provides value to customers, which you might not need to do with the innovators or early adopters. For instance, selling your CAD/CAM software to the majority of the market could require you to show how your customers will speed up their design process and save money by adopting your product, rather than just demonstrating how cool the software is.

Moreover, you need to develop a complete solution to your customers' problems—things like manuals, customer service, set-up assistance, and so on—because the majority of the market seeks solutions to its problems, rather than just pieces of technology. If you don't provide a complete solution, you are likely to fail to cross the chasm. For instance, Nokia experienced problems crossing the chasm when it introduced the N800 Internet Tablet, a small computer. The company was relying on open source software developers to provide a calendar application, a way to synch data to a desktop computer, and the software necessary to record pictures, so the company didn't develop these features. But mainstream customers were expecting Nokia to provide a complete product with all of these features already present, and so didn't adopt the product.[28]

Furthermore, you will need to pursue a **vertical marketing strategy** (focusing on customers in a single industry) rather than a **horizontal marketing strategy** (serving customers in multiple industries at the same time). The majority of the market seeks references from adopters they know, and this is not likely if those customers are in different industries. It is also less expensive and easier to create these references if you focus on a single industry at a time because of the links that exist between customers in the same industry.[29] Lastly, if you adopt a horizontal marketing strategy, your resources will be spread too thinly for you to develop a complete solution to your customers' problems and demonstrate the value of that solution to them.[30]

FIGURE 3.4 Crossing the Chasm

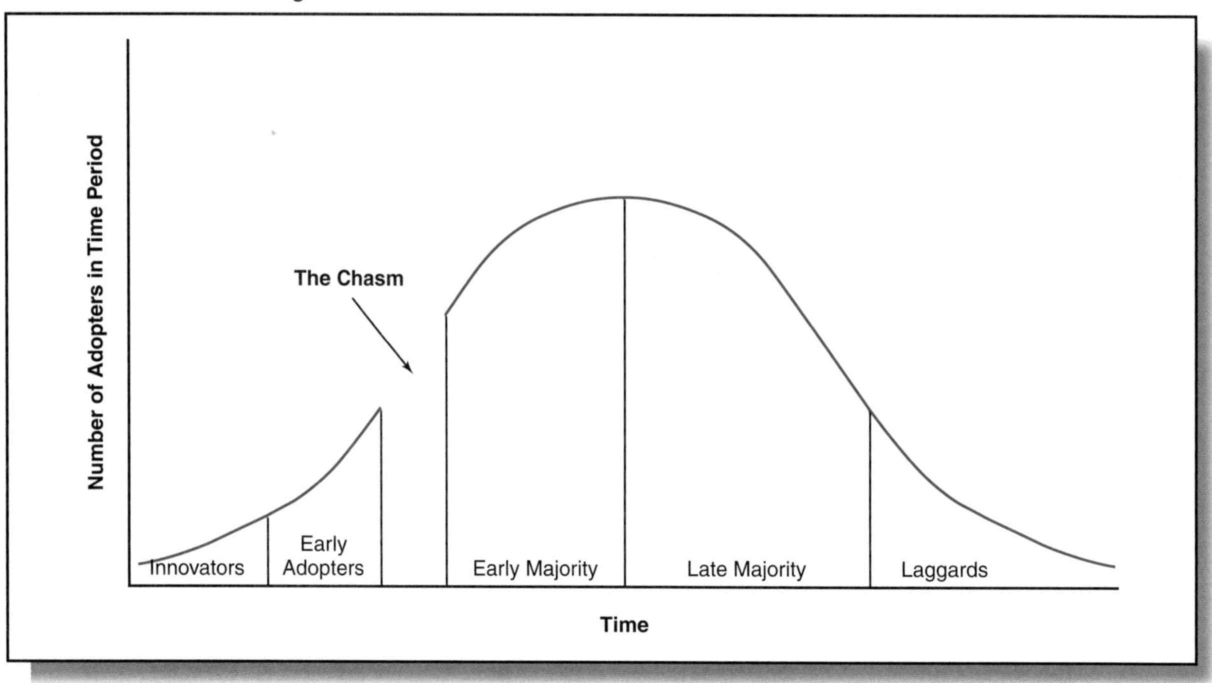

There is a chasm between early adopters and the early majority that results from differences in the demands of marketing to the two groups of customers.

Source: Adapted from Moore, G. 1991. *Crossing the Chasm*. New York: Harper Collins.

Choosing the Customer

Because you need to concentrate on a single vertical market to transition to the early majority, you need to figure out which vertical market to focus on first. You should target the one that has the greatest need for your new product because your ability to demonstrate the value of that product will be greatest in this market.[31] For example, Dell first chose to sell computers to business customers because they had a greater need than consumers for low-priced high-end computers with little tech support.

A vertical market has a great need for your new product if the new product would improve your customers' productivity, cut their costs, or allow them to do something that they otherwise could not do.[32] For example, UPS has a great need for a computerized system to track the location of its vehicles because such a system would allow the company to identify the most precise routes to destinations, thus cutting the cost of delivering packages.

To determine which vertical market has the greatest need for your new product, you need to estimate its value to different markets. The best way to do this is to estimate the time it takes to pay back the cost of the product in each market. Take, for example, a product that makes roof shingles reflect light. The benefit of this product can be estimated because reflecting light reduces the cost of cooling a building. Therefore, value of the product can be measured by the payback from reduced energy costs. Given the cost of the shingles, the average payback time is three years.[33] But this is the average. You can examine different vertical markets—new home builders, commercial builders, and roofing companies—to see which one has the shortest payback time.

If you cannot estimate the payback from purchasing the product, you might be able to quantify the value to the different vertical markets by estimating the amount of savings that customers achieve from using the product. For instance, the developers of online sales systems were able to figure out which type of retailers—clothing, books, insurance, airline tickets, and so on—had the greatest need for their technology because they could measure the labor cost savings in different vertical markets.

Unfortunately, estimating the value of a product to different vertical markets is not always this straightforward. Sometimes, the benefits of the new product cannot be captured in numbers. For example, the value of a new service that helps companies to improve labor relations may be hard to quantify because it lies in having "happier" employees. In general, when it is difficult to quantify the value of a new product to potential vertical markets, it is also difficult to identify the vertical market with the greatest need to adopt that product.

Beachhead Strategy

For new companies to successfully cross the chasm, it is sometimes necessary for them to segment the early majority of the market, looking for a niche whose needs are not well met. By focusing on this segment of the market, the company can make headway.[34]

An example of this process at work is the effort to introduce VOIP telephone service. Most customers were reluctant to adopt this new technology initially because it did not provide many of the features of traditional telephone service. However, some segments of the market were not well served by the traditional telephone companies and saw VOIP as a solution to their problems.[35] These firms became the initial customers of new companies that provided VOIP telephone service. Once the new companies established this beachhead in the market, they were able to expand to other market segments by improving the technology underlying their products and making it useful to a wider range of the market.[36]

Key Points

- Because the early majority adopts products for different reasons from early adopters, companies often find it difficult to transition to the mainstream of the market.
- Providing a complete solution, demonstrating the value proposition, and focusing on a single vertical market are all things companies need to do to transition to the early majority.
- Firms are most successful at transitioning to the early majority if they focus on the vertical market with the greatest need for their product.
- The vertical market with the greatest need for a new product is the one in which the introduction of the product would lead to the largest increase in customers' productivity, the largest decline in customers' costs, or the greatest opportunity for customers to do something that they otherwise could not do.
- To determine which vertical market has the greatest need for a new product, you can estimate the time it takes customers to pay back the cost of the product or the amount of savings the customer gets from using it; however, the value of a new product to customers cannot always be quantified.
- To successfully cross the chasm, it is sometimes necessary for new companies to segment the early majority of the market and focus on the portion that is underserved by existing products.

Although it is very difficult to accurately forecast anything that happens in the future, you still need to estimate what demand will be for your new products for several reasons:

- First, demand forecasts help you to determine how much to produce, allowing you to avoid underproduction and the loss of potential customers to your competitors, as well as overproduction, and the corresponding need to cut your price and profits.
- Second, demand forecasts help you to project your future costs in businesses based on economies of scale. In businesses with high fixed costs, unit costs decrease as the volume of production increases, making your costs a function of the size and rate of sales growth.[37]
- Third, demand forecasts help you to determine the payback on your investment in product development. Because the costs of developing a new product tend to be fixed, the payback on investments in product development depends a great deal on amount and timing of sales.
- Fourth, demand forecasts help you to make pricing and advertising decisions[38] because they provide information about customers' price sensitivity, and the effect of advertising on sales.[39]
- Fifth, demand forecasts help you to determine the competitiveness of your market.[40] Stagnating and shrinking markets are more competitive than growing markets because sales in the former have to come at the expense of competitors, but sales in the latter can come from new customers.

Forecasting Demand

As Figure 3.5 illustrates, predicting anything that will happen in the future isn't easy. This holds true for market demand as well as anything else. Because markets are dynamic, you can't estimate future demand solely on the basis of the current market size. In fact, for many new technology products and services, a huge market tomorrow may not even exist today. Take, for example, the market for Internet search, which is enormous, but did not even exist in the early 1990s.

Moreover, forecasting demand depends a lot on the timing of customer adoption. To understand why, just think about the difference in two forecasts of a billion dollar

FIGURE 3.5
Predicting the Future

While it's easy to predict past success, it isn't easy to predict future winners.
Source: The Piranna Club, August 19, 2006.

market—one which indicates that this market will emerge in 5 years, and the other, which indicates that it will emerge in 150 years. The market size at different points in time over the next 150 years will be completely different in the two forecasts.

Furthermore, the rate of growth of demand over time is not linear. (Remember the S-shaped curve of adoption from the first section of the chapter?) Because demand growth accelerates when markets shift from the early adopters to the early majority, a key to forecasting demand accurately lies in identifying when this point of acceleration will occur.

Finally, forecasting demand depends a lot on the accuracy of information about the factors that influence diffusion patterns. (**Diffusion** is the rate at which a new technology product becomes adopted by potential users.[41]) As we will see later in this chapter, both the functional form and the rate of diffusion can be affected by a variety of factors, including the external environment, the proportion of innovators and imitators in a market, and the nature of the product.

Information Diffusion Models

As the previous section indicated, you need to know something about the factors that affect the diffusion of your new products to forecast demand accurately. Information diffusion models explain that the functional form of diffusion is primarily a result of the distribution of **innovators** (customers who learn about new products from sources other than previous adopters, such as from advertising)[42] and **imitators** (customers who learn about new products from previous adopters).[43]

In all markets, some people are innovators and adopt new products independently of learning information from others, while most people are imitators, and adopt new products when they learn that others have adopted them.[44] The distribution of innovators and imitators varies across markets and affects the shape of the diffusion curve. As was explained in the beginning of the chapter, when there are few innovators and many imitators, diffusion follows an S-shaped pattern.[45] In the beginning, only small numbers of innovators purchase new technology products and services, and the rate of diffusion is slow. However, as imitators learn about the new product from the innovators, the rate of diffusion accelerates. As the market becomes saturated, the rate of diffusion slows, and ultimately reaches an asymptote.[46]

(While you might start to think that the only shape that professors of technology strategy can draw is an "S," the diffusion S-curve is different from the technology S-curve described in Chapter 2 and the adoption S-curve described earlier in this chapter. The diffusion S-curve measures the number of adopters as a function of time. It is based on the idea that some parts of the market adopt independently from the decisions of other users, while other parts of the market do not. The adoption S-curve measures the percent of the market adopting a new product as a function of time. It is based on the idea that the shape of cumulative adoption is related to the shape of the distribution of adopters, with a normal distribution of adopters yielding an S-shaped curve of cumulative adoption. The technology S-curve described in Chapter 2 measures the level of performance of a technology as a function of the amount of effort put into its development. It is based on the concept that performance improvements require a lot of effort initially and later run into diminishing returns.)

The diffusion of new technology products and services is not always S-shaped. For instance, diffusion patterns look more like a convex curve than an "S" if there are many innovators and few imitators. This happens when potential adopters are not

FIGURE 3.6
The Diffusion of the CT Scanner and Cable Television

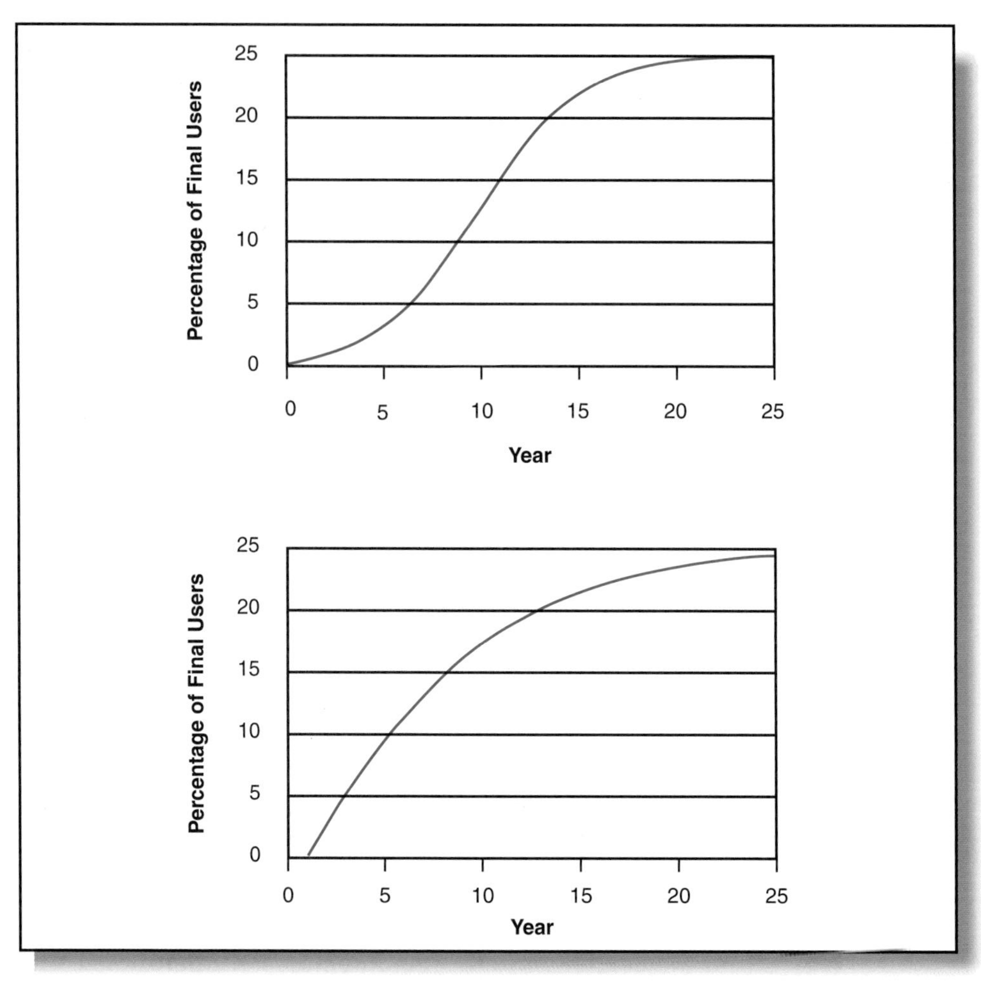

The diffusion of the CT scanner took the standard S-shape, whereas the diffusion of cable television was more of a convex curve.

Source: http://andorraweb.com/bass/index.=?show[examples]=1.

closely connected, leading the decisions of earlier adopters to have little influence on the decisions of later ones.[47] Figure 3.6 provides the example of the diffusion of the CT scanner, which followed an S-shaped pattern, and cable television, which followed a pattern best described as a convex curve.

How Not to Do It

The previous section indicates that you need some method for estimating the size of the market for your new products. Unfortunately, the most common way that entrepreneurs and managers do this is not very accurate. The typical approach involves looking up the size of a similar market and using that number as the estimate of the size of the target market. For instance, take a company that has developed a new communications device called e-mail. At the time that the company has developed the first e-mail communications system, no one has a way to send messages from computer to computer. So the manager responsible for analyzing the market size estimates it by looking up how many telephone calls people make each day, and uses that information to estimate the number of e-mail messages that people will send.

However, to use information on the telephone market to estimate the market for e-mail, you have to assume that the purpose of e-mail is to *replace* the telephone, making e-mail a substitute for that product, rather than a substitute for regular mail, faxes, face-to-face conversations, or other forms of communication. If the true purpose of e-mail is to substitute for other products like face-to-face conversations, regular mail, or faxes, then market size estimates for e-mail that are based solely on the size of the market for telephone calls are not going to be very accurate. Moreover, e-mail is often a *complement* to telephone communication because people often send e-mail messages to clarify telephone calls, and vice versa. As a result, by assuming that e-mail substitutes for the telephone in the evaluation of market size, you will likely get a number that has nothing to do with the size of the actual market for e-mail communication.

The Bass Model

So what should you do to estimate demand if just looking at an analogous market won't work? You should use quantitative or qualitative methods for forecasting the diffusion of new technology products. One quantitative tool for forecasting the diffusion of new technology products that many companies use is the **Bass diffusion model**.[48] In this model, diffusion patterns are a function of the size of the market, the rate of adoption by innovators and imitators, and the proportion of adopters in the previous time period.[49] They can be measured by the following formulas:[50]

The likelihood of purchase by a new adopter at time period t is

$$p+(q/m)n_{t-1},$$

where
p is the likelihood that an innovator will adopt,
q is the likelihood that an imitator will adopt,
m is the total number of adopters that will never be exceeded,
n_{t-1} is the cumulative number of adopters of the product through the previous time period.

The number of new adopters during time period t, $S(t)$, equals

$$[p+(q/m)n_{t-1}][(m-n_{t-1})]$$

That is, in the Bass model, the number of adopters in a period is measured by the sum of the probability that an innovator adopts the new product plus the probability that an imitator adopts, divided by the size of the market, and multiplied by the proportion of the market that has already adopted the product, all multiplied by the size of the market, less the amount of the market that has already adopted. For instance, suppose the rate of adoption by innovators is 0.10 and the rate of adoption by imitators is 0.50, the market size is 10 million, and no one has yet adopted the product. The number of adopters in the year would be predicted to be: $(0.10 + (0.5/10) \times (0)) \times (10 - 0) = 0.1 \times 10 = 1.0$ million customers.

So where do you get the numbers to plug into the Bass model? Typically, users estimate the market size by conducting a survey of potential customers or by examining secondary data. And they estimate the likelihood that innovators and imitators will adopt by extrapolating from historical data on the diffusion patterns of similar products.[51] For example, you could create estimates of the rate of innovator and imitator adoption of HDTV by extrapolating from published data on the rates for those two groups taken from studies of color televisions, VCRs, and other consumer electronics.[52]

The basic Bass diffusion model can be modified to include a variety of factors that affect the diffusion of new technology products, such as price,[53] and changes in the size of the potential market.[54] You can estimate the diffusion of complementary products (like washing machines and dryers) by using a variant of the model that makes the diffusion of one product contingent on the diffusion of another.[55] Similarly, you can capture the effect of substitution by future generations of products, and accommodate repeat purchases with other variants.[56] You can even estimate market share by linking the Bass model to discrete choice models that compare the likelihood that a customer will choose between your product and your competitors' products, using information from a conjoint analysis on different product attributes.[57]

How effective are these models at predicting the adoption of new technology products and services? They are pretty good, but not outstanding. And they are better for some products than for others. The Bass model is most accurate at predicting the diffusion of **consumer durables** (consumer products, like appliances, whose value is not exhausted in a year), than other types of products.[58]

Moreover, the Bass model has important limitations, which you need to consider if you are going to use it.

1. You cannot use the Bass model to estimate diffusion at the very beginning of a product's life because the model uses information from previous periods to estimate diffusion in subsequent periods.
2. The accuracy of the Bass model's predictions depends on the accuracy of your assessments of size of the potential market. If you cannot estimate the size of the potential market, or if your estimate is wildly inaccurate, then the Bass model isn't going to be much help to you.
3. The Bass model assumes that the diffusion of a technology product depends on only demand-side factors.[59] If the size of a market, or the rate of adoption of a product, are determined by supply-side factors, such as the availability of raw materials or production capacity, then the results of a Bass model analysis will be very inaccurate.
4. The accuracy of the Bass model is much lower when competing technologies are being introduced, or when economic shocks are occurring, because the model has no mechanism for estimating the effect of competitive factors on new product adoption, or ways to incorporate the effects of external economic shocks.
5. The accuracy of the Bass model declines as you get further away in time from the initial adoption point.

GETTING DOWN TO BUSINESS

Diffusion of MP3 Players[60]

Let's look at an example of how a company might use the Bass Diffusion Model. Suppose you are an entrepreneur who has founded a company to make MP3 players at the start of that industry. You need to create estimates of market size and growth to select a revenue model and to justify your financing strategy to potential investors. So you decide to use the Bass model to create those estimates.

To calculate the diffusion of your product, you first need to estimate the total size of the market. Let's suppose that you conducted a survey of potential customers and found out that this number is 40 million people. You then need to estimate the rate of adoption by innovators and by imitators. You can create these estimates by looking at analogous products to the MP3 player—things like the portable CD radio and the cellular telephone. If you find the average rate of adoption for innovators and imitators of these kinds of products is 0.01 and 0.40, respectively, you can generate the diffusion pattern shown in Figure 3.7.

FIGURE 3.7 A Demand Forecast for MP3 Players

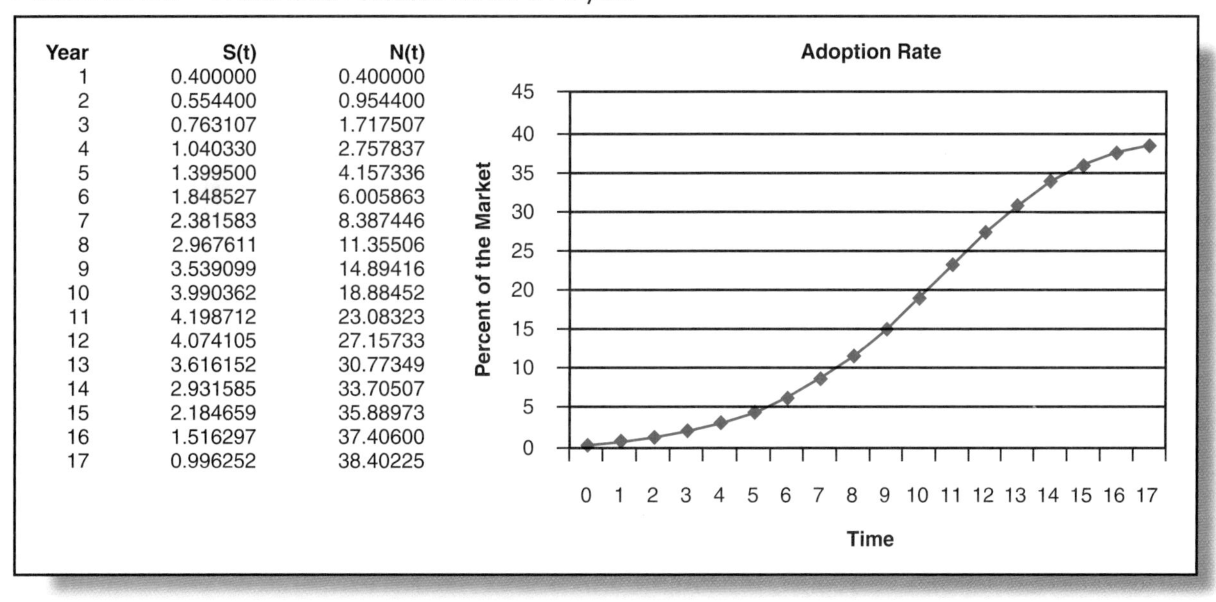

Year	S(t)	N(t)
1	0.400000	0.400000
2	0.554400	0.954400
3	0.763107	1.717507
4	1.040330	2.757837
5	1.399500	4.157336
6	1.848527	6.005863
7	2.381583	8.387446
8	2.967611	11.35506
9	3.539099	14.89416
10	3.990362	18.88452
11	4.198712	23.08323
12	4.074105	27.15733
13	3.616152	30.77349
14	2.931585	33.70507
15	2.184659	35.88973
16	1.516297	37.40600
17	0.996252	38.40225

The Bass model allows you to forecast demand for new products.

The Delphi Technique

Because of these limitations of the Bass model, some experts believe that you are better off making your organization flexible so that it can respond to changes in market demand, rather than trying to predict how demand will change over time. However, you do need to have some idea of where demand is going to figure out how much to produce when first introducing a new technology product.

For this reason, some experts suggest that you use more qualitative methods to forecast demand. One important qualitative method that many companies use is called the Delphi Technique. As Table 3.1 shows, with the Delphi Technique, experts

TABLE 3.1
The Process for Using the Delphi Technique
The use of the Delphi Technique involves the simple 10-step process, outlined here.

Step	Action
1	Pick a facilitator
2	Identify an expert group
3	Create an initial list of criteria
4	Have the experts rank criteria
5	Estimate the mean and standard deviation
6	Have panel re-rank the newly ordered criteria
7	Identify preferences and constraints
8	Have the panel rank alternatives by constraints and preferences
9	Analyze the results and return them to the panel
10	Repeat the ranking process until you achieve stability on the rankings

Source: Adapted from: Cline, A. *Prioritization Process Using Delphi Technique,* downloaded from http://www.carolla.com/wp-delph.html.

are selected and asked anonymously for their estimates of the likelihood of particular outcomes occurring. The participants return their estimates to a coordinator, who compiles them. The summary data outlining the mean and range of viewpoints is then returned to the respondents who are asked for new estimates in light of the information presented by the other experts.[61]

The Delphi Technique has been used with success in a variety of settings, and it is an important tool for you to use to identify potential technological trends that might impact the development of your new products and services. However, you need to be aware of several major weaknesses of the technique.

First, the method is sensitive to the precision of the questions that you ask. If you ask imprecise questions, the respondents might not reach a consensus, and, if they do, that consensus might not be about the major trends that you are seeking to understand.

Second, the Delphi Technique is sensitive to variance in the expertise of the respondents. If the respondents are not experts on the topic about which they are being queried, then the technique provides little more than a consensus of the ignorant, which is of little use to you as a technology strategist.[62]

Third, the validity of the technique is limited by the intervention of unexpected events that the experts do not incorporate into their analyses. Because the Delphi Technique is built on consensus, it is only accurate at predicting things that most people expect to occur.

Product Diffusion Models

Some researchers propose product diffusion models that are different from the information diffusion models, like the Bass model, described previously. Product diffusion models look at product characteristics to explain their rate of diffusion and have found a variety of different characteristics to be important. First, the greater the benefit that a new product provides to customers and the lower its cost, the faster it will diffuse because customers have more motivation to adopt a product that benefits them more and costs them less.[63] For instance, asymmetric DSL technology diffused slowly because the cost to deploy the technology was high in comparison to alternative technologies for transmitting digital data.[64]

The cost of a new technology product is affected by the degree of change that customers have to make to adopt it, leading diffusion to slow as the needed amount of change increases. For example, 3G mobile phones diffused to wireless carriers more slowly than previous generations of mobile phone technology because the adoption of 3G required the carriers to purchase a new wireless spectrum, and to create new signal stations, whereas previous mobile phone technologies did not require such significant upgrades.[65]

The effect of the degree of change on the rate of diffusion is not limited to industrial products. The diffusion of CDs was slower than the companies producing them had expected because customers saw replacing or copying their cassette tapes as a cost to adopting the new technology.[66]

Sometimes, companies need to produce several generations of new products before those products offer enough benefit to customers to spur diffusion. For example, digital cameras did not begin to diffuse quickly until after manufacturers had produced several generations of them, and their level of resolution had reached that of chemical film cameras.

Second, the amount of learning necessary to understand the benefits of a product will slow its rate of diffusion because people must understand the value of new products and services to adopt them.[67] Thus, the provision of information about a new product (including advertising) and the opportunity to test it enhance the rate of diffusion.[68]

Third, the perceived risk of a new product lowers its rate of diffusion.[69] If new products are perceived as risky because competing technical standards exist[70] or because the products are novel, potential adopters often delay adoption to gather more information.[71] For example, diffusion of high-definition DVDs has been slow because manufacturers have failed to agree on a common technical standard, and consumers have been unwilling to risk selecting a format that is not the industry standard.[72] (Don't worry if you don't get this point yet; Chapter 12 will discuss technical standards in greater detail and will consider this issue again.)

Fourth, the characteristics of the adopters affect the rate of diffusion of new products. Products diffuse faster when adopters are tolerant of uncertainty because this attribute makes them more open to new product ideas. They also diffuse faster when adopters are wealthier because wealth provides a cushion against poor adoption decisions.[73]

Moreover, new products diffuse more quickly when adopters are geographically proximate, or tightly connected through social networks because these attributes accelerate information transfer.[74] For instance, pharmaceutical firms, like Merck, try to get their new drugs into the hands of well-connected doctors because the interconnectedness of the medical community makes diffusion of medically related products very rapid.

Fifth, aspects of the environment in which the adoption decision is being made affect the rate of diffusion of new products.[75] For instance, new products that involve capital investment diffuse faster as interest rates decline because the cost of adoption decreases as interest rates go down. Moreover, this effect is larger for more expensive products because the use of financing increases with the cost of products.

Social factors also affect the diffusion of new products. When new products are socially acceptable, they diffuse more quickly than when they face social opposition. For example, a product that allows people to use a Web cam and Internet technology to record their prayers and send them over the Internet to a religious site has diffused slowly because it has faced opposition from religious leaders who think that the use of high technology for prayer demeans religion.[76]

Political factors, too, influence the rate of diffusion of new products. For instance, when an important political group opposes the adoption of a new product, its diffusion is slowed. FM radio, for instance, diffused slowly because the AM radio industry got the Federal Communications Commission (FCC) to earmark the 42 to 50 MHz radio frequency range to television, to require a new frequency for FM radio, and to allow programs to be broadcast on both AM and FM radio.[77]

Political opposition can even stop the diffusion of a new product. Consider what happened to the diffusion of nuclear power in the United States after the Three Mile Island disaster led to a dramatic rise in political action against the nuclear power industry.[78] Because people became afraid of a nuclear disaster and public support for nuclear power dropped dramatically, the diffusion of nuclear power stopped.[79]

The Importance of Complementary Technologies

The nature of an innovation affects its rate of diffusion. In general, new products based on discrete technologies diffuse faster than ones based on systemic technologies because new products based on systemic technologies can only diffuse as fast as their slowest diffusing component. For example, the diffusion of music downloads was slow until MP3 players were developed and the downloaded music could be made portable.

For systemic technologies, an important part of managing a strategy of technology diffusion involves coming up with ways to ensure that complementary technologies will develop, and figuring out when those technologies will come to fruition. For example, General Motors and Ford have prototype fuel cell vehicles that they are planning to introduce to the market in 2011. However, both companies recognize that the adoption of these vehicles depends a great deal on the ability to refuel them, making the complementary technology of a system of hydrogen refueling stations of central importance. For this reason, Ford and General Motors, among other automakers, are working with the major oil companies to establish a network of hydrogen refueling stations throughout the country so that they can introduce their new vehicles in the next five years.[80]

Substitution

As was described in the opening vignette to this chapter, the diffusion of new products and services is often affected by substitution. (**Substitution** is the replacement of one technology by another that can achieve the same objective.[81]) While some observers argue that all new products and services are substitutes for some existing product or service, substitution is more direct in some cases than in others. For example, fiber optic cable is a very clear substitute for coaxial cable because telecommunications firms now use the former for almost all purposes that they previously used the latter. Similarly, DVD players have substituted for laser disk players, and effectively stopped their diffusion.[82]

Substitution can have a profound effect on how your company competes. In general, it causes great hardship for large, established firms because these firms tend to have major investments in older technology. When a new technology substitutes for an old technology, these companies need to spend hundreds of millions and sometimes billions of dollars to replace their technology. For example, the substitution of VOIP technology for traditional circuit switch telephone technologies required the large, established telephone companies, like British Telecom and AT&T, to spend billions of dollars to replace their circuit switch technology with VOIP. This substitution puts the established telecom companies at a disadvantage when they compete with small, new firms founded on the basis of the new technology.

Because technology substitution makes the investments of established firms obsolete, it is a very important part of the strategy that successful technology entrepreneurs use to compete with established firms, particularly in industries that face significant economies of scale. As new technology substitutes for old technology, the scale economies at the businesses producing products based on the old technology erode, leading their costs to increase. When the established firms' costs rise above those of the new firms, the established firms become uncompetitive. If the loss of scale economies increases the cost structure of the established firms enough, new firms can drive the established firms out of business.

For example, when the minimills entered the steel industry with a new technology that substituted for traditional steel casting, the integrated steel mills initially suffered only modest declines in the volume of steel that they produced. However, these modest declines in production volume set off a downward spiral that ultimately resulted in the bankruptcy of many integrated steel companies. Because economies of scale are very large in the steel industry, the initially small substitution of new steel making technology that resulted in small declines in the volume of production at the integrated steel mills led to large increases in those mills' cost of production. Those cost increases encouraged further substitution to the new minimill technology, which further reduced integrated steel mill production volumes, and further increased their production costs, creating a downward cycle for the integrated steel mills.

While the steel mill example illustrates the value of substitution to the formulation of an effective technology strategy at start-up companies, the formulation of such a strategy is difficult for several reasons. First, substitution is very much affected by political actions to deter change taken by the providers of products based on the old technology. For example, the producers of glass windshields have taken action to keep auto insurers from authorizing repairs to cars made with plastic windshields because plastic could substitute for glass in windshield repair. Because auto insurers pay for most of the repairs to automobiles damaged in collisions, this political action has kept the substitution from occurring.

Established companies that exploit the old technology usually have an advantage in the realm of political action because they employ large numbers of people whose jobs would be lost if the technology substitution occurs, creating a strong incentive for policy makers to support their efforts to deter substitution. Moreover, the mechanisms to discourage policy makers from taking actions that slow technology substitution involve lobbying and political influence, activities with which many entrepreneurs are unfamiliar.

A second reason why a substitution-based technology strategy is difficult to implement is that substitution is typically partial. As Figure 3.8. shows somewhat humorously, when one technology substitutes for another, the new technology usually does not completely replace the old technology. For example, the use of trucks did not completely substitute for the use of railroads to transport goods. Because substitution is usually partial, you need to figure out how much substitution will occur to determine if it will be enough for you to compete successfully against established firms who are exploiting the old technology.

A final reason why a substitution-based technology strategy is difficult to pull off is that substitution takes very different amounts of time in different industries. For example, as Table 3.2 shows, it took synthetic rubber 58 years to drive natural rubber down to only 10 percent of the market, but detergent soap took only 9 years to push natural soap down to the same percent market share.[83] As a result, with

FIGURE 3.8
Complete Substitution Is Not Always Possible

Few technology products are ever complete substitutes for the products used before them.
Source: New Yorker Magazine. 1975. *The New York Album of Drawings, 1925–1975,* New York: Penguin Books.

TABLE 3.2
Time to 90 Percent Substitution
The time it takes for a new technology product to substitute for 90 percent of the uses of an old technology product varies widely, making it difficult for technology strategists to plan for substitution.

Substitutes	Years to 90 Percent Substitution
Synthetic/natural rubber	58
Synthetic/natural fibers	58
Plastic/natural leather	57
Margarine/butter	56
Water/oil house paint	43
Open hearth/Bessemer steel	42
Sulfate/tree sapped turpentine	42
Plastic/hardwood home floors	25
Organic/inorganic pesticides	19
Synthetic/natural tire fibers	17
Plastics/metals in cars	16
Animal/detergent soap	9

Source: Adapted from Girifalco, L. 1991. *Dynamics of Technical Change.* New York: Van Nostrand Reinhold.

soap, the companies introducing products based on the new technology were able to exploit the rapid deterioration of the cost structure of established firms in ways that companies introducing products based on synthetic rubber could not.

Without accurate estimates of the time horizon of substitution, you cannot balance supply and demand for products based on the new technology. Take, for example, an automobile company that wants to develop electric vehicles as a substitute for gasoline-powered vehicles. If the company bets that the substitution will take 20 years and it occurs more quickly, the company will not have enough cars to satisfy demand and will allow competitors to enter. On the other hand, if the company bets that the substitution will take five years and it occurs more slowly, it will run out of cash before the new technology begins to replace the old one.

Established companies using the old technology also have to estimate the time horizon of substitution accurately because that horizon affects how quickly they need to transition to the new technology. Unfortunately, they often believe that this time horizon will be longer than it actually is, leaving them flat-footed in the face of changing demand and competition. For instance, Kodak's senior management realized that digital technology ultimately would replace traditional film technology, but they assumed that this substitution would occur much more slowly than it did, leaving Kodak without adequate digital camera alternatives to its traditional products, and forcing it to undertake a painful restructuring.

Sometimes established companies don't wait for competitors to develop substitutes for their products and deliberately develop new products based on new technologies to substitute for their own products. Examples of these efforts include the development of 3G telephones in place of GSM phones, high-definition DVDs in place of standard DVD technology, and digital televisions in place of analog ones.

The deliberate development of new products based on new technologies requires you to consider another important aspect of substitution, product cannibalization. This is a difficult and risky dimension of technology strategy. If you mismanage the transition to the new technology, your company will lose the sales of its old product without having a replacement for it, after having incurred the costs of unnecessary product development. So, if you are going to substitute a new technology for an old one, you need to balance the advantages that you will derive from deterring competitors who would otherwise introduce the new technology, against the costs of product cannibalization.

Key Points

- Forecasting demand helps you to determine how much of your product to produce, to calculate the payback on your investment in product development, to project your costs, to set a pricing and advertising plan, and to formulate the right competitive strategy.
- Forecasting demand is difficult because markets are dynamic, demand growth is nonlinear, and the accuracy of forecasts depends on the accuracy of estimates about the timing of customer adoption and the factors that affect the diffusion of technology.
- Information diffusion models measure the functional form of diffusion, using data on the proportion of innovators and imitators in the customer population.
- The Bass model is a tool to estimate the diffusion of technology products that is based on the rate of adoption by innovators and imitators, and the size of the potential market.
- Versions of the Bass model can incorporate the effects of the marketing mix, changes in the size of the potential market, the effect of diffusion of complementary products, substitution by future generations of products, and competition between products with different features.
- The Delphi Technique is a methodology for gathering data from experts about technology trends.
- Product diffusion models measure the rate of diffusion, which depends on the costs and benefits of the new product, the amount of learning necessary to understand its benefits, the perceived risks of the product, the nature of the innovation, the characteristics of the target market, and the nature of the environment in which the adoption decision is being made.
- Substitution is the replacement of one technology for another that can achieve the same objective; it is an important part of technology strategy because it influences the competition between incumbent firms and new entrants.
- The implementation of a substitution strategy is difficult because substitution can be multilevel, face political opposition, be partial, and take a long or a short time.
- Companies sometimes deliberately develop new products to substitute for old products; this process is risky and can severely damage a firm's competitive position if done incorrectly.

Discussion Questions

1. What shapes can the adoption curve for new technology products take? What is the relationship between the distribution of adopters and the shape of the adoption curve?
2. Why do innovators, early adopters, early majority, late majority, and laggards adopt products? What effects do differences in their motivations have on technology strategy?
3. What's the difference between the adoption S-curve described in this chapter and the technology S-curve described in Chapter 2?
4. What is the typical pattern of diffusion of new technology products? What factors affect this pattern? How do these factors affect this pattern?
5. What are the similarities and differences between product and information diffusion models?
6. Is the Delphi Technique better than the Bass model at forecasting demand or vice versa? Why?
7. How should you manage technology substitution? Why do you make these recommendations?

Key Terms

Adoption: The decision by customers to purchase a new product or service.

Bass Diffusion Model: A mathematical tool for estimating the rate at which a new product or service diffuses to a potential market.

Consumer Durables: Consumer products whose value is not exhausted in a year.

Crossing the Chasm: The process of transitioning from the early adopters of a product to the early majority.

Diffusion: The rate at which a new technology product or service is adopted by potential users.

Early Adopters: The segment of the market that follows the innovators in the adoption of a new technology product or service.

Early Majority: The segment of the market that adopts a product slightly before the market average.

Horizontal Marketing Strategy: A strategy of serving customers in multiple industries at the same time.

Imitators: Customers who learn about new products from previous adopters.

Innovators: (1) The group of customers that adopt a new technology product or service immediately after it is introduced; (2) customers who learn about new products from sources other than previous adopters.

Laggards: The segment of the market that adopts new technology products or services last.

Late Majority: The segment of the market that adopts new technology products or services after the early majority.

Normally Distributed: A bell-shaped distribution in which a small portion of the population falls at each end, and the majority falls in the middle.

Substitution: The replacement of one technology for another that can achieve the same objective.

Take Off: The period in time when the S-curve of adoption accelerates because of the transition from the early adopters to the majority of the market.

Vertical Marketing Strategy: A strategy of focusing on serving customers in a single industry at a time.

Putting Ideas into Practice

1. **Technology Adoption** Select a new technology product or service that you know well. Describe the distribution of adopters. (e.g., Is it normal or does it take another shape? If it takes another shape, what is that shape?) What proportion of adopters fall into each group? Describe the characteristics of the different adopter groups? (e.g., innovators, early adopters, early majority, late majority, and laggards). Identify the factors that you think will influence adoption of this product or service by the different groups of adopters. Explain how to "cross the chasm" to gain adoption of the product or service by the majority of the market.
2. **Information Diffusion Models** Go to http://andorraweb.com/bass/index.=?show[examples]=1. Click on the diffusion models for the different products listed. What does the diffusion pattern look like for each of the products? (For example, are they S-shaped? Are they linear? Are they upward sloping curves?) Why are the diffusion patterns different? What conditions must be present to have each of the diffusion patterns?
3. **Forecasting Technology Diffusion** The purpose of this exercise is for you to use the Bass model to predict the adoption of a new technology product, in this case the e-book. Your assignment is to develop a forecast for the diffusion of that product or service over the next 20 years. Assume that the maximum number of e-books that could ever be sold in a year in the United States is 500 million. In addition, assume the rate of innovator adoption to be 0.082 and the rate of imitator adoption to be 0.416. In an Excel spreadsheet, calculate the number of customers adopting annually, as well as the cumulative number of adopters in each year. Then plot the expected pattern of diffusion over the next 20 years. Now change the rate of innovator adoption to 0.20 and the rate of imitator adoption to 0.05. How does the adoption pattern change? What if you change the rate of innovator adoption to 0.05 and the rate of imitator adoption to 0.20? What happens if you leave the rate of imitator and innovator adoption the same but change the maximum market size never to be exceeded to 100 million? Why is the adoption pattern different under these different scenarios?

NOTES

1. http://www.taipeitimes.com/News/worldbiz/archives/2004/01/15/2003087725
2. http://www.taipeitimes.com/News/worldbiz/archives/2004/01/15/2003087725
3. http://www.usatoday.com/money/industries/manufacturing/2004–01–13-kodak-cameras_x.htm
4. http://www.boston.com/business/articles/2005/07/21
5. Mahajan, V., and E. Muller. 1998. When is it worthwhile targeting the majority instead of the innovators in a new product launch? *Journal of Marketing Research*, 35(4): 488–295.
6. Rogers, E. 1983. *Diffusion of Innovations*. New York: Free Press.
7. Calculated from information downloaded from http://andorraweb.com/bass/index.=?show[examples]=1.
8. Gatignon, H., J. Eliashberg, and T. Robertson. 1989. Modeling multinational diffusion patterns: An efficient methodology. *Marketing Science*, 8(3): 231–247.
9. Moore, G. 1991. *Crossing the Chasm*. New York: Harper Collins.
10. Mohr, J., S. Sengupta, and S. Slater. 2005. *Marketing of High-Technology Products and Innovations* (2nd edition). Upper Saddle River, NJ: Prentice Hall.
11. Vanac, M. 2007. Vision for the future. *The Plain Dealer*, February 14: C1, C3.
12. Brown, S., and V. Venkatesh. 2003. Bringing non-adopters along: The challenge facing the PC industry. *Communications of the ACM*, 46(4): 76–80.
13. Burt, R. 1973. The differential impact of social integration on the participation in the diffusion of innovations. *Social Science Research*, 2: 125–144.
14. Schilling, M. 2005. *Strategic Management of Technological Innovation*. New York: McGraw-Hill.
15. Burgelman, R., C. Christiansen, and S. Wheelwright. 2004. *Strategic Management of Technology and Innovation*. New York: McGraw-Hill Irwin.
16. Moore, *Crossing the Chasm*.
17. Brown and Venkatesh, Bringing non-adopters along.
18. Day, G. 2000. Assessing future markets for new technologies. In G. Day and P. Schoemaker (eds.), *Wharton on Managing Emerging Technologies*. New York: John Wiley.
19. Brown and Venkatesh, Bringing non-adopters along.
20. Rogers, *Diffusion of Innovations*.
21. Brown, R. 1992. Managing the "S" curves of innovation. *Journal of Consumer Marketing*, 9(1): 61–72.
22. Ibid.
23. Vara, V., and M. Mangalindan. 2006. Web pioneers eBay and Amazon face a threat from older retailers. *Wall Street Journal*, November 16: A1, A12.
24. Brown, Managing the "S" curves of innovation.
25. Moore, *Crossing the Chasm*.
26. Afuah, A. 2003. *Innovation Management*. New York: Oxford University Press.
27. Moore, *Crossing the Chasm*.
28. Mossberg, W. 2007. Nokia's marriage to small computers still has its problems. *Wall Street Journal*, February 22: B1.
29. Lal, R. 2002. Documentum Inc., *Harvard Business School Case*, Number 9–502–026.
30. Moore, *Crossing the Chasm*.
31. Ibid.
32. Ibid.
33. Rosch, W. 2004. New roofing shingles mimic nature. *Cleveland Plain Dealer*, June 17: D3.
34. Christiansen, C. 1997. *The Innovator's Dilemma*. Cambridge: Harvard Business School Press.
35. Grant, P., and A. Latour. 2003. Battered telecoms face new challenge: Internet calling. *Wall Street Journal*, October 9: A1, A9.
36. Christiansen, *The Innovator's Dilemma*.
37. Ofek, E. 2005. Forecasting the adoption of a new product. *Harvard Business School Note*, Number 9–505–062.
38. Van den Bulte, C., and S. Stremersch. 2004. Social contagion and income heterogeneity in new product diffusion: A meta-analytic test. *Marketing Science*, 23(4): 530–544.
39. Mahajan, V., E. Muller, and F. Bass. 1990. New product diffusion models in marketing: A review and directions for research. *Journal of Marketing*, 54(1): 1–26.
40. Cooper, R., and E. Kleinschmidt. 1987. New products: What separates winners from losers? *Journal of Product Innovation Management*, 4: 169–184.
41. Girfalco, L. 1991. *Dynamics of Technological Change*. New York: Van Nostrand Reinhold.
42. Dodson, J., and E. Muller. 1978. Models of new product diffusion through advertising and word of mouth. *Management Science*, 24: 1568–1578.
43. Bass, F. 1969. A new product growth model for consumer durables. *Management Science*, 15: 215–227.
44. Ofek, Forecasting the adoption of a new product.
45. Van den Bulte and Stremersch, Social contagion and income heterogeneity in new product diffusion.
46. Girfalco, *Dynamics of Technological Change*.
47. Rogers, *Diffusion of Innovations*.
48. Bass, A new product growth model for consumer durables.
49. Ibid.
50. Ofek, Forecasting the adoption of a new product.

51. Mahajan, Muller, and Bass, New product diffusion models in marketing.
52. Sultan F. 2001. *Marketing Research for High Definition Television*. Boston: Harvard Business School Publishing.
53. Kalish, S. 1985. A new product adoption model with pricing, advertising, and uncertainty. *Management Science*, 31(12): 1569–1585.
54. Mahajan, Muller, and Bass, New product diffusion models in marketing.
55. Ibid.
56. Norton, J., and F. Bass. 1987. A diffusion theory model of adoption and substitution for successive generations of high technology products. *Management Science*, 33 (9): 1069–1086.
57. Ofek, Forecasting the adoption of a new product.
58. Ibid.
59. Day, Assessing future markets for new technologies.
60. Adapted from Ofek, Forecasting the adoption of a new product.
61. Goldfisher, K. 1992–1993. Modified Delphi: A concept for new product forecasting. *Journal of Business Forecasting*, 11(4): 10–11.
62. Day, Assessing future markets for new technologies.
63. Wejnert, B. 2002. Integrating models of diffusion of innovations: A conceptual framework. *Annual Review of Sociology*, 28: 297–326.
64. Tuzo, T. 1997. Asymmetric digital subscriber line: Prospects in 1997. *Harvard Business School Case*, Number SM-69.
65. Eisenmann, T., and F. Suarez. 2003. Symbian: Setting the mobility standard, *Harvard Business School Case*, Number 9–804–076.
66. Greenstein, S., and V. Stango. Forthcoming. The economics of standards and standardization. In S. Shane (ed.), *Blackwell Handbook on Technology and Innovation Management*. Oxford: Blackwell.
67. Rogers, *Diffusion of Innovations*.
68. Day, Assessing future markets for new technologies.
69. Ibid.
70. Van den Bulte and Stremersch, Social contagion and income heterogeneity in new product diffusion.
71. Wejnert, Integrating models of diffusion of innovations.
72. Sandoval, G. 2005. Dueling DVD formats put buyers in the middle. *The Plain Dealer*, August 24: C3.
73. Rogers, *Diffusion of Innovations*.
74. Wejnert, Integrating models of diffusion of innovations.
75. Ibid.
76. Rhoads, C. 2007. Web site to holy site: Israeli firm broadcasts prayers for a fee. *Wall Street Journal*, January 25: B1–2.
77. Dhebar, A. 1995. The introduction of FM radio (A), (B), and (C), *Harvard Business School Teaching Note*, Number 5–594–072.
78. Girfalco, *Dynamics of Technological Change*.
79. Three Mile Island Accident, http://en.wikipedia.org/wiki/Three_Mile_Island_accident.
80. Boudette, N. 2006. GM hopes engine of the future sells cars now. *Wall Street Journal*, November 29: B1, B2.
81. Rogers, *Diffusion of Innovations*.
82. Dranove, D., and N. Gandal. 2003. The DVD-vs.-DIVX standard war: Empirical evidence of network effects and preannouncement effects. *Journal of Economics and Management Strategy*, 12(3): 363–386.
83. Girfalco, Dynamics of Technological Change.

Chapter 4

Sources of Innovation

Learning Objectives

After reading this chapter, you should be able to:

1. Define a technological opportunity, and explain why such opportunities exist.
2. Spell out how technological, political, regulatory, social, and demographic changes generate opportunities for technological innovation.
3. Identify the different loci of innovation, and explain how different institutions contribute to technological innovation.
4. Describe the roles played by the public sector in the national innovation system, and explain how universities help private firms to innovate.
5. Identify the different components of R&D, and figure out the pros and cons of investing in R&D.
6. Describe the different ways that organizations link R&D activities, and account for the trade-offs between the different approaches.

7. List the different forms of innovation, and explain why technological innovation takes these different forms.
8. Interpret the effects of industry on the forms of innovation, and figure out the right strategy for exploiting the different forms of innovation.

Technological Opportunities: A Vignette[1]

Howard Becker, the CEO of a Mayfield Heights, Ohio, company called Comet Video Technologies LLC, took advantage of two sources of innovation in developing his company's first product, Comet Vision. This device can send streaming video to other computers, cell phones, or personal digital assistants when it is connected to a video camera, a computer, and the Internet. Becker is targeting this product at parents of older children who want to know what the children are doing when they are home alone. By attaching up to six video cameras to a computer, parents can see what is happening inside their homes by accessing their office computers, cell phones, or personal digital assistants.

Becker's company took advantage of an important technological change, the development of a wireless compression technology that allows video to be streamed to telephones and personal digital assistants at a low bandwidth. Most video is transmitted electronically through mathematical algorithms, such as JPEG and MPEG. However, these algorithms require a high bandwidth. Comet Video Technologies LLC developed a new algorithm that transmits video at a low bandwidth, making possible video transmission to a wider variety of electronic devices.

The company also took advantage of an important social change, the concern of working parents about what their children are doing when the children are not with them. With a greater number of children who are home alone after school, getting into trouble, many parents want a way to keep an eye on what their children are doing when they cannot be at home.

Comet Vision shows how entrepreneurs—and managers of established companies—often develop technological innovations in response to social and technological changes that create opportunities for new products and services. It also demonstrates how many opportunities result from a combination of different changes, in this case social and technological.

INTRODUCTION

Opportunities to create new products and processes emerge from change—in technology, in politics and regulation, and in social and demographic factors. Effective technology strategists either create or identify those opportunities and figure out the best way to take advantage of them—by creating new products, establishing new production processes, organizing businesses in new ways, introducing new inputs, or tapping new markets.

The first section of the chapter examines the sources of opportunities, focusing on technological, political-regulatory, and social-demographic changes. The second section focuses on the entities that engage in technology development and explains how each of them contributes to the innovation process. The third section explains why firms invest in research and development, despite an inability to capture all of the returns to investment in it. The final section examines the different forms that innovation can take, and the key relationships between the types of changes that occur and the form of innovation.

SOURCES OF OPPORTUNITIES

The technological innovation that firms undertake is often triggered by some kind of change, whether that change is the result of organized human action or is accidental. For example, the invention of the laser was a source of opportunity because that invention made it possible to develop a new product for storing music (the compact disk), a new product for ringing up groceries (the checkout scanner), and a new service for correcting vision problems (laser eye surgery) among other things. In the absence of the invention of the laser, the creation of these products and services would not have been possible.

Researchers have identified three major sources of opportunity for innovation: technological change, political and regulatory change, and social and demographic change, which are discussed next.

Technological Change

As you might expect, technological change is one of the most important triggers of technological innovation, largely because technological change allows people to do things that could not be done before or only could be done in a less efficient manner.[2] Take, for example, the invention of the computer software behind e-mail. This software made it possible to communicate in ways that people find "better" for many purposes than communicating by telephone, fax, or letter. The technological change here—the advance in computer software—opened up an opportunity for innovation: the creation of a new communication product called e-mail.

It's important for you to understand the connection between technological change and opportunities for innovation. Unfortunately, it isn't a straightforward, one-to-one relationship. Many technological changes do not make any innovations possible, while others generate a multitude of opportunities. For example, less than 5 percent of all newly patented inventions ever lead to a commercial product or service.[3] On the other hand, some technological changes, like the microchip, have led to the creation of a very large number of new products.[4]

Sometimes, the introduction of a new technology creates demand for new products to counteract its negative side effects, as well as to exploit its benefits. For instance, the development of the Internet has helped companies by allowing their employees to get easier access to information that facilitates purchasing, evaluation of potential products and employees, information search, and a variety of other uses. However, Internet connections at the office also allow employees access to pornographic, gambling, hateful, and other types of Internet sites that could potentially open up companies to liability.[5] Thus, the development of the Internet has created the opportunity for companies to produce Web-filtering software that allows Internet search while blocking objectionable sites.

Technological change also does not immediately lead to opportunities for innovation, making the time from invention to innovation uncertain and potentially very long. Take, for example, the case of the photocopy machine. Chester Carlson invented xerography in 1937, but it was not until 1949 that the first Xerox photocopying machines were introduced to the market.

Many times this delay occurs because **complementary technologies**, or technologies that are used along with the focal technology, need to be invented before an innovation can be developed. For example, the development of high-speed radio devices, like cellular telephones, first required advances in complementary metal oxide semiconductor (CMOS) technology, which made it possible to use CMOS chips rather than just using ones made from gallium arsenide or silicon germanium. As a result, cellular telephones were not immediately developed once they were technically possible.[6]

What makes predicting the timing of the relationship between technological change and opportunities for innovation so complicated is that technological change is rarely linear (as we discussed in the Chapter 2). Take fuel cells as an example. The invention of fuel cells did not immediately open up the opportunity to change automobile engines from internal combustion to fuel cells. Why? Because the cost effectiveness of producing fuel cells has grown very slowly. An automobile needs an engine that generates 100 kilowatts of power. Because it costs about $5,000 to produce the power plant of a car, that's a price of about $50 per kilowatt. However, fuel cells currently cost about $3,000 per kilowatt. Consequently,

FIGURE 4.1
Technological Change Is an Important Source of Innovation

Even the problems created by the introduction of new technologies create opportunities for other innovations.

Source: http://www.speedbump.com, January 23, 2007.

right now, there is no commercial opportunity to build cars powered by fuel cells. As improvements in the technology lower the per kilowatt cost of fuel cells, the opportunity for people to make fuel cell cars should open up.

Important Attributes of Technological Change

Because there is not a simple, one-to-one relationship between technological change and innovation, technology strategists need to evaluate technological changes to determine if they are likely to make valuable innovations emerge in the future. So what attributes of technological change are associated with subsequent innovation? The magnitude of the technological change is one important factor. Changes of greater magnitude open up more opportunities for innovation because they allow new technology to be used in more ways than changes of small magnitude permit.

Take, for instance, the creation of a new type of electrical circuit. If that new circuit is only 10 percent faster than an older one, it will replace the older circuit in only a small number of products. Only those products in which a 10 percent improvement would exceed the cost of the change will use the new circuit. In contrast, if the new circuit is 500 percent faster than the older one, its benefits will exceed the cost of the change in a much larger range of products.

Then there is the generality of the change. Some new technologies, like genetic engineering or nanotechnology, are **general purpose technologies**. They make possible the creation of a wide range of new products. For instance, the invention of genetic engineering has led to the creation of new products in human and animal health-care, agriculture, and industrial chemicals, while the invention of nanotechnology has made possible new products in televisions, sensors, transistors, batteries, resins, and clothing.[7]

General purpose technologies are particularly important to innovation by new firms because these technologies permit strategic flexibility. If a new company is exploiting a general purpose technology and one market application proves to be inappropriate, the company can shift to other applications, minimizing the likelihood that the company will reach a dead end and will have to shut down.

General purpose technologies do not provide the same benefit to established firms because these firms see general purpose technologies through the lens of their existing businesses. Consequently, they can rarely take advantage of the flexibility that these technologies provide, given their unwillingness to pursue technology development when the value of a technology lies in another industry.

The commercial viability of the change also matters. Some new technologies do not offer much commercial potential—at least not for a long time. For instance, the space shuttle is a very large change over alternatives for getting into space because it saves a huge amount of money over rockets that cannot be reused. However, the commercial benefits of the space shuttle are rather limited because there are only a few commercial applications to which it can be put.

Of course, you need to be careful in thinking about the commercial viability of the change. Sometimes, a change will lead to a large number of commercial outcomes, but people don't recognize that immediately. This is what happened with the laser. At the time it was invented, no one thought that it would have any commercial use; but just the opposite happened. The laser turned out to have commercial viability in a wide range of areas. The lesson here is that the commercial viability of a technological change affects subsequent innovation, but it is often very difficult for entrepreneurs and managers to see those opportunities at the time that the change occurs.

Political and Regulatory Change

Political and regulatory changes also make innovation possible in a variety of ways. Regulation sometimes makes innovation possible by providing **subsidies** that pass off the cost of innovation to the government, thus reducing the cost of innovation to the companies undertaking it.[8] (A subsidy is a payment by the government that makes up the difference between what customers will pay for a product or service and the cost of producing it.) For example, the governments of several European countries created Airbus, an aerospace rival to Boeing, by subsidizing the cost of R&D investments by that company. Similarly, changes in federal tax credits for hybrid vehicles in the United States have increased the incentives for auto companies to produce those cars.

Another way that political and regulatory change can make innovation possible is—perhaps paradoxically—by spurring firms to come up with new products and processes to solve problems created by regulation.[9] For example, European Union restrictions on the importation of hazardous materials led Corning to develop a new type of glass for liquid crystal displays that is free of heavy metals, and can be used in televisions and laptop screens sold in Europe.[10] Similarly, the Sarbanes-Oxley Act, which requires publicly traded companies in the United States to ensure that their accounting and finance systems can detect and deter fraud, has created an opportunity for software companies to create new products to store audit-related materials electronically.[11]

Political and regulatory change can also make innovation possible by spurring competition between firms. As firms seek to generate competitive advantages, they often innovate. For instance, the deregulation of telecommunications led firms in that industry to compete to introduce less expensive ways to transmit voice and data that gave them a price advantage over their competitors.[12] Similarly, regulations in France that required the national telephone company to make its phone lines available to other Internet service providers has spurred competition that has led those providers to offer download speeds over 20 times faster than in the United States, where federal courts have blocked regulators from implementing similar policies.[13]

Political and regulatory change can make innovation possible by providing access to resources that permit the development of new products and services. This

FIGURE 4.2 Political and Regulatory Change

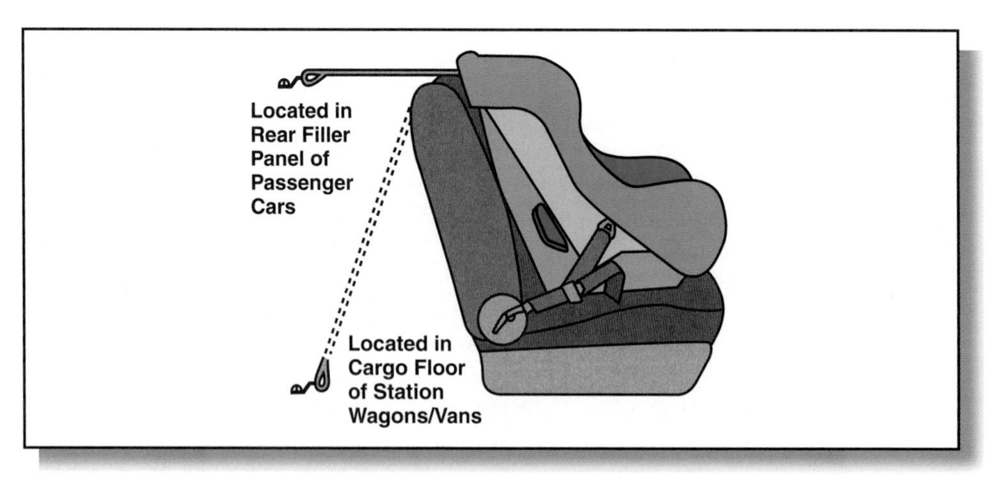

The child's car seat shown here is a product that resulted from changes in government regulations about such things as the mechanism to latch the seat to the car and the stability of the base.

Source: National Highway Traffic Safety Administration, http://www.nhtsa.dot.gov/CPS/safetycheck/TypeSeats/index.htm.

often occurs when the government controls the access to key resources and changes the regulations about how companies can access them. For instance, in the United States, the Federal Communications Commission's decision in 1997 to open up 300 MHz of the radio spectrum in the 5 GHz frequency for short-range high-speed digital communications created opportunities for companies to develop an innovative product called wireless local area network (LAN).[14]

Social and Demographic Change

Another important category of change that generates opportunities for innovation is social and demographic change, which opens up opportunities by altering people's preferences and by creating demand for products that had not existed before. For instance, the shift of women into the workforce, and the corresponding increase in demand for speed in food preparation, created the opportunity to introduce many types of frozen food into the marketplace. Similarly, as Figure 4.3 shows,

FIGURE 4.3
Social and Demographic Change

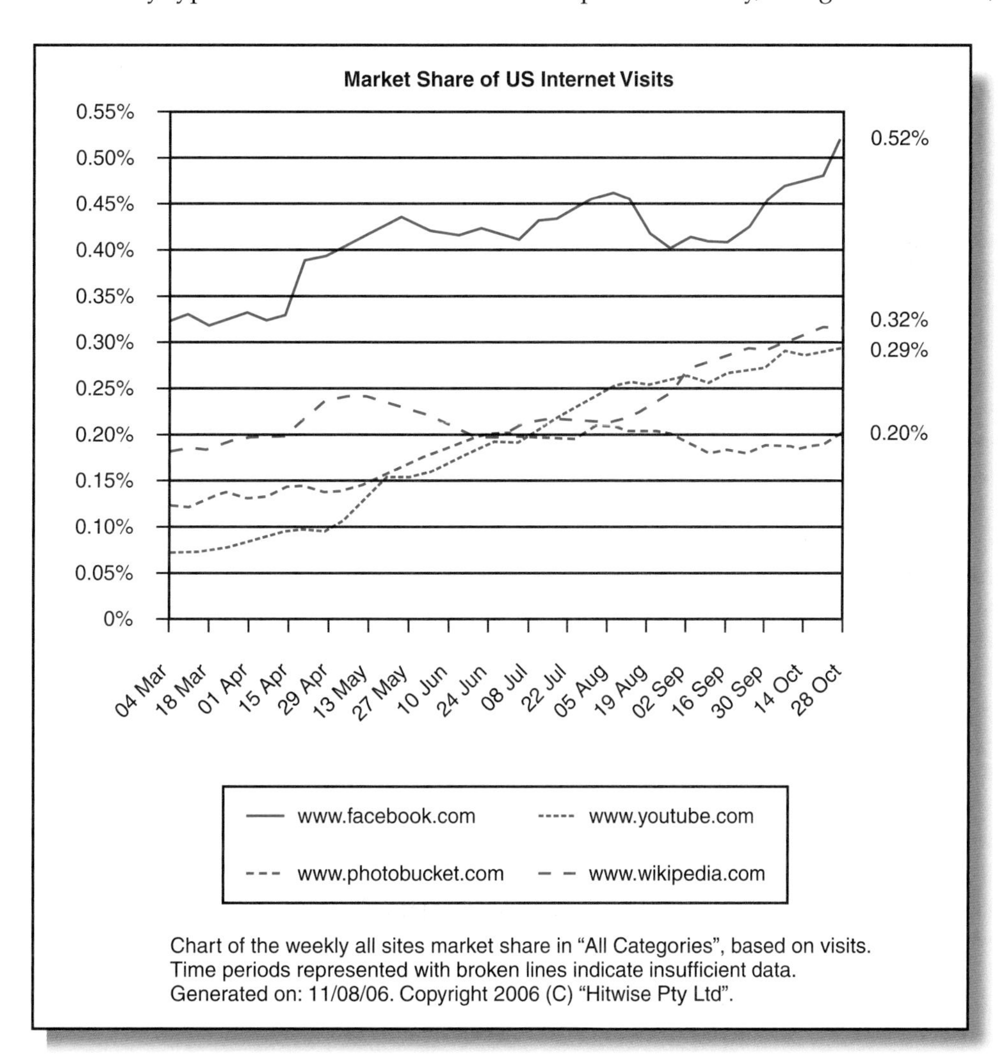

Social trends toward the use of technology to keep in touch with friends created the opportunity for social networking Web sites like Facebook, which, as the figure shows, has become one of the more heavily viewed Internet sites.

Source: http://weblogs.hitwise.com/leeann-prescott/2006/11/social_networking_sites_recove.html.

a desire of people to be more closely connected and in greater touch with their friends led to the development of social networking Web sites, like Facebook and MySpace, which have quickly become some of the most used Internet sites.

One important type of opportunity-generating social and demographic change is a social trend, which leads potential customers to change their needs and preferences. For example, the opportunity to produce deodorant was the result of a social trend that led people to believe that body odor was offensive. While there is actually no medical or health need for people to mask their body odor, the trend in believing that body odor is offensive opened up the opportunity for companies to create products that mask that odor.

Demographic trends are another important source of opportunity to introduce innovative new products and services. For example, as birth rates decline and people live longer, the population of many developed countries is aging. This demographic trend makes it possible to introduce products and services targeted at the elderly, for which sufficient demand did not exist 25 years ago. In Japan, Toyota is responding to this demographic trend by producing Welcabs—specially designed cars with wheelchair ramps and special seats that make it easier for the elderly to get in and out.[15]

A third type of opportunity-creating social or demographic change is a shift in perception. Sometimes, an opportunity to innovate occurs because people perceive things in a new way. Take, for example, the use of vaporized hydrogen peroxide to clean facilities contaminated with anthrax and other biological agents. Steris Corporation, a Cleveland, Ohio, medical device firm, adapted this technology, which it had originally developed for decontaminating pharmaceutical clean rooms, to work against biological agents. Why? The anthrax terrorism scare in the Hart Senate Office Building and U.S. post offices changed people's perception about the need for government offices to be clean of biological contaminants.[16]

Combination

The previous discussion of the effects of technological, political, regulatory, social, and demographic change on innovation described these sources of opportunity separately from one another. However, this approach is purely an artifact of exposition. (The relationship between the change and the creation of opportunities to innovate would not be as clear if the changes were not discussed separately.) In reality, opportunities for innovation are often the result of many different types of changes.

Take, for example, the efforts of companies like Net Nanny and Cyber Patrol to introduce new products to protect children against harmful Internet content. The need to protect children against certain Internet content came about both because of a technological change (the invention of the Internet) and because of a social trend (the tendency for children to be home alone after school).

Similarly, in the United States, a start-up company called iFly has developed an air taxi service that will let corporate travelers fly in small planes between local airports, much like a limousine service takes passengers between locations within a city. The opportunity for iFly's innovation came about as a result of both technological, and political and regulatory changes. On the technological side, the development of four-person micro jets with fuel efficient engines, something that previously had not been technically feasible, was an important change that made this innovation possible. On the political and regulatory side, the time-consuming post-September 11 security procedures at U.S. airports was an important source of the opportunity.[17]

Key Points

- Technological, regulatory, political, social, and demographic changes are all sources of innovation and can operate separately or in combination.
- Technological changes vary in their magnitude, generality, and commercial viability, all of which influence opportunities for innovation.
- Predicting the relationship between technological change and the opportunity for innovation is difficult because the relationship is not always one-to-one, leads to the creation of additional opportunities, and because the effect of technological change on the creation of opportunity is rarely immediate.
- Political and regulatory change affects opportunities for innovation by changing the costs of, or rewards to, innovation; by spurring firms to respond to problems created by regulation; by providing potential innovators with access to resources and subsidies; and by spurring competition between firms.
- Social and demographic changes affect opportunities for innovation by altering people's preferences and by creating demand for things that had not existed before.

Locus of Innovation

A variety of different entities undertake technological innovation, including companies, individuals, universities, and government agencies. Of these four sources, companies account for the most technological innovations. Sometimes company innovation occurs because businesses are trying to come up with new products or services to sell to customers or to improve their processes for making and distributing those products or services. Other times, however, companies develop technological innovations because no supplier offers a product or service that they need.[18] For example, Texas Instruments, a customer of semiconductor manufacturing equipment, developed the planar process for making semiconductors on a silicon substrate because no other companies could provide it with that technology.[19]

Another important source of technological innovation is the government. Federal government agencies in many countries come up with their own technological innovations. For instance, researchers at Los Alamos National Laboratory developed a radio frequency identification system for locating nuclear materials, which led to the development of automated toll payment systems, like the E-ZPass.[20]

The government also helps develop technological innovation by sponsoring basic research undertaken by universities and industry,[21] particularly in the life sciences and engineering (see Figure 4.4). For instance, the Defense Advanced Research Projects Agency (DARPA) of the Defense Department has funded much of the basic research in computer science in the United States, including basic research on the Internet that began in the 1960s.[22]

FIGURE 4.4
Share of Federal Funding of Different Technical Fields

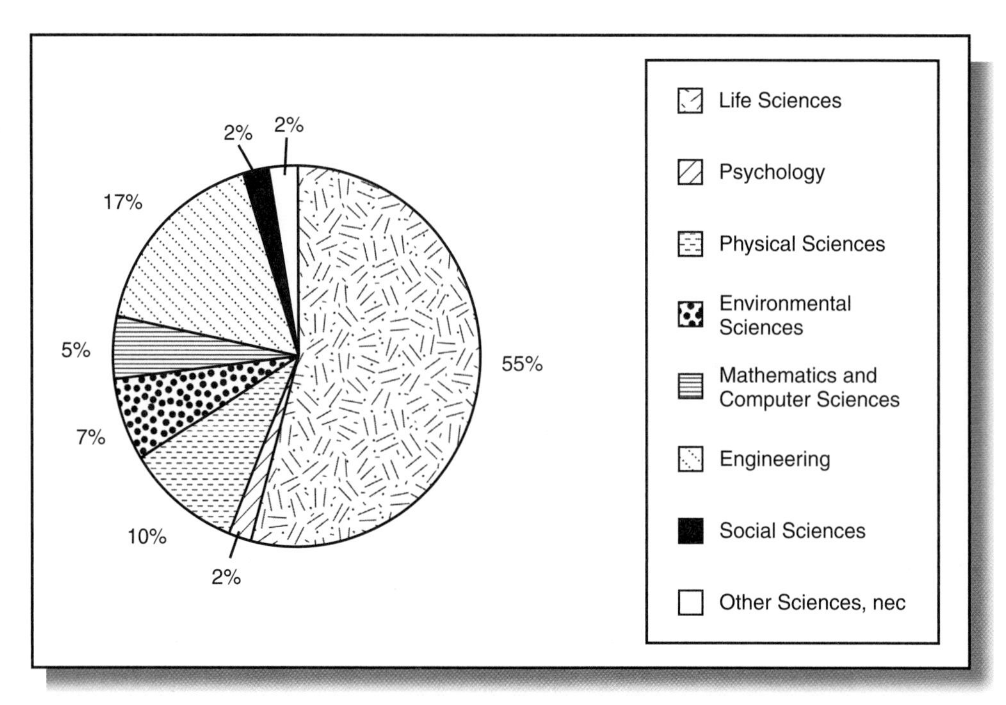

Federal funding of research is not equal across fields; life sciences account for over half of all federal funding of R&D in the United States.

Source: National Science Foundation/Division of Science Resources Statistics, *Survey of Federal Funds for Research and Development*, Fiscal Years 2002, 2003, and 2004.

The government also sponsors corporate R&D. By paying for some of the R&D costs of private companies, governments increase the willingness of firms to develop technological innovations.[23] For instance, General Motors Corp., Ford Motor Co., Hyundai Motor Co., and DaimlerChrysler are all recipients of U.S. Department of Energy funding under a program to develop vehicles that run on fuel cells, rather than on gasoline.[24]

To help industry develop new technology, in 1986, the U.S. federal government created a new arrangement to allow joint business-government research efforts called Cooperative Research and Development Agreements (CRADA). Under these agreements, government laboratories can participate in cooperative research with industry with the government matching the financial contributions of industry with contributions of personnel and equipment.[25] As Figure 4.5 shows, while these agreements were once very popular, they are now being created at a much lower rate.

In addition to its funding of research, the government also supports technological innovation by serving as a lead customer, and paying a high price, for the initial versions of new technology products.[26] The classic example of a government agency that serves as a lead customer is NASA, the U.S. space agency, which pays a high price for the initial versions of many new technology products, which then are developed for other customers.

Individuals are a third locus of innovation. While individuals once accounted for a much larger number of technological innovations than they do today, they are still an important locus, particularly for innovations that lead to the creation of new

FIGURE 4.5
CRADA Alliances

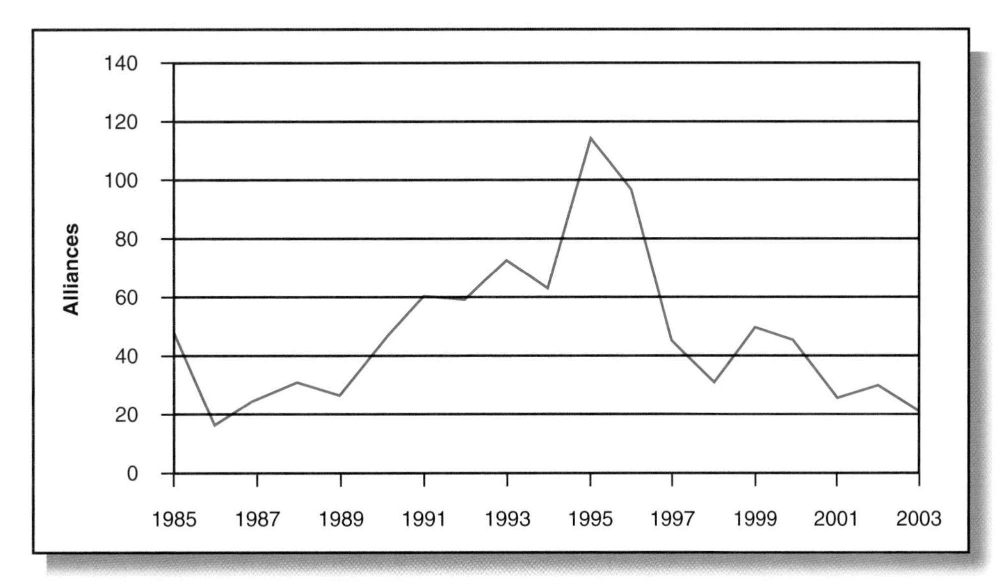

This figure shows that the number of new CRADA alliances formed each year grew substantially from the mid 1980s to the mid 1990s, but then returned to the level of the late 1980s.

Source: Science and Engineering Indicators, 2006, http://www.nsf.gov/statistics/seind06/c4/fig04–18.htm.

companies. For example, Steve Jobs and Steve Wozniak created the Apple I personal computer in the garage of Steve Jobs's parents before starting a company, making the original Apple computer an individual innovation.[27]

Academic institutions are a fourth important locus of innovation. In fact, one study showed that 11 percent of all new products and 9 percent of all new industrial processes would not have been developed in a timely manner if it were not for academic research.[28] Moreover, Figure 4.6 shows that the amount of research conducted at universities is growing. Since the 1960s, two forces have led to the increasing amount of real university research funding: the financing of defense-related R&D during the cold war and later the war on terrorism; and the financing of biomedical research and the rise of National Institutes of Health (NIH) funding. In 1998, academic research funding further accelerated because economic growth and increases in tax revenues led to a federal budget surplus and a relaxation of restrictions on federal spending on university research.[29] R&D spending in industry has declined some in recent years (but then rebounded) because of the movement of much R&D activity out of the United States to other countries, a shift of many large American companies out of basic research, the bursting of the information technology bubble in 2000, and the decreasing demand of the federal government for military-related innovations.

Universities often conduct basic scientific research that leads to applied R&D, which results in technological innovation. For example, Ben & Jerry's has sought to build an acoustic freezer based on the results of U.S. Navy funded research at Penn State that uses sound waves instead of chlorofluorocarbons and hydrofluorocarbons as a cooling mechanism.

In addition, many universities license the technological innovations developed by their faculty, staff, and students to the private sector and share the royalties from those licenses with the inventors.[30] While this activity has been

FIGURE 4.6 The Rise of Academic Research

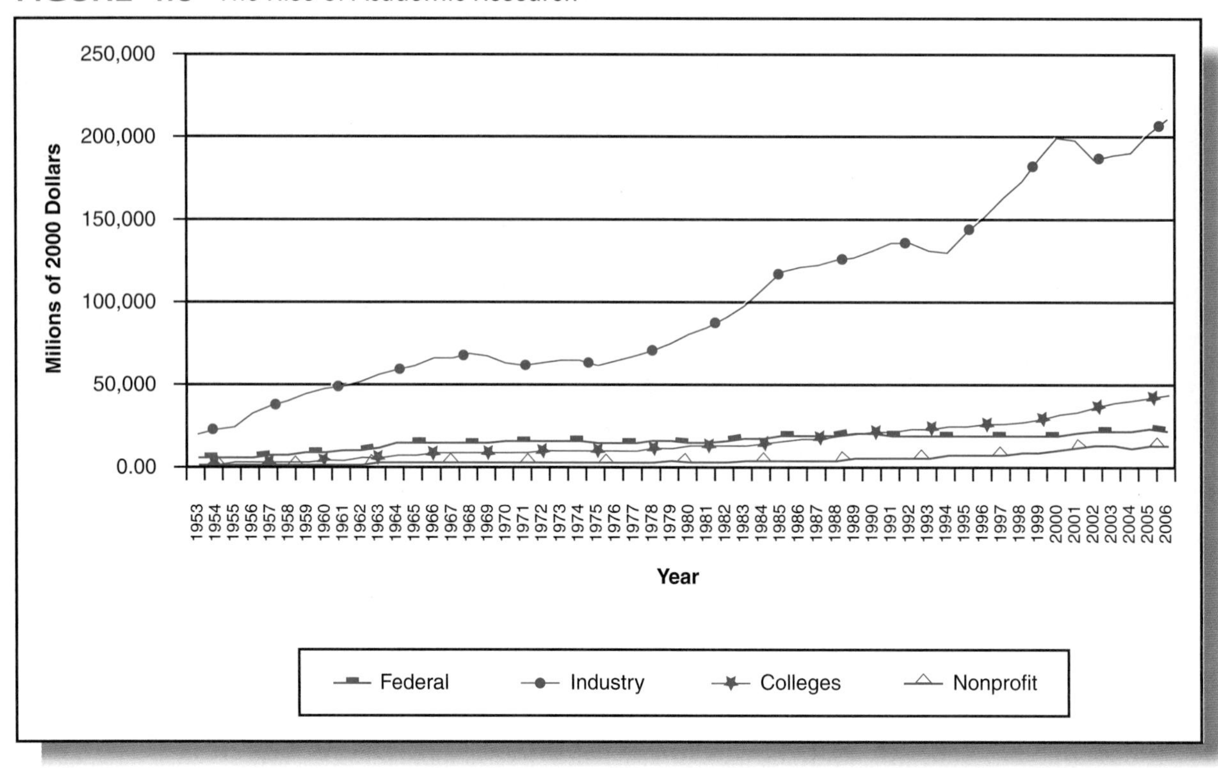

The amount of research undertaken at universities and colleges has grown dramatically over the past five decades, and it is now above the level undertaken by the federal government.

Source: Created from data contained in National Science Foundation, Division of Science Resources Statistics, National Patterns of R&D Resources, annual series. See appendix table B-1.

occurring since the 1920s, the number of universities engaged in it has grown dramatically since the **Bayh-Dole Act** was passed in 1980. This law gave universities the rights to federally funded inventions, which account for about two-thirds of all university inventions, thus creating a strong incentive for universities to license their inventions to industry. Gatorade, the sports beverage, and Taxol, the cancer drug, are both examples of new products developed from licenses to university inventions.

Universities also train students, who bring cutting-edge knowledge to the businesses that hire them.[31] This role of universities is so important that Intel funds semiconductor researchers at several major U.S. universities just so that they can hire the students of those researchers as development engineers.[32]

Because universities are an important locus of technological innovation, many companies develop strategies to leverage the value of university research. For instance, Intel has established several small research labs just outside the campuses of the University of California at Berkeley, Carnegie Mellon University, the University of Washington, and Cambridge University. In these labs, 20 Intel and 20 university researchers work side-by-side to learn about university research and figure out how to transfer it to Intel.[33]

All of the previous not withstanding, universities are a more important locus of innovation for some technologies than others. For instance, university research is particularly important in semiconductors, measuring and controlling devices, pulp and paper, drugs, petroleum refining, aircraft, surgical and medical instruments, and computing.[34] And the academic fields of chemistry, computer science, electrical engineering, materials science, and mechanical engineering are more important than other disciplines to technological innovation.[35] Therefore, the degree to which technology strategists focus on universities as a locus of innovation depends a great deal about what businesses their companies are in.

Key Points

- Business undertakes most of the R&D conducted in the United States.
- Individuals conduct much less of the technological innovation than they once did.
- The government plays an important role in the technology innovation process by conducting research, paying for research done by others, and by serving as a lead customer.
- Universities help firms innovate by training students, by conducting research, and by licensing technology developed by their faculty, staff, and students.

Research and Development

As the previous section indicated, not all innovation is reactive. Companies can, and do, create innovations through deliberate investment in R&D efforts designed to create new products and services, and new processes for providing them. The implicit model behind corporate R&D is shown in Figure 4.7. Investments are made in R&D to create technological inventions. Those inventions make possible innovations, which, in turn, are embodied in new products and services that meet previously unmet needs or allow existing products and services to be produced more efficiently. All of this is done with the goal of capturing the commercial value generated from the effort.

Types of R&D

R&D is composed of three different activities: Basic research, applied research, and development. **Basic research** is the effort to understand the technical or scientific

FIGURE 4.7 Model Underlying Investment in R&D

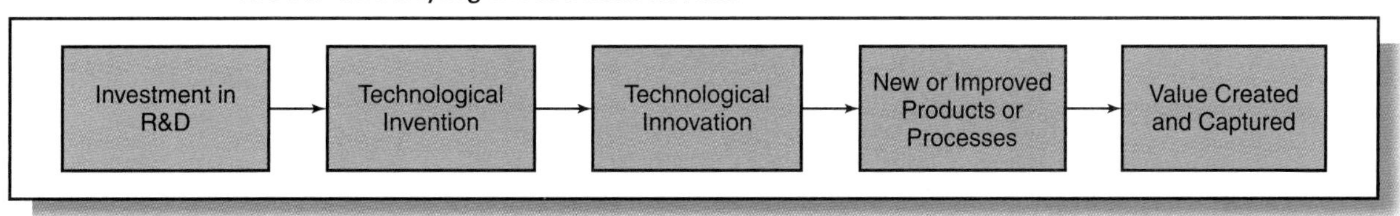

This figure shows the process model underlying investment in R&D; companies invest in R&D to create technological inventions that make possible innovations, which they can use to produce new products and services whose commercial value they try to capture.

principles in a field. Usually, it is undertaken without commercial goals in mind. For example, scientists conducted basic research in physics that led to the discovery of the transistor.[36]

Applied research is the effort to understand technical or scientific principles with a specific commercial goal in mind. For example, scientists and engineers conducted applied research on the transistor to develop a new type of radio receiver that could be used with it.[37]

Development is the effort to use technical knowledge to produce something of commercial use. For example, engineers used knowledge of transistors and radio receivers to design a transistor radio that could be sold to customers.

Why Firms Invest in R&D

Of the approximately $300 billion spent annually on R&D in the United States, industry accounts for approximately 71 percent, the federal government accounts for 8 percent, colleges and universities account for 16 percent, and nonprofit organizations account for 5 percent.[38]

However, as Table 4.1 shows, private sector R&D expenditure is not evenly distributed across firms. Large firms account for most of the R&D spending (though not as a percentage of sales); and, even among them, R&D expenditure is highly skewed,

TABLE 4.1
Top 20 R&D-Spending Corporations in 2003
R&D spending by companies is highly concentrated in a small number of businesses; just three companies—Microsoft, Ford, and Pfizer—accounted for 11 percent of all U.S. R&D.

Company (country)	R&D (millions) 2003	R&D as Percent of Sales 2003
Microsoft (United States)	7,779	21.1
Ford Motor (United States)	7,500	4.6
Pfizer (United States)	7,131	15.8
DaimlerChrysler (Germany)	6,689	4.1
Toyota Motor (Japan)	6,210	3.9
Siemens (Germany)	6,084	6.8
General Motors (United States)	5,700	3.1
Matsushita Electric Industrial (Japan)	5,272	7.7
International Business Machines (United States)	5,068	5.7
GlaxoSmithKline (United Kingdom)	4,910	13.0
Johnson & Johnson (United States)	4,684	11.2
Sony (Japan)	4,683	6.9
Nokia (Finland)	4,514	12.8
Intel (United States)	4,360	14.5
Volkswagen (Germany)	4,233	4.0
Honda Motor (Japan)	4,086	5.5
Motorola (United States)	3,771	13.9
Novartis (Switzerland)	3,756	15.1
Roche Holding (Switzerland)	3,694	15.3
Hewlett-Packard (United States)	3,652	5.0

Source: Adapted from information in Science and Engineering Indicators 2006, http://www.nsf.gov/statistics/seind06/c4/tt04–06.htm.

TABLE 4.2
R&D Intensity of Different Industries
There is wide variation in the R&D intensity of the different industry sectors, with professional, scientific, and technical services investing over 10 percent of their sales in R&D, and utilities investing only 0.10 percent.

Industry	R&D Intensity
Beverage and tobacco products	0.5
Chemicals	5.6
Computer and electronic products	9.3
Construction	1.2
Electrical equipment, appliances, and components	2.2
Fabricated metal products	1.5
Finance, insurance, and real estate	0.3
Food	0.6
Furniture and related products	0.8
Health-care services	2.2
Information	5.7
Machinery	4.2
Management of companies and enterprises	4.1
Mining, extraction, and support activities	3.3
Miscellaneous manufacturing	5.6
Nonmetallic mineral products	1.0
Other nonmanufacturing	4.9
Paper, printing, and support activities	1.1
Petroleum and coal products	0.3
Plastics and rubber products	2.1
Primary metals	0.7
Professional, scientific, and technical services	10.2
Retail trade	0.8
Textiles, apparel, and leather	1.0
Transportation and warehousing	0.4
Transportation equipment	2.7
Utilities	0.1
Wholesale trade	3.6
Wood products	0.7
All industries	**3.2**

Source: Adapted from Science and Engineering Indicators, 2006, http://www.nsf.gov/statistics/seind06/append/c4/at04–22.xls.

with only eight companies accounting for one-quarter of all R&D conducted by the private sector in the United States.[39]

Moreover, as Table 4.2 shows, private sector R&D expenditures vary greatly across industries, with some industries spending a much higher percentage of their sales on this activity than others.

This variation across companies and industries in investment in R&D suggests that companies have both reasons to invest and reasons not to invest in R&D, and the relative balance between the two depends on the industry and the firm. So why do firms invest in R&D? There are basically five reasons:

1. To create new technologies that can serve as the basis for new products and services.[40] For example, DuPont invested in research in polymers to create nylon,[41] and Monsanto invested in research in genetic engineering to create new agricultural pest control products.[42]

2. To develop products to replace those threatened by substitutes. For example, AMP invested in research to replace its coaxial cable products because the company's managers believed that fiber optics would substitute for coaxial cable and make its products obsolete.[43]
3. To differentiate products from those of competitors. For example, many machine tool makers in the United States and Germany invest in R&D to develop versions of their products that have features not present in the products made by their lower-cost Chinese competitors.
4. To create strong intellectual property positions by making fundamental technological discoveries on which pioneering patents can be obtained. For example, investment in R&D allowed Texas Instruments to develop the pioneering patent for the integrated circuit, which positioned the company as a leader in semiconductor technology for many decades.
5. To create **absorptive capacity**, which is the capability to recognize and use knowledge from elsewhere. This capacity allows companies to use information from customers, suppliers, and even competitors to develop new products and services, or new ways of making them.[44] For example, IBM conducts R&D on X-ray lithography even though it doesn't use that technology to make semiconductors so that it can recognize and take advantage of any developments made outside the company to use that technology for the purpose of making semiconductors.[45]

Costs of R&D

While investing in R&D provides the benefits just described, they come at a cost. First, investments in basic research are uncertain, leading many investments to fail to generate a positive return. R&D projects often require years of investment before they reach the point of commercial viability, and only a small portion of the projects ever yield commercial outcomes. For instance, to develop a single cholesterol drug, Pfizer Inc. had to examine 400,000 chemicals, looking for possible drug leads. It then had to test those drug leads on animals to determine efficacy. Following that, Pfizer had to go through several stages of FDA approval before it had a cholesterol drug that it could sell on the market. Until the company had an approved cholesterol drug on the market, it could not capture any value from the hundreds of millions of dollars that it had invested in the development of the drug.[46]

As a result, almost all of the financial returns from R&D projects are generated by a handful of projects, many years after the investments are first made, leaving average returns poor. For example, investors in public biotechnology companies have put more than $100 billion into those companies since they were first founded, but the cumulative returns for all public biotechnology companies since the formation of the industry has been negative $40 billion.[47]

Second, investments in basic research often lock companies into strategies that are hard to change. Because the returns to investment in R&D only occur after years of investment, companies have to stick with particular development paths over long periods of time to benefit financially from investments in basic research. Unless a company can accurately predict the direction of technological change far into the future, it might become locked into the wrong strategic focus by selecting the wrong research projects and hiring the wrong researchers.

For example, to conduct basic research on semiconductors, IBM must hire researchers with expertise on particular classes of molecules. These researchers cannot easily change their research focus to other classes of molecules, let alone to investigate Internet search engines. So if IBM is wrong about the direction of technological development, it will be stuck with semiconductor researchers whose work leads to dead ends, while lacking the expertise necessary to develop technology in a growth area like search engines.[48]

Third, the returns from investments in R&D are difficult to appropriate. (**Appropriability** is the degree to which the value that results from an investment in the development of an innovation can be captured by the company that made the investment.) This lack of appropriability stems from two factors:

- R&D has positive externalities, which make it possible for companies that did not pay for the R&D to benefit from the technological innovations of other firms.[49] For example, Xerox's discoveries of the Ethernet, graphical user interface, and computer mouse helped other firms figure out how to develop products from these technologies.
- Knowledge leaks, or spills over, from one firm to another. For example, when one company hires employees away from another company, those employees bring with them knowledge of what their prior employer was doing, creating **knowledge spillovers**.[50] (As a result, some companies, such as Samsung[51] deliberately try to capture knowledge from other companies by systematically hiring their employees, particularly in technical areas that are new to the hiring company.[52])

Linking Research and Development

Part of technology strategy is concerned with linking research and development. The failure to link basic research with subsequent development will keep you from earning financial returns on your investments in basic research. For example, Xerox lost many billions of dollars of potential returns on its investment in basic research on the computer mouse, the graphical user interface, and the Ethernet because the company had no mechanism for linking these inventions to the activities of its operating units, which controlled development projects.

However, creating effective links between research and development is not easy because the former is undermined when it is closely tied to the latter. If researchers become too heavily involved in short-term problem solving on behalf of operating units, then they do not have time for research projects.[53] Consequently, many

TABLE 4.3
The Benefits and Costs of R&D
There are advantages and disadvantages to conducting basic research, which firms must balance when deciding whether or not to undertake it.

Benefits	Costs
Creates products and services that meet customer needs	Is difficult to appropriate the returns
Replaces products and services threatened by substitutes	Is highly uncertain and does not yield high average financial returns
Develops the absorptive capacity to recognize the value of externally developed technology	Locks companies into particular strategies
Creates a strong intellectual property position	
Differentiates products from those of competitors	

companies disconnect research from development. For example, 3M recently centralized its 12 R&D groups to keep its researchers from becoming too involved in product development at the expense of basic research.[54]

Moreover, companies often lose their best researchers and find it difficult to attract new ones if they require researchers to spend too much time on development. Leading researchers want the freedom to select research problems, publish their results, and work on cutting-edge projects. These goals are much more easily accomplished when researchers focus on research than when they conduct development work.

Furthermore, transitioning technology from basic research to development is difficult and time consuming. Engineers (who conduct most development work) and scientists (who conduct most basic research) often work in different locations with different equipment, design rules, and materials. Therefore, the transition from basic research results to products often involves changing the materials from which a product is made, redesigning it, and changing the process through which it is created.

Take, for example, the problems that IBM faces when it wants to transition its semiconductor research to development projects. At IBM, researchers use germanium and gallium arsenide to produce semiconductors in low volume, while the development group uses silicon to produce semiconductors in high volume. Thus, IBM has to redesign its semiconductors to work in silicon when new technologies are transitioned to development from research.[55]

Finally, research scientists often have difficulty transferring information about their discoveries to development engineers. Scientists often believe that they are superior to engineers because they "do science," which leads to low levels of interaction and poor communication between basic researchers and development personnel.[56] And engineers, who create new technology products, tend to have very little understanding of the basic science underlying technology.[57] As a result, engineers often need scientific discoveries "translated" for them, which is problematic because scientists and engineers often do not have a common language or format for communication.[58]

Several policies encourage productive links between research and development.

1. **Giving researchers financial incentives to work on projects that are aligned with development goals.** Because people respond to financial incentives, these policies help to align research and development. For example, IBM has developed a program to give more funding to researchers who are willing to work more closely with development personnel than to those who will not work closely with them.[59]
2. **Requiring your research laboratories to use the same equipment and materials as your product development laboratories.** For instance, Intel requires its researchers to use the same equipment as development personnel, and to operate that equipment inside of an existing fabrication plant, which minimizes the need for redesign and retesting when discoveries move from research to product development.[60]
3. **Exposing basic researchers to development, perhaps for a temporary period when they first join your company.** For example, Intel requires all newly hired research scientists to first work in product development, reducing the interpersonal tension between research scientists and development engineers without detracting too much from the scientists' goal of conducting research.
4. **Making fit with company strategy a criterion in the evaluation of proposals for research funding.** By funding only R&D projects related to the capabilities of your firm, as opposed to everything that shows technical merit or feasibility, then you can better tie together research and development activities.[61]

GETTING DOWN TO BUSINESS
Investing to Create Opportunities[62]

Even in a product as low tech as panty hose, companies invest in R&D to come up with innovations that will increase demand and create competitive advantage over rivals. Some companies are using nanotechnology to design panty hose that have baked-in microcapsules filled with aloe. Because the microcapsules break as the wearer moves, these panty hose help the wearer to combat dry skin. Other companies are adding a derivative of menthol to the yarn that they use to make panty hose. The addition of menthol helps the user feel cooler when she puts on the panty hose, thus making the product more attractive to customers.

Corporate investment in R&D in panty hose occurs both in company laboratories and through funded research at universities. For example, at the industry-funded Textile Protection and Comfort Center at North Carolina State University, researchers investigate new technologies for the textile industry. In fact, the ideas behind the two panty hose innovations just described—nanocapsules filled with aloe and the addition of menthol to yarn—were developed there. Thus, even in as mundane a product as panty hose, companies conduct R&D to improve their products, and use universities to help them with their innovation efforts.

Key Points

- R&D is composed of basic research, applied research, and development.
- Industry performs more than two-thirds of all R&D in the United States.
- Companies invest in R&D to create new technologies that provide the basis for new products and services, to improve their products and differentiate them from those of competitors, to create strong intellectual property positions, and to absorb externally generated ideas.
- However, investment in R&D is uncertain; it rarely generates short term financial returns; it locks companies into strategies that are difficult to change; and its benefits are difficult to appropriate.
- Companies need to link research and development to capture the returns to basic research; however, this is not easy because too strong ties divert researchers to short-term problem solving, make employment unappealing to them, and make transitioning technology from basic research to applied development difficult and time consuming.

FORMS OF INNOVATION

When formulating a technology strategy, you need to consider the different forms that technological innovation can take. While people intuitively think of new technology making possible innovations that take the form of new products or services, those innovations can also take the form of new production processes that allow older products and services to be made in new ways, new raw materials that change the composition of products or services, new ways of organizing that change the way that existing products are produced and sold, and new markets where the products or services can be offered.[63] Table 4.4 shows some examples of different forms that innovations have taken.

The creation of the compact disk is an example of innovation, which occurred in response to the development of a new technology—the laser—and took the form of a new product.

TABLE 4.4 Relationship Between Technology and Form of Innovation
This table shows examples of business ideas that emerged when innovation took different forms in response to technological change.

TECHNOLOGICAL CHANGE	EXAMPLE OF A BUSINESS IDEA	FORM OF THE INNOVATION
Genetic engineering	Synthetic insulin	New product or service
Internet	Online pet food sales	New way of organizing
Refrigeration	Refrigerated ship	New market
Computer	Factory automation	New method of production
Nanotechnology	Smart paints	New raw material

Amazon.com's effort to sell books on the Internet is an example of an innovation, which occurred in response to the development of a new technology—the Internet—that took the form of a new way of organizing a firm. Amazon.com's product, the book, has been around a long time and is exactly the same when sold online as in a bookstore. The company's innovation lies in its way of organizing the business as a retail bookstore with no retail outlets.

The export of frozen meat by ship to distant locations is an example of an innovation that took the form of a new market. The invention of the refrigerated ship in the 1880s didn't change the product—the filets were still filets—and the business was still organized the same way. The difference was that Argentinean beef could now be sold in Europe as opposed to just in Argentina.

The substitution of ceramic composites for metals in the production of motor vehicles is an example of an innovation that took the form of new materials. The product, the car, and the market, you and me, are the same whether ceramics are used or not. Moreover, GM is still organized the same way whether ceramics are used in place of metals or not. Even the process of making cars doesn't change in response to this innovation.

Finally, continuous strip production is an example of a new production process used by steel minimills to make an existing product, steel, in a more efficient way. The product, market, and way of organizing the business are the same at Nucor as at U.S. Steel. The difference is that Nucor can make steel out of scrap, which the integrated steel mills' continuous casting process doesn't allow.

Forms of Innovation and Technology Strategy

The different forms that innovation can take matter for your company's technology strategy in two important ways. First, the value that your company can generate and capture from innovating depends on the form that its innovation takes. Sometimes producing old products and services by organizing in new ways, by using new raw materials, or by developing new production processes is more beneficial to a company than developing new products and services. For example, in many industries, the gains from applying nanotechnology lie in the development of new materials or new production processes, not in the development of new products.

Moreover, even if value that your company could generate was equal across forms of innovation, the amount of that value it could capture might be different for different forms of innovation. Successful innovation requires companies to develop products and services that competitors will not immediately imitate. The ability to deter imitation is often higher when the innovation takes the form of a new process or a new material than when it takes the form of a new product or service. As we will discuss in greater detail in Chapter 9, secrecy is an important barrier to imitation. Because new products and services are sold to customers, their workings cannot easily be kept secret from competitors who can buy a sample and reverse engineer it. However, production processes and input materials can be kept hidden from customers and competitors. Thus, by producing an old product with a new production process, your company will be better able to deter imitation and may capture more of the value from innovating than it will by producing a new product.[64]

Second, the form that your innovation takes affects your company's ability to get customers. It is easier to attract customers by targeting a new market with an existing product or service than it is to create a new product or service for an old market. Your company's efforts to innovate only matter to customers if they affect the attributes of the products or services that customers buy. If innovating does not affect those attributes—for example, books are the same whether bought online or at a bookstore—then customers will be more willing to purchase a product than if innovating changes those attributes. When your innovation changes product attributes, you also need to persuade customers that the new product is better than the old one.

Industry and the Nature of Innovation

As a technology strategist, you also need to be aware that the nature of innovation depends on the production process in an industry. Some products, like aluminum, petroleum, and steel, are nonassembled. Lacking components, these products are created through steps or subprocesses that are chemical, thermal, or machined. Other products are simple assembled products, made up of linked subsystems, such as stoves, guns, and skis.[65] Still other products are assembled systems, composed of distinct subsystems that interact and must be linked together, such as automobiles, televisions, and airplanes.

The differences in these production processes affect the nature of innovation in different industries. For instance, technological innovation in nonassembled products is most likely to take the form of improved materials, product performance, processes, or production scale.[66] Improving input materials is a common way to innovate in simple assembled products. And mechanization, automation, and standardization, and changes in products' dimensions, are very common ways to innovate in assembled systems. [67]

The locus of innovation also varies across industries. In some industries, like machinery, electrical equipment, and medical instruments, customers are an important locus of innovation. Because customers in these industries have an understanding of their unfulfilled needs, they often innovate to meet those needs, or work closely with their suppliers to develop innovations that would benefit them.[68] For example, Singer Sewing Machine Company developed the electronic cash register when it could not get any provider of mechanical cash registers to produce one for it.[69]

In other industries, such as food products, lumber, metal working, drugs, soap, and semiconductors, suppliers are an important source of innovation, pushing technological advance to increase demand for their products. For instance, Alcoa and Reynolds Aluminum, two metals companies, developed the two-piece aluminum can and gave the technology to produce it to can manufacturers as a way to increase demand for aluminum.[70]

Key Points

- Innovations can take the form of new methods of production, new raw materials, new ways of organizing, and new markets, as well as new products and services.
- Understanding the form that innovation takes is important because it affects the value that a firm can generate and capture from innovation, as well as the ease of attracting customers.
- The nature of innovation varies across industries.

Discussion Questions

1. What are some major sources of innovation? Why are these things sources of innovation?
2. Should companies invest in research? Why or why not?
3. Why is most research conducted by large corporations and not small ones?
4. Why do small firms have a greater R&D intensity than large firms?
5. What role does the government play in fostering innovation? Is it necessary to have the government involved to have innovation? Why or why not?
6. Is it a good or a bad thing that universities are engaging in a greater share of R&D in the United States? Why do you take this position?
7. What are some different forms that innovation can take? Why do different innovations take different forms?

Key Terms

Absorptive Capacity: The capability to recognize and use knowledge generated elsewhere.

Applied Research: The effort to understand technical or scientific principles with a specific commercial goal in mind.

Appropriability: The degree to which the value that results from an investment in the development of an innovation can be captured by the company that made the investment.

Basic Research: The effort to understand the technical or scientific principles in a field without particular commercial goals in mind.

Bayh-Dole Act: A law that gave universities the rights to federally funded inventions, thus creating a strong incentive for universities to license the inventions of their faculty, staff, and students.

Complementary Technology: A technology that is used along with a focal technology.

Development: The use of technical knowledge to produce something of commercial value.

General Purpose Technology: A technology that can be applied in multiple markets.

Knowledge Spillovers: The leakage of a firm's knowledge to another firm.

Subsidy: A payment by the government that makes up the difference between what customers will pay for a product or service and the cost of producing it.

Putting Ideas into Practice

1. **Finding Opportunities for Innovation** Many technological innovations occur in response to some kind of change. You can use some of the concepts that were discussed in this chapter to identify an opportunity for technological innovation by identifying the sources of that innovation. Please follow the steps below to do that:

   Step 1: Identify some changes that you have noticed in (1) technology, (2) demographics, (3) social trends, (4) regulation, and (5) politics. List the changes below.

   Changes You Observe

   Technology:
   1.
   2.
   3.

Demographics:
1.
2.
3.

Social trends:
1.
2.
3.

Regulation:
1.
2.
3.

Politics:
1.
2.
3.

Step 2: After you have made this list, think of the following types of technological innovations that could occur in response to them: (1) new products or services, (2) new markets, (3) new production processes, (4) new raw materials, and (5) new ways of organizing. Please match the source of the opportunity with the type of innovation that the change would lead to.

Example: The invention of fuel cells (a technological change) creates the opportunity to make cars that do not have internal combustion engines (a new product).
1.
2.
3.
4.
5.

2. **Form of Innovation** The purpose of this exercise is to get you thinking about the different forms that innovation can take. Using what you have learned from this chapter, fill in the chart at bottom of page using examples that were not contained in Table 4.4.
3. **Corporate R&D** Pick a company. Go to its Web site or conduct a Google search to find information about its R&D. How much money does the firm spend on R&D? What proportion is spent on basic research and what portion is spent on applied development? Explain how the firm manages the balance between "basic" and "applied" research. Why does it make the choices that it does? Are the choices "correct"? Why?

NOTES

1. Adapted from Gomez, H. 2006. Tech gadget lets parents watch kids home alone. *The Plain Dealer*, January 6: C1, C3.
2. Shane, S. 1996. Explaining variation in rates of entrepreneurship in the United States: 1899–1988, *Journal of Management*, 22(5): 747–781.
3. Intellectual Property Professional Information Center. 2006. Perception gap hindering efforts to improve patent system, Dudas says. *Patent Trademark and Copyright Journal*, 71(1756): February 10, http://ipcenter.bna.com/pic2/ip.nsf/id/BNAP6LVKRG?.
4. http://inventors.about.com/library/weekly/aa122999a.htm
5. Vara, V. 2007. How firms' web-filter use is changing. *Wall Street Journal*, January 16: B3.
6. Eisenmann, T., and N. Tempest. 2001. Atheros Communications. *Harvard Business School Case*, Number 9–802–073.
7. Regalado, A. 2004. Nanotechnology patents surge as companies vie to stake claim. *Wall Street Journal*, June 18: A1, A2.
8. Afuah, A. 2003. *Innovation Management*. New York: Oxford University Press.

Form of the Innovation	Technological Change That Made It Possible	Business Idea
New product or service		
New way of organizing		
New market		
New method of production		
New raw material		

9. Sathe, V. 2003. *Corporate Entrepreneurship*. Cambridge, UK: Cambridge University Press.
10. Ramstad, E. 2006. Corning develops a glass without heavy metals. *Wall Street Journal*, March 22: B2.
11. Clark, D., and C. Forelle. 2005. Sun Microsystems to buy StorageTek for $4.1 billion. *Wall Street Journal*, June 3: A3, A6.
12. Holmes, T., and J. Schmitz. 2001. A gain from trade: From unproductive to productive entrepreneurship. *Journal of Monetary Economics*, 47: 417–446.
13. Abboud, L. 2006. How France became a leader in offering faster broadband. *Wall Street Journal*, March 28: B1, B4.
14. Eisenmann and Tempest, Atheros Communications.
15. Sapsford, J. 2005. As Japan's elderly ranks swell Toyota sees new path to growth. *Wall Street Journal*, December 21: A1, A12.
16. Metzger, R. 2004. Steris making strides in post-terror technology. *The Plain Dealer*, March 28: G1, G5.
17. McCartney, S. 2004. Taxi! Fly me to Cleveland. *Wall Street Journal*, May 19: D1, D10.
18. Von Hippel, E. 2001. Innovation by user communities: Learning from open source software. *Sloan Management Review*, 42(4): 82–86.
19. Von Hippel, E. 1988. *The Sources of Innovation*. New York: Oxford University Press.
20. The history of RFID Technology, http://www.rfidjournal.com/article/articleview/1338/1/129.
21. Afuah, *Innovation Management*.
22. Markoff, J. 2005. A blow to computer science research. *New York Times*, April 2: B1–2.
23. Afuah, *Innovation Management*.
24. Thomas, K. 2005. U.S. funds fuel cell vehicle development. *The Plain Dealer*, March 31: B3.
25. Adams, J. 2005. Industrial R&D laboratories: Windows on black boxes? *Journal of Technology Transfer*, 30(1/2): 129–137.
26. Afuah, *Innovation Management*.
27. http://en.wikipedia.org/wiki/Apple_computer#1975_to_1980:_The_early_years
28. Mansfield, E. 1991. Academic research and industrial innovation. *Research Policy*, 20: 1–12.
29. Koizumi, K. 2001. 25 years of the AAAS Report: Historical perspectives of R&D in the federal budget, *AASS Report XXV*, http://www.aaas.org/spp/rd/chap2.htm.
30. Afuah, *Innovation Management*.
31. Adams, Industrial R&D laboratories.
32. Chesbrough, H. 1999. Intel Labs (A): Photolithography strategy in crisis. *Harvard Business School Case*, Number 9–600–032.
33. Tennenhouse, D. 2004. Intel's open collaborative model of industry-university research. *Research Technology Management*, 47(4): 19–26.
34. Klevorick, A., R. Levin, R. Nelson, and S. Winter. 1995. On the sources of significance of inter-industry differences in technological opportunities, *Research Policy*, 24: 185–205.
35. Ibid.
36. Afuah, *Innovation Management*.
37. Ibid.
38. *Science and Engineering Indicators 2006*, http://www.nsf.gov/statistics/seind06/c4/tt04–01.htm.
39. Jankowski, J. 1998. R&D: Foundation for Innovation. *Research Technology Management*, 41(2): 14–20.
40. Roberts, E. 2001. Benchmarking global strategic management of technology. *Research Technology Management*, March–April: 25–36.
41. Narayanan, V. 2001. *Managing Technology and Innovation for Competitive Advantage*. Upper Saddle River, NJ: Prentice Hall.
42. Burgelman, R., C. Christiansen, and S. Wheelwright. 2004. *Strategic Management of Technology and Innovation*. New York: McGraw-Hill Irwin.
43. Sathe, *Corporate Entrepreneurship*.
44. Afuah, *Innovation Management*.
45. Chesbrough, Intel Labs (A).
46. Hensley, S., and R. Winslow. 2004. Blood work. *Wall Street Journal*, April 8: A1, A12.
47. Ibid.
48. Chesbrough, H. 2001. Managing research in Internet time, *Harvard Business School Teaching Note*, Number 5–601–122.
49. Cohen, W., and D. Levinthal. 1990. Absorptive capacity: a new perspective on learning and innovation. *Administrative Science Quarterly*, 35: 128–152.
50. Song, J., P. Almeida, and G. Wu. 2003. Learning-by-hiring: When is mobility more likely to facilitate interfirm knowledge transfer? *Management Science*, 49(4): 351–365.
51. Kim, L. 1997. The dynamics of Samsung's technological learning in semiconductors. *California Management Review*, 39: 86–100.
52. Song, Almeida, and Wu, Learning-by-hiring.
53. Argyres, N., and B. Silverman. 2004. R&D, organization structure, and the development of corporate technological knowledge. *Strategic Management Journal*, 25: 929–958.
54. Stevens, T. 2004. 3M reinvents its innovation process. *Research Technology Management*, March–April: 3–5.
55. Chesbrough, Managing research in Internet time.
56. Chesbrough, Intel Labs (A).
57. Shapero, A. 1997. Managing creative professionals. In R. Katz (ed.), *The Human Side of Managing Technological Innovation*. New York: Oxford University Press, 307–319.
58. Shapero, Managing creative professionals.

59. Chesbrough, Managing research in Internet time.
60. Chesbrough, Intel Labs (A).
61. Maidique, M., and Hayes, R. 1984. The art of high technology management. *Sloan Management Review*, 25: 18–31.
62. Adapted from Dodes, R. 2005. Weird science of pantyhose. *Wall Street Journal*, November 19–20: P4.
63. Schumpeter, J. A. 1934. *The Theory of Economic Development: An Inquiry into Profits, Capital Credit, Interest, and the Business Cycle*. Harvard University Press: Cambridge, MA.
64. Mansfield, E. 1985. How rapidly does technology leak out? *Journal of Industrial Economics*, 34(2): 217–223.
65. Tushman, M., and L. Rosenkopf. 1992. Organizational determinants of technological change: Toward a sociology of technological evolution. *Research in Organizational Behavior*, 14: 311–347.
66. Ibid.
67. Klevorick, Levin, Nelson, and Winter, On the sources of significance of inter-industry differences in technological opportunities.
68. Von Hippel, Innovation by user communities.
69. Afuah, *Innovation Management*.
70. Ibid.

Chapter 5

Selecting Innovation Projects

Learning Objectives

After reading this chapter, you should be able to:

1. Identify the different ways that firms manage the uncertainty of innovation, and understand the pros and cons of the different approaches.
2. Describe the different tools for making decisions about innovation, spell out how each of them works, and explain the pros and cons of each.
3. Use the analytical hierarchy process to evaluate innovation projects.
4. Calculate the internal rate of return, payback period, and net present value of innovation projects.
5. Conduct a real options analysis of an innovation project.
6. Explain why real options analysis is often better than net present value calculation for evaluating innovation projects.
7. Use a decision tree to make decisions about an innovation project.
8. Explain why companies need to manage product portfolios, describe the different portfolio management tools that they use, and understand the value of these tools.

Real Options: A Vignette[1]

Gemplus International SA, the world's largest provider of smart cards, is researching wearable, ultrathin, secure, wireless communication devices because the company believes that, through R&D, it will identify new products and services that will provide it with a competitive advantage. However, Gemplus has limited resources and needs to invest its funds wisely, lest it use up its limited resources on poor investments.[2]

Gemplus's management takes a real-options approach to making R&D decisions. **Real options** are investments that give a firm the right, but not the obligation, to make an investment. For example, the company has invested in reducing the power requirements for wearable wireless communication devices because lower power requirements could provide the basis for a new family of products based on location services. Moreover, the low-power-technology investments could help the company improve its ability to manage bandwidth if improvements are made in peer-to-peer network architectures. The company is also investing in location technology because such technology would position the company well to compete in service delivery if the market for content-rich service delivery develops.[3]

By making these initial investments, Gemplus preserves the right to make future investments in the technology should uncertainty be resolved in a favorable way over time, but does obligate itself to make additional investments. By treating investments as real options, Gemplus minimizes investments in dead ends and preserves flexibility, while still undertaking the projects necessary to create future products and services. As a result, Gemplus can position itself for the future without straining its financial resources.[4]

INTRODUCTION

Companies often need to choose between different innovation projects under considerable uncertainty about whether the technological developments are feasible (technical uncertainty), the market needs the innovation (market uncertainty), the resources to put the innovation into place are available (financial uncertainty), and the value from successful innovation can be captured (competitive uncertainty). Doing this successfully requires the use of decision-making tools that are effective for making choices about novel activities when the future is unknown.

This chapter explains how to make decisions about innovation projects. The first section examines the strategic approaches that you can take to manage uncertainty. The second section describes several tools that you can use to evaluate innovation projects. The final section discusses the management of an innovation project portfolio, and the tools that support such an activity.

Innovation is a highly uncertain process. However, researchers have identified four strategic actions that you can take to manage this uncertainty: seeking high returns, minimizing investment, maintaining flexibility, and reallocating uncertainty to others.

First, you should focus on high return opportunities or you will not be able to generate enough of a financial return from innovation to justify bearing the uncertainty of undertaking it. This means that many innovation opportunities that generate positive financial returns may not be worth pursuing once uncertainty is taken into consideration. Take, for example, a technology with only a one-in-ten chance of success. If you need to generate an average rate of return on your invested capital of 10 percent to satisfy your shareholders, you should only pursue this project if a successful outcome will generate a 100 percent rate of return. Because you will make nothing on nine out of ten projects, a 100 percent rate of return on the successful project is necessary to get a 10 percent average rate of return across all projects.[5]

Second, you should minimize your investment in nonsalvageable assets. Minimizing your investment in these assets will help you to manage uncertainty because the cost of bearing uncertainty is very much affected by the size of your potential loss. Because you can recoup part of your investment in assets with high salvage value if your innovation effort is unsuccessful (but you cannot recoup your investment in nonsalvageable assets), your potential loss is higher with investments in nonsalvageable assets.

So what can you do to minimize your investment in assets with low salvage value? You can use standard inputs, like generic machinery, rather than customized inputs. That way, if your innovation effort is unsuccessful, you can sell the generic machinery to another business and salvage some of the value of your investment. In contrast, if you use customized inputs, you will lose the entire value of your investment in them if your innovation effort fails because those assets cannot be sold to others.[6]

Of course, you will need to use some customized inputs because generic inputs cannot provide the basis of a competitive advantage. Therefore, you need to consider which assets are the source of your business's competitive advantage when you invest in innovation. Only those inputs that are the source of competitive advantage should be customized. For example, back office accounting software in an accounting firm might make sense to customize, but that same software should not be customized in a medical device firm where it will not create a competitive advantage.

You can also minimize your investment by beginning your operations on a small scale. For example, if you start with a single manufacturing plant and expand only if you are successful, you can minimize the size of your investment in fixed assets, and, consequently, the uncertainty that you need to bear when innovating.[7]

Finally, you can minimize your investment by turning fixed costs into variable costs. (**Variable costs** are costs that are dependent on the level of production of your good or service, whereas **fixed costs** are costs that are independent of your level of production.) By reducing the amount of your up-front expenditures, you will reduce the size of your loss in the event that your innovation effort is unsuccessful. For example, you might contract with software engineers to write code for you in return for a royalty on the sales of your product instead of paying them a salary. As Table 5.1 shows, this approach will reduce your fixed costs and, consequently, your breakeven level of sales. As a result, you will bear less risk in the development of your software.[8]

TABLE 5.1 Reducing Fixed Costs

By turning the fixed cost of writing the software code into a variable cost, the company in this example can reduce its level of initial fixed investment and its breakeven level of sales, decreasing the risk it needs to bear.

COST	VERSION 1	VERSION 2
	Software developers on salary	Software developers getting royalty
Writing software code	$100,000	$0.50/copy sold
Debugging and preparing manual, and packaging	$50,000	$50,000
Distributing software	$0.25/copy	$0.25/copy
Fixed costs	$150,000	$50,000
Variable costs	$0.25/copy	$0.75/copy
Sales price	$10.00/copy	$10.00/copy
Contribution margin per unit	=$10.00 − $0.25 = $9.75	=$10.00 − $0.75 = $9.25
Breakeven level of sales	=$150,000/$9.75 = $15,385	=$50,000/$9.25 = $5,405

Third, you should maintain flexibility so that your company can change direction rapidly if doing so becomes necessary. Because you can't know in advance the outcome of technical, market, financial, and competitive uncertainty, your ability to change strategic direction when unexpected events occur allows you to avoid a downside loss, and thus to manage uncertainty.

Take, for example, the history of computer disk drives. Many manufacturers of these devices managed the uncertainty of introducing a new product into an unknown market by being flexible about the target market that they were pursuing. When the manufacturers found that the customers that they initially targeted did not want the disk drives that they were offering, they shifted target markets until they found one interested in adopting the drives.[9]

Fourth, you should reallocate uncertainty to other parties who are better able or more willing to bear it. For example, DoMoCo, the Japanese phone company, managed the uncertainty of identifying desirable content for its mobile telephone ringtones by creating a competition among content providers. If the content proved to be popular, then the company paid the content provider a royalty, but if it was not, then DoMoCo bore none of the uncertainty of developing it.[10]

Why are other parties willing to bear uncertainty for you? One reason is that they like it. For example, business angels (individuals who use their own money to finance start-ups) often bear some uncertainty for entrepreneurs by investing in their start-ups. They make these investments because they enjoy the entrepreneurial process and see uncertainty bearing as the price of admission to this activity.

Another reason that others will bear your uncertainty is that they are better able to do so than you. Diversified investors, specialists, and companies operating at less than full capacity are all better able to bear uncertainty than others. Diversified investors can bear a lot of uncertainty because they face a low likelihood that their entire investment will be lost in the event of a single adverse outcome. (Diversifying involves investing in a portfolio of uncertain projects with outcomes that are independent of each other.[11]) For example, venture capital firms can often bear more uncertainty than the entrepreneurs that they finance because their simultaneous investment in several uncertain ventures with uncorrelated outcomes ensures that they will not lose all of their money if one of the start-ups fails. By contrast, each of the entrepreneurs will lose his or her entire investment if his or her venture fails.

Specialized investors can better bear uncertainty than unspecialized ones because their specialization gives them information that reduces the amount of uncertainty that they face. For example, a factor (a company that purchases accounts receivable) can collect your bills with greater certainty than you because the factor knows how to collect debt. So specialized investors, like factors, will take on uncertainty that you seek to shed.[12]

Companies that are currently engaged in an activity at less than full capacity can better bear uncertainty than companies that want to start an activity from scratch because those operating at less than full capacity can undertake the activity in question with a lower potential downside loss. Take, for example, the biotechnology companies with manufacturing plants whose product approvals have been delayed by the FDA. These companies often produce products for other biotechnology companies in their idle plants because the downside loss from an adverse outcome is less for them than for a company that creates a plant from scratch.[13]

Key Points

- You can manage uncertainty in innovation by seeking high enough returns to justify the cost of bearing it.
- You can manage uncertainty in innovation by minimizing the magnitude of your investments in nonsalvageable assets through the use of generic inputs, small scale production, and the transformation of fixed assets into variable ones.
- You can manage uncertainty in innovation by maintaining the flexibility to change strategic direction.
- You can manage uncertainty in innovation by reallocating it to those parties more willing, or able, to bear it.

Decision-Making Tools

As a technology entrepreneur or manager, you should also use decision-making tools for making choices about innovation projects. These tools take a wide variety of forms. Some tools are **qualitative,** and compare projects on the basis of scales or words, while others are **quantitative,** and evaluate projects based on the basis of numerical calculations. Tools can also be **comparative,** and pit projects against each other, or **scoring,** and compare projects against standard scales.

Different types of decision-making tools have different advantages and disadvantages. Quantitative tools provide a way to get more precise estimates of the contributions of various factors that affect a decision, but can often result in false precision when these factors are not easily quantified. Scoring models incorporate nonfinancial criteria relatively easily, which facilitates thinking about project attributes and strategic factors, but often results in poor decisions when the scores are not precise estimates of the underlying concepts being evaluated.[14] Comparative models allow you to consider different projects in relationship to each other but lead to inaccurate results when you lack information about project quality or need to compare many projects.[15]

Checklists

The **checklist** is one example of a scoring model. With a checklist, projects are evaluated on whether or not they meet specific criteria. Checklists are a useful tool whenever the

FIGURE 5.1 The Value of Decision-Making Tools

Decision-making tools help you figure out whether an innovation project is worth doing or not.
Source: Washington Post, 2002.

presence or absence of key factors, such as the existing or future capabilities of a firm and its competitors, affects a decision to move forward on a project.[16]

For instance, as Table 5.2 shows, a business that is considering the purchase of software might seek a package that is inexpensive to obtain, support, staff, and use; with a high level of scalability, security, interoperability, and reliability; a long expected life; and easy customization. By using a checklist, the business can determine if these attributes are present or absent in the different software packages that it is considering.[17]

Analytical Hierarchy Process

Analytic hierarchy process (AHP) is a decision-making tool in which a problem, like the adoption of a new technology, is broken down into a hierarchy of different criteria and choices. When this technique is used, a decision maker starts by identifying the objective, the alternatives, and the criteria on which the decision will be made. For instance, suppose you are considering a new computer system for a factory, and you are trying to choose between four different alternatives. You know that there are three dimensions of the computer system that you care about: compatibility with the existing system, difficulty in learning how to use it, and cost. But the dimensions are not equally important to you.

TABLE 5.2
A Checklist for Selecting Software
Companies often use checklists like this one to make decisions about the adoption of new computer systems.

Dimension	Present?
Low cost to purchase	
Low cost to staff	
Low cost to support	
Low cost to train employees to use	
Easy to customize	
High level of reliability	
High level of interoperability with other systems	
High degree of scalability	
High level of security	
Long expected life	

Source: Adapted from Tuma, D. 2005. Open source software: Opportunities and challenges. *Cross Talk*, June, http://www.stsc.hill.af.mil/. . ./2005/01/0501tuma.html.

The first step in using AHP is to figure out how much you care about different criteria. To do this, you start by making pair-wise comparisons of the importance of each dimension versus another on a scale of 0 to 10. In this case, suppose that you found that the difficulty of learning the system is four times more important to you than the compatibility with the existing system; cost is nine times more important than the difficulty of learning the system; and cost is three times more important than compatibility with the existing system.[18]

You can use these values to create a matrix like the one shown in Table 5.3. To do this, you enter the values in the cells that show the relationship between the two dimensions.[19] You then enter the values for the diagonal (which are always equal to one because each dimension is equally important as itself). Finally, you

TABLE 5.3 Analytical Hierarchy Process
This table shows the analysis for the analytical hierarchy process for the new computer system described in the text.

	Compatibility with the Existing System	Difficulty in Learning the System	Cost
Compatibility with the Existing System	1.00	4.00	3.00
Difficulty in Learning the System	0.25	1.00	9.00
Cost	0.33	0.11	1.00
	Compatibility with the Existing System	**Difficulty in Learning the System**	**Cost**
Compatibility with the Existing System	=1.00/1.58 = **0.63**	=4.00/5.11 = **0.78**	=3.00/13.00 = **0.23**
Difficulty in Learning the System	=0.25/1.58 = **0.16**	=1.00/5.11 = **0.20**	=9.00/13.00 = **0.69**
Cost	=0.33/1.58 = **0.21**	=0.11/5.11 = **0.02**	=1.00/13.00 = **0.08**

put in the remaining values, which are the inverse of the first set of values that you put in. (For instance, because the difficulty of learning the system is four times as important to you as compatibility with the existing system, the compatibility with the existing system is one-fourth as important as difficulty of learning the system.)[20]

Next, you need to calculate the "priority vector," which is done as follows: First, you calculate the sum of each column. In the example, these calculations are 1.58 (e.g., 1.00 + 0.25 + 0.33), 5.11 (e.g., 4.00 + 1.00 + 0.11), and 13.00 (e.g., 3.00 + 9.00 + 1.00). Second, you divide each entry in the table by its column sum, which is shown in the second panel of Table 5.3. Third, you take the averages for each row, which are 0.55 [e.g., = (0.63 + 0.78 + 0.23)/3], 0.35 [e.g., = (0.16 + 0.2 + 0.69)/3], and 0.10 [e.g., = (0.21 + 0.02 + 0.08)/3].

The row averages make up the "priority vector," or the weighted priority of each attribute.[21] The priority vector shows that compatibility with the existing system is the most important dimension, followed by difficulty in learning the system, followed by cost.

When making decisions about innovation projects, people are often inconsistent because they do not see the elements of the project as a coherent whole when answering questions about choices between different dimensions. AHP can be used to see how consistent or inconsistent the decision makers are. If the comparisons are consistent, then the values of the normalized columns in the lower panel of an AHP matrix, such as the one shown in Table 5.3, would all be the same. If the comparisons are not consistent, you can see how inconsistent they are by calculating the priorities scaled against each other, a number that is called the inconsistency ratio.[22] If your calculations show that the inconsistency in decision making is very large, it is important for you to understand why before making decisions about the innovation project. Otherwise, you will not be able to produce an internally consistent innovation.[23]

The reason why AHP is called a hierarchy model is that you can use the values that you calculate to weight decisions at different levels of analysis. For instance, you could go through similar calculations for each of the four alternatives that you are considering in the example we have been discussing, and then weight those values by the priority vector values that you calculated to make your overall decision about the innovation.

Net Present Value

Many of the quantitative methods used to evaluate innovative projects are based on the analysis of discounted cash flows. The two most common of these methods are **net present value (NPV)** calculation, which estimates the value of a project today, given the amount and timing of cash outflows and inflows and the discount rate; and **internal rate of return** calculation, which estimates the rate of return on a project, given the level of expenditure and the timing and amount of cash inflows and outflows.[24]

Table 5.4 shows an example of a net present value calculation for an innovation project. To calculate the net present value, the cash inflows for each year are divided by one over the discount rate to the power of the number of years in the future that the inflow occurs. This example shows that, at a 10 percent per year interest rate, a project that generates $1,000,000 per year for five years is worth $3,790,786 in today's dollars.

TABLE 5.4 Present Value of Future Cash Flows

This table shows an example of the calculation of the net present value of future cash flows when the discount rate is 10 percent and there are five years of annual cash flows of $1 million per year.

YEAR	CASH INFLOW	PRESENT VALUE	COMPUTATION
1	$1,000,000	$ 909,091	PV = CI/(1.10)
2	$1,000,000	$ 826,446	PV = CI/(1.10^2)
3	$1,000,000	$ 751,315	PV = CI/(1.10^3)
4	$1,000,000	$ 683,013	PV = CI/(1.10^4)
5	$1,000,000	$ 620,921	PV = CI/(1.10^5)
Total	$5,000,000	$3,790,786	Sum of PV for Years 1–5

The ability to measure the returns from innovation in today's dollars is important because the investment that you make to generate those returns might have to be made today, and net present value lets you compare both the expenditures and returns in current dollars. For instance, if this project cost $4 million to undertake and the full $4 million had to be spent immediately, the project might not be worth undertaking. While at first glance the project might look worthwhile—it would return $5 million in cash inflows in comparison to the $4 million in cash outflows—the timing of the cash inflows and the time value of money is such that this project would actually yield less in today's dollars than it would cost to undertake.

You can also use net present value calculations to figure out how long it will take you to pay back your initial investment. For instance, if we look at the same example that we have been examining—an up-front investment of $4 million and cash inflows of $1 million per year for five years, with a discount rate of 10 percent—we can figure out the cumulative cash flows by year, which are shown in Table 5.5. This table shows that at the $4 million investment cost, the project costs are never recouped. However, at a $1.7 million investment cost, the project costs would be recouped by the end of the second year.

Internal Rate of Return

Another calculation that you might want to make is for the internal rate of return on a project, which is the discount rate that yields a net present value of zero. The internal rate of return gives you an indication of the financial rate of return that a project

TABLE 5.5 Payback Period

This table shows the payback period for the discounted cash flows shown in Table 5.4; if the initial investment cost $1.7 million, then the cost of the project would be paid back in two years.

YEAR	DISCOUNTED CASH INFLOW	CUMULATIVE DISCOUNTED CASH INFLOW
1	$909,091	$ 909,091
2	$826,446	$1,735,537
3	$751,315	$2,486,852
4	$683,013	$3,169,865
5	$620,921	$3,790,786

TABLE 5.6 Internal Rate of Return

This table shows the first iterations of the calculation of the internal rate of return for a project that costs $850,000 to implement and generates $200,000 per year in cash flow over five years.

YEAR	0	1	2	3	4	5	TOTAL
Income stream	($850,000)	$200,000	$200,000	$200,000	$200,000	$200,000	$150,000
Present value, at a 5% discount rate	($850,000)	$190,476	$181,406	$172,768	$164,541	$156,705	$ 15,895
Present value, at a 6% discount rate	($850,000)	$188,679	$177,999	$167,924	$158,419	$149,452	($ 7,527)
Present value, at a 5.675% discount rate	($850,000)	$189,260	$179,096	$169,478	$160,377	$151,764	($ 26)

could generate. Because you know the rate of return that you can earn on risk-free investments—those in U.S. government securities—you can estimate whether the project provides you with enough of a premium to justify bearing uncertainty to undertake it.

The internal rate of return is calculated in an iterative manner. As Table 5.6 shows, you first identify the magnitude and timing of your incoming and outgoing cash flows from the project. Then you discount those cash flows to create a rate of return where the present value of the sum of the initial investment and the annual incoming cash flows is zero. In this example, a 5 percent internal rate of return yields a positive net present value, while a 6 percent internal rate of return yields a negative one. Therefore, the actual internal rate of return must be between 5 percent and 6 percent. By iterating, you will see that 5.675 percent is as close as you can get to a net present value of zero at the three-decimal-point level.

At the time that this book was written, five-year U.S. treasury notes were providing a rate of return of 4.750 percent. Because the innovation project in the example has a five-year internal rate of return of only 5.675 percent, it would make sense as an investment only if it involved slightly more risk than buying the securities of the U.S. government.

For most innovation projects, the internal rate of return that is necessary for someone to undertake an investment needs to be much higher. For example, a venture capitalist typically expects an internal rate of return of more than 50 percent to make an investment in a technology start-up at the seed stage—when the entrepreneur has no more than a business plan. That's because the risk that the investor will lose what he or she invested in the start-up is so much higher than the risk that the investor will lose what he or she lent to the U.S. government.

While discounted cash flow analysis is an important tool for making decisions about innovation projects, it faces several important limitations. First, it cannot incorporate many important nonfinancial factors that influence decisions, such as the reaction of competitors, the relationship of one part of the business to another, or organizational learning.[25] Second, discounted cash flow calculations do not consider the option value of doing something, like R&D, that could benefit a variety of different projects or make highly profitable future investments possible.[26] Third, discounted cash flow calculations rely on point estimates, which require decision makers to assume (with questionable accuracy) the probability of that outcome occurring when a range of possibilities exist.[27] Fourth, the accuracy of discounted cash flow calculations depends very much on the accuracy of assumptions about costs, revenues, and time horizons, which is problematic because such assumptions are often very inaccurate for innovation projects.

As a result, discounted cash flow calculations favor short-term, high-probability events over long-term, low-probability ones. In fact, discounted cash flow calculations often suggest that firms not make investments in innovation, particularly if the firm already has products with high cash flow.[28]

Real Options

Real-options analysis is a tool that can be used to overcome the limitations of discounted cash flow analysis. As the opening vignette explained, real options provide the right, but not the obligation, to make a future investment. They involve an "option price," which is the cost of developing a new technology, and an "exercise price," which is the cost of exploiting the technology. Options are "exercised" when the decision is made to exploit the technology.[29] Options have positive value if the expected cash flow from making the investment is less than the value of maintaining the option.

Real options are useful for making decisions about investments in innovation for several reasons. First, technology development occurs in an evolutionary fashion that is largely staged. In the typical technology development scenario, initial research results in the invention of a new technology, which is followed by product development, manufacturing, and marketing, in that order. Because several things are unknown at each stage of this process, and will only become known if the venture passes through the previous stage successfully, you cannot accurately estimate the value of technology development through all of the stages. Under these circumstances, real options are useful because they limit decision making to information known at the stage at which the decision is being made.[30]

Second, real options help you to maintain flexibility and avoid committing valuable resources to infeasible alternatives. For example, you might not know if your biotechnology start-up should pursue the human or veterinary market for a new drug until you've conducted tests on its efficacy on both groups. By using real options to make decisions, you can evaluate the drug's efficacy on the two groups before making a decision about developing a sales force. As a result, you can avoid spending money on a sales force for a market in which the drug doesn't work.

Third, real options permit you to postpone decisions until uncertainty is reduced.[31] Over time, you might gather information about market needs, customer demand, or ways to reduce costs or appropriate returns.[32] Because this additional information helps you to make accurate decisions, delaying your investment until after it has been gathered has value if your investment is irreversible.[33] (For instance, expenditures on the construction of a plant that can do nothing but smelt aluminum are irreversible because the costs cannot be recovered or used for anything else.) By allowing expenditures to be delayed until after information about their value can be gathered, real-options analysis facilitates more accurate decision making.

Figure 5.2 illustrates how real-options analysis can lead to more accurate decisions than discounted cash flow analysis. The figure shows the evaluation of a wireless PDA product for which you first need to invest $12 million in research. This investment can result in one of three possible outcomes: excellent performance, which has a 30 percent chance of occurring; good performance, which has a 60 percent chance of occurring; or poor performance, which has a 10 percent chance of occurring. After research is completed, the company undertaking the project will need to invest

FIGURE 5.2
Real Options Evaluation for a Wireless PDA

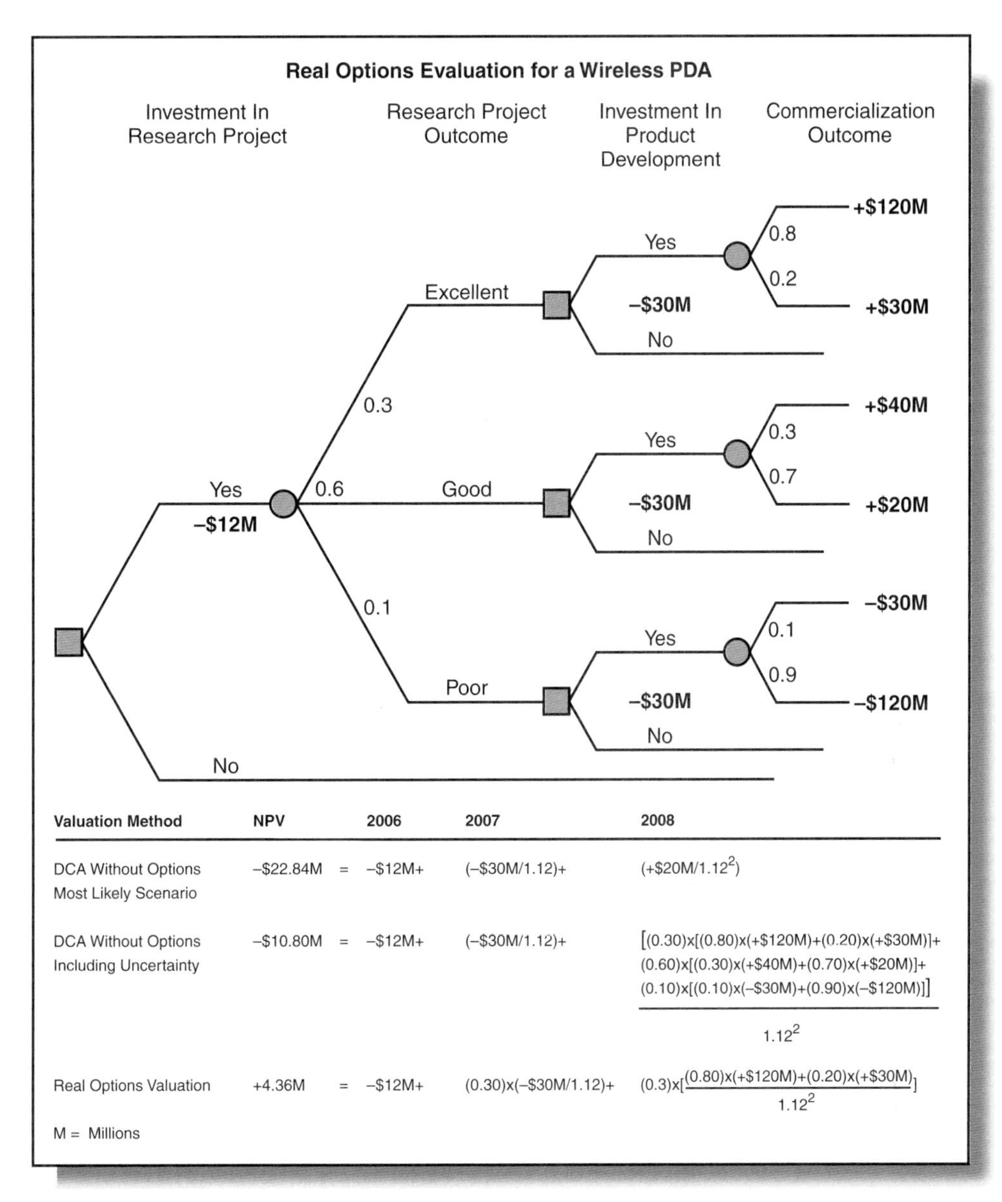

Using real options leads to the calculation of a positive net present value for the wireless PDA project, while not using real options leads to the calculation of a negative net present value.

Source: Adapted from Faulkner, T. 1996. Applying "options thinking" to R&D valuation. *Research Technology Management*, May–June: 52.

$30 million in product development, which will either yield a viable product or it won't. If the product is viable, then it can have different potential commercialization outcomes. If the project has an excellent research outcome, the product will have an 80 percent chance of earning $120 million and a 20 percent chance of earning only $30 million. If the project has a good research outcome, the product will have a 30 percent chance of earning $40 million and a 70 percent chance of earning $20 million. And if the project has a poor research outcome, the product will have a 10 percent chance of losing $30 million and a 90 percent chance of losing $120 million. The company has a discount rate of 12 percent per year.

As the figure shows, the net present value of the most likely scenario—a good research outcome (60 percent chance) and $20 million of commercial value (70 percent chance)—is negative $22.84 million. That is, in net present value terms the company spends $12 million in 2006 and $26.79 million in 2007 ($30 million divided by 1.12) but earns only $15.94 million in net present value terms in 2008 ($20 million divided by [1.12 squared]). Clearly, this net present value is not going to motivate the company to undertake the wireless PDA project.

If you calculate the net present value to include uncertainty by weighting the stream of payments for probability of the different scenarios occurring, then the net present value of the project is not as bad, but is still negative $10.80 million. The company will spend $12 million in 2006. In addition, in net present value terms, it will spend $26.79 million in 2007 ($30 million divided by 1.12). But the net present value of earnings in 2008 is only $27.98 million because the payments are weighted for the probability of occurring before they are discounted. That is, the expected payment is (0.30) × [(0.80) × (+$120M) + (0.20) × (+$30M)] + (0.60) × [(0.30) × (+$40M) + (0.70) × (+$20M)] + (0.10) × [0.10) × (−$30M) + (0.90) × (−$120M)]. Because the net present value for this approach is still negative, the company again will not undertake the project.

But if we examine the project using real options, the net present value is positive $4.36 million, and the project will be undertaken. Why is the analysis different with real options? Although there is no difference in the net present value of the initial research investment when looked at from a real-options perspective as compared to a discounted cash flow perspective, there is a difference in the net present value of the product development expenditure and the commercialization outcome. Because the company knows that it is foolish to persist with its PDA product if the outcome of research is not "excellent," it makes that investment, if, and only if, the results of the research are excellent. Because the odds that the research results will be excellent are only 30 percent, it only has a 30 percent chance of incurring the expense of the $30 million. As a result, the net present value of that investment is only 30 percent of what it was under the two discounted cash flow scenarios [(0.30) × (−$30M/1.12)].

In addition, the net present value of payments in 2008 is also different because the company will only make the product development investment if the outcome of research is excellent. Thus, there are only two possible scenarios that need to be calculated: the 80 percent chance of making $120 million and the 20 percent chance of making $30 million. As a result, the amount that is discounted to present value for 2008 is (0.80) × (+$120M) + (0.20) × (+$30M). Of course, this path only has a 30 percent chance of occurring, so this payment for year 2008 needs to be multiplied by 0.30. However, the potential outcomes on the excellent research path are so high that the net present value for the project is positive even though there is only a 30 percent chance that the money will be earned.

In general, real-options analysis generates a larger value than discounted cash flow analysis if the magnitude of the outcome is large relative to the magnitude of the investment, when the magnitude of the future outcome is uncertain, when the time horizon is long, and when some information can be gathered in the future that will eliminate uncertainties.[34] Thus, real-options analysis is more appropriate than discounted cash flow analysis for decisions about innovation because the costs, revenues, and time horizons of these projects cannot be measured accurately,[35] there is uncertainty about the value of future payoffs, and project costs are irreversible.[36]

FIGURE 5.3
Scenario Analysis

Scenario analysis generally examines best, worst, and intermediate scenarios.
Source: http://www.andertoons.com.

Scenario Analysis

Another tool for making decisions about innovation projects is **scenario analysis,** which is the representation of investments under different assumptions about key factors that influence those investments (see Figure 5.3). Scenario analysis is useful because it helps you to identify the sources of uncertainty rather than assuming them away.[37] By making different assumptions about key variables affecting an innovation project, you can figure what uncertainties the project faces, what factors might lead things to go wrong, and what factors might drive desired outcomes.[38] As a result, you can partial out uncertain from certain outcomes and make decisions about the project more accurately.[39] For example, an entrepreneur writing a business plan for a business selling electronic books might create different scenarios for the business that considered whether complementary technology in electronic book readers could be developed at different points in time, and whether customers were as willing or less willing to read fiction and nonfiction books in electronic form as compared to paper.

Scenario analysis can be made more sophisticated by using **Monte Carlo simulation,** which is a way to create a probability distribution of outcomes through the use of computer software that experiments with randomly selected values of inputs.[40] With Monte Carlo simulation, you can look at the effect of tens of thousands of possible combinations of inputs on outcomes, rather than the effect of only a handful, thus generating more precise results from scenario analysis.

Decision Trees

Successful innovation also requires you to make accurate decisions about investments at different stages of the innovation process. The **decision tree**—a visual representation of decisions and their effects on outcomes, costs, and risks—is a tool that helps you to do that. For example, Figure 5.4 shows the decision tree for a choice between two possible innovation projects. The figure shows that project one will take twice as much R&D expense as project two and will have a 30 percent chance of failing, as compared to no chance for project two. Both projects then require a $1 million investment in manufacturing. The first project has an expected value of only $7.2 million. That is, −$4 million + (0.7 × −$1 million) + 0.7 × [(0.1 × $30 million) + (0.70 × $20 million) + (0.20 × 0)],

FIGURE 5.4
A Decision Tree

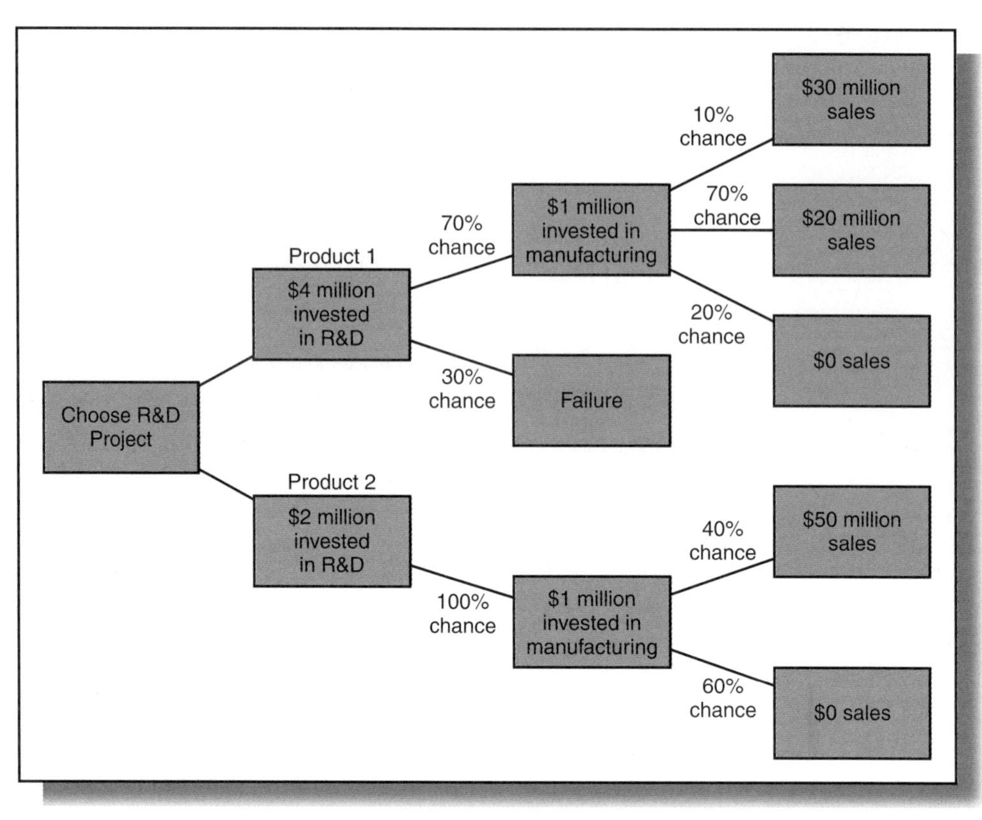

This figure shows a decision tree for the choice between two possible innovation projects.

while the second project has an expected value of $17 million. That is, −$2 million + −$1 million + [(0.4 × $50 million) + (0.6 × $0)]. The decision tree clearly shows that the second project is a better choice.

Decision trees provide a quantitative evaluation of a choice that is based on the value and probability of outcomes, and that accounts for the influence of staged decision making on risk. However, decision trees are limited by the use of discounted cash flows to calculate the branches on the trees, which makes them subject to all of the weaknesses of the use of discounted cash flows as a decision-making tool.[41]

Key Points

- Companies use quantitative and qualitative, and comparative and scoring, tools to select innovation projects to pursue; the different types of tools have different advantages and disadvantages.
- The checklist is a scoring model that evaluates projects on whether or not they meet specific criteria.
- The analytic hierarchy process is a decision-making tool in which a problem, like the adoption of a new technology, is broken down into a hierarchy of different criteria and choices.
- Many quantitative decision-making tools are based on the analysis of discounted cash flows; the two most common are net present value calculation, which estimates the value of a project today given the amount and timing of cash outflows and inflows and the discount rate; and internal rate of return

calculation, which estimates the rate of return on a project given the level of expenditure and the timing and amount of cash inflows and outflows.
- Real options are an important tool for selecting innovation projects that overcomes the weaknesses of decision making based on discounted cash flow calculations; they give the right, but not the obligation, to make future investments.
- Scenario analysis is a quantitative tool for making decisions about innovation projects that examines the effect of different assumptions about key factors.
- The decision tree is a visual representation of decisions and their effects on outcomes, costs, and risks; it helps you to make accurate decisions about investments at different stages of the innovation process.

Portfolio Management

When companies develop multiple products, they often use portfolio management tools to make decisions about innovation.[42] These tools help them to coordinate the different parts of the innovation process, set up the right order for those activities, and determine what resources are needed at different points in the process.[43] They also help decision makers to decide the order in which to pursue projects and the allocation of resources between them, given resource constraints.[44]

Portfolio management tools include any method that allows you to compare a set of projects against your strategic goals and to allocate scarce product development resources across the different projects.[45] They typically rely on some type of scoring system that allows you to compare and prioritize development projects along some dimension, such as financial return, strategic fit, or resource demands.[46]

Portfolio models take a wide variety of forms, including pie charts and bubble diagrams, and can be based on either quantitative or qualitative analyses. However, **project maps** are the most important type.

As Figure 5.5 shows, project maps show the placement of projects into three different categories[47]—derivative projects (efforts that extend existing projects, like the development of Liquid Tide), platform projects (efforts to create new product families, like the Toyota Camry), and breakthrough projects (efforts based on fundamentally new ideas, like the development of the first digital camera).[48]

Project maps help you in several different ways. First, they link product development efforts to strategy.[49] Because project mapping helps to allocate resources across different types of projects, you can use it to manage your company's growth. If you want to accelerate growth, you can allocate more resources to platform and breakthrough projects; whereas if you want to maximize profits from your R&D investments, you can allocate more resources to derivative projects.[50]

Moreover, project maps help you to formulate the right product development plan, given your business strategy. You can use project maps to ensure that you have the resources and capabilities for the product development goals that you seek to achieve.[51] By analyzing the distribution of your product development projects across types, you can also use project maps to determine the number and mix of projects that you need to achieve your strategic goals.[52] For instance, a project map might reveal gaps in your efforts to develop breakthrough projects and lead you to allocate more human and financial resources to the development of those types of projects, rather than to derivative ones.

FIGURE 5.5
Mapping Product Development Projects

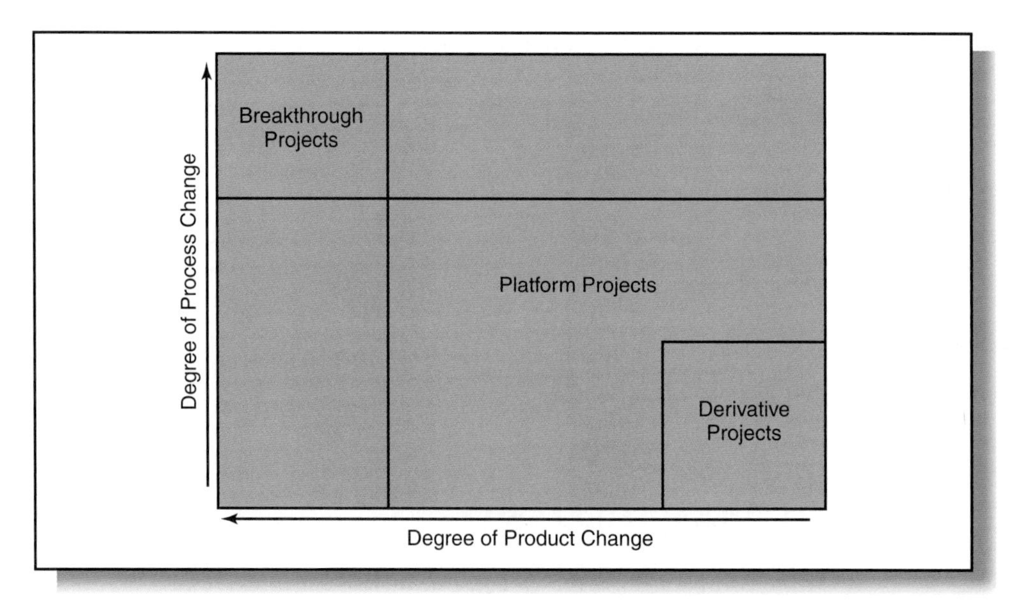

Mapping product development projects is a good way to ensure a balance between derivative, platform, and breakthrough projects because it helps technology strategists to visualize the allocation of resources across different types of innovation projects.

Source: Adapted from Wheelwright, S., and K. Clark. 2003. Creating project plans to focus product development. *Harvard Business Review,* September: 2–15.

This latter benefit of project maps is particularly important because you need to invest sufficient resources in platform and breakthrough projects for your company to achieve long-term growth.[53] However, your employees will be much more likely to propose derivative projects than platform or breakthrough projects. Without a mechanism to preserve your resources for other projects, derivative projects will use up all of your company's resources for innovation. Because project maps help you to define your company's allocation of human and financial resources across different types of projects, they force you to evaluate project ideas against an overall plan, which helps you to maintain nonderivative projects in your company's portfolio.

Second, project maps help you to avoid overcommitting your organization to innovation.[54] Companies often undertake too many innovation projects simultaneously, achieving less innovation than they would have achieved had they undertaken fewer projects. For example, in a study of its own innovation efforts, Exxon Chemical found that it would have had more and better innovation if it had cut in half the total number of projects it undertook.[55] Project maps allow you to measure how many projects you have underway at your company at any point in time, which helps you to avoid overcommitting your company to innovation.

Third, project maps help you to sequence your innovation projects to utilize your human resources effectively.[56] Research has shown that the productivity of engineers drops dramatically if they have to manage too many projects simultaneously.[57] Project maps provide information about your company's project allocation, which helps you to make better decisions about hiring product development personnel and assigning them to projects, thereby improving your human resource utilization.

Fourth, project maps help you to better manage your R&D personnel. Because engineers learn by doing, they can undertake more complicated innovation projects

GETTING DOWN TO BUSINESS
Medtronic's Use of Portfolio Management Tools[58]

Medtronic is a good example of a company that uses portfolio management tools. This Minnesota-based medical device company develops its products off of common platforms. For example, the company creates its cardiac pacemakers off of a hybrid integrated circuit platform, which allows it to make a variety of derivative products at low cost.[59]

It wasn't always that way. The company used to develop one product variant at a time, with no derivative products being offered. For instance, the company produced only one version of its very popular Activitrax pacemaker, even though different segments of the market were willing to pay different prices for versions with different features.[60]

However, the company's poor performance at product development led its management to require that all new products be designed so that derivative models could be created from a single platform.[61] The company then used the platform to reach all ends of the market, designing fully featured products to hit the high end of the market and products without some of the features and functionality for the middle and lower end. This approach allowed Medtronic to reach the low end of the market without doing what many companies have to do: cut prices on older models.[62]

The end result of this effort to create a platform strategy? An increase in the number of derivative products that were developed off of a single platform from 1 to 41.[63]

as they become more seasoned. As a result, you should first allocate new engineers to derivative projects, then move them to platform projects, and finally transfer them to breakthrough projects. Because project maps provide a layout of your company's projects, they facilitate the assignment of engineers to projects appropriate to their level of experience.[64]

Key Points

- Companies with multiple product lines and multiple products within those lines often use portfolio management tools to align their product development efforts with their technology strategies.
- A project map is a portfolio selection tool that helps you to manage the allocation of resources across platform, derivative, and breakthrough projects.
- Project mapping links product development efforts to firm strategy, helps companies to avoid overcommitment to innovation, facilitates the sequencing of product development efforts, and improves the management of product development personnel.

DISCUSSION QUESTIONS

1. What strategies can firms adopt to manage uncertainty? What are the pros and cons of these strategies?
2. What are the advantages and disadvantages of different tools for evaluating innovation projects? When might qualitative tools be more appropriate than quantitative ones, and vice versa?
3. What is the analytical hierarchy process? When do you want to use it? How does it help you to make decisions about innovation projects?
4. What problems occur when discounted cash flow analysis is used to make decisions about innovation projects? What decision-making tools should be used if those problems are present? Why?
5. How does scenario analysis help you to make decisions about innovation projects? Why does scenario analysis help?

KEY TERMS

Analytic Hierarchy Process (AHP): A comparative model of project evaluation in which managers create a hierarchy of evaluation criteria.

Checklist: The most basic form of scoring model; projects are evaluated on whether or not they meet specific criteria.

Comparative Models: Decision-making tools that compare projects against each other.

Decision Tree: A visual representation of decisions and their effects on outcomes, costs, and risks that people can use to make decisions about ways to achieve a goal.

Fixed Costs: The costs that are not dependent on the volume produced.

Internal Rate of Return: A calculation of the payoff of a project, given the level of expenditure and the timing and amount of cash inflows and outflows.

Monte Carlo Simulation: A use of computer software to create a probability distribution of financial outcomes based on randomly selected values of inputs.

Net Present Value (NPV): An estimate of the financial value of a project today, given the expenditures, timing, amount of cash inflows and outflows, and the discount rate.

Project Map: A portfolio management tool that allocates projects into three types: derivative, platform, and breakthrough.

Qualitative Methods: Decision-making tools that compare projects on the basis of scales or words.

Quantitative Methods: Decision-making tools that select projects based on the basis of numerical calculations.

Real Options: A decision-making tool that is based on the idea that investments give the right, but not the obligation, to make future investments.

Scenario Analysis: A decision-making tool that represents investments under different assumptions about key factors that influence those investments.

Scoring Models: Decision-making tools that compare projects against standard scales.

Variable Costs: The costs that are dependent on the volume produced.

PUTTING IDEAS INTO PRACTICE

1. **Net Present Value Analysis** Your company is thinking of establishing a new semiconductor plant. It will cost you $1 billion to build the plant, all of which will be incurred in the first year. From the second through the fifth year, the plant will generate $250 million per year in revenue. At the end of the fifth year, the plant will become obsolete and will no longer generate any revenue. Your company pays 11 percent to borrow money. What is the net present value of this project? Does the net present value calculation indicate that you should pursue the project or not? Why?
2. **Real Options Analysis** Assume that you are a manager at a pharmaceutical company where researchers are working on a drug to treat heart disease. The cost of making the drug and the revenues that you will earn from selling an effective drug are uncertain because you don't know how effective the drug will be or what conditions the drug will treat. You need to conduct $1 million of R&D to develop the drug. If you are successful, you will have to undertake clinical trials that could cost either $25 million or $80 million, with a 50 percent chance of each type of trial occurring. If you succeed at clinical trials, there are two indications for the drug, each of which has a 50 percent chance of occurring—one which will generate $40 million revenue, and the other which will generate $60 million in revenue. Your company's cost of capital is 10 percent. Use the net present and real-options approaches to calculate the discounted cash flows for the project. What does each approach suggest that you do? Why?[65]
3. **Developing Project Maps** The purpose of this exercise is to develop a project map for General Motors. (For information on the company and the innovations that it is developing, go to www.gm.com.) Place the following projects on a project map for General Motors, identifying the breakthrough projects, platform projects, and derivative projects: hybrid diesel-electric busses, Chevy Tahoe and GMC Yukon hybrid cars, E85 flex-fuel vehicles, Chevy Volt electric car, Chevy Equinox fuel cell; engine power trains (e.g., LS7-V8 for the Corvette), marine power trains (e.g., Vortec 8100), Sit-N-Lift power seat; Onstar navigation system, active fuel management, variable valve timing, six-speed transmission, 1.8 Ecotec engine, and displacement on demand. Given the firm's strategy, is the allocation of projects across the three types correct? Why or why not? Should the company undertake additional projects? If so, what? Should the company allocate more resources to any of the existing projects? Why or why not? How does this project allocation influence what the firm will be able to do?

Notes

1. Adapted from Mun, J. 2002. *Real Options Analysis: Tools and Techniques for Valuing Strategic Investments and Decisions*. New York; Wiley and Sons.
2. Ibid.
3. Ibid.
4. Ibid.
5. Jaffe, A., and J. Lerner. 2004. *Innovation and Its Discontents*. Princeton, NJ: Princeton University Press.
6. Bhide, A., and H. Stevenson. 1992. Attracting stakeholders. In W. Sahlman and H. Stevenson (eds.), *The Entrepreneurial Venture*. Boston: Harvard Business School Press, 149–159.
7. Caves, R. 1998. Industrial organization and new findings on the turnover and mobility of firms. *Journal of Economic Literature*, 36: 1947–1982.
8. Roberts, E. 1991. *Entrepreneurs in High Technology*. New York: Oxford University Press.
9. Christiansen, C., and J. Bower. 1996. Customer power, strategic investment, and the failure of leading firms. *Strategic Management Journal*, 17: 197–218.
10. Moon, Y. 2004. NTT DoCoMO: Marketing i-mode. *Harvard Business School Note,* Number 5–5–3–097.
11. Morris, P., E. Teisberg, and A. Kolbe. 1991. When choosing R&D projects, go with long shots. *Research Technology Management*, January–February: 35–40.
12. Bhide and Stevenson, Attracting stakeholders.
13. Pisano, G. 1995. Nuclean Inc. *Harvard Business School Case,* Number 9–692–041.
14. Brunner, D., L. Fleming, A. MacCormack, and D. Zinner. Forthcoming. R&D project selection and portfolio management: A review of the past, a description of the present, and a sketch of the future. In S. Shane (ed.), *Handbook of Technology and Innovation Management*. Cambridge, UK: Blackwell.
15. Ibid.
16. Schilling, M. 2005. *Strategic Management of Technological Innovation*. New York: McGraw-Hill.
17. Tuma, D. 2005. Open source software: Opportunities and challenges. *Cross Talk,* June, http://www.stsc.hill.af.mil/. . . /2005/01/0501tuma.html.
18. Hallowell, D. Analytical hierarchy process (AHP)—Getting oriented, http://www.isixsigma.com/library/content/c050105a.asp.
19. Ibid.
20. McCaffrey, J. The analytical hierarchy process, http://msdn.microsoft.com/msdnmag/issues/05/06/TestRun.
21. Ibid.
22. Hallowell, Analytical hierarchy process (AHP)—Getting oriented.
23. Ibid.
24. Schilling, *Strategic Management of Technological Innovation*.
25. Schilling, M., and C. Hill. 1998. Managing the new product development process: Strategic imperatives. *Academy of Management Executive*, August: 67–81.
26. Ibid.
27. Narayanan, V. 2001. *Managing Technology and Innovation for Competitive Advantage*. Upper Saddle River, NJ: Prentice Hall.
28. Ibid.
29. Hamilton, W. 2000. Managing real options. In G. Day and P. Schoemaker (eds.), *Wharton on Managing Emerging Technologies*. New York: John Wiley.
30. Ibid.
31. Dixit, A., and R. Pindyck. 1995. The options approach to capital investment. *Harvard Business Review*, May–June: 105–115.
32. McGrath, R. 1997. A real options logic for initiating technology positioning investments. *Academy of Management Review*, 22(4): 974–996.
33. Dixit and Pindyck, The options approach to capital investment.
34. Faulkner, T. 1996. Applying "options thinking" to R&D valuation. *Research Technology Management*, May–June: 50–56.
35. Utterback, J. 1994. *Mastering the Dynamics of Innovation*. Boston: Harvard Business School Press.
36. Fichman, R. 2004. Real options and IT platform adoption: Implications for theory and practice. *Information Systems Research*, 15(2): 132–154.
37. Schwartz, P. 1996. *The Art of the Long View: Planning for the Future in an Uncertain World*. New York: Currency Doubleday.
38. Ahn, J., and A. Skudlark. 2002. Managing risk in a new telecommunications service development process through a scenario planning approach. *Journal of Information Technology*, 17: 103–118.
39. Wright, A. 2000. Scenario planning: A continuous improvement approach to strategy. *Total Quality Control*, 11(4): 433–438.
40. Brunner, Fleming, MacCormack, and Zinner, R&D project selection and portfolio management.
41. Ibid.
42. Rothwell, R. 1994. Industrial innovation: Success, strategies, trends. In M. Dodgson and R. Rothwell (eds.), *The Handbook of Industrial Innovation*. Aldershot, UK: Edward Elgar, 33–53.
43. Afuah, A. 2003. *Innovation Management*. New York: Oxford University Press.

44. Brunner, Fleming, MacCormack, and Zinner, R&D project selection and portfolio management.
45. MacMillan, I., and R. McGrath. 2002. Crafting R&D project portfolios. *Research Technology Management*, 45(5): 48–59.
46. Cooper, R., S. Edgett, and E. Kleinschmidt. 1997. Portfolio management in new product development: Lessons from the leaders—II. *Research Technology Management*, 40(6): 43–52.
47. Cooper, R., S. Edgett, and E. Kleinschmidt. 2002. Optimizing the stage gate process: What best-practice companies do—II. *Research Technology Management*, 45(6): 43–49.
48. Schilling, *Strategic Management of Technological Innovation*.
49. Schilling and Hill, Managing the new product development process.
50. Burgelman, R., C/ Christiansen, and S. Wheelwright. 2004. *Strategic Management of Technology and Innovation* New York: McGraw-Hill/Irwin.
51. Schilling and Hill, Managing the new product development process.
52. Wheelwright, S., and K. Clark. 2003. Creating project plans to focus product development. *Harvard Business Review*, September: 2–15.
53. Schilling and Hill, Managing the new product development process.
54. Burgelman, Christiansen, and Wheelwright, *Strategic Management of Technology and Innovation*.
55. Canner, N., and N. Mass. 2005. Turn R&D upside down. *Research Technology Management*, 48(2): 17–21.
56. Burgelman, Christiansen, and Wheelwright, *Strategic Management of Technology and Innovation*.
57. Pelz, D., and F. Andrews. 1976. *Scientists in Organizations*. Ann Arbor, MI: University of Michigan.
58. Adapted from Christiansen, C. 1998. We've got rhythm! Medtronic Corporation's cardiac pacemaker business. *Harvard Business School Teaching Note*, Number 5–698–056.
59. Ibid.
60. Ibid.
61. Ibid.
62. Ibid.
63. Ibid.
64. Wheelwright and Clark, Creating project plans to focus product development.
65. Dixit and Pindyck, The options approach to capital investment.

Chapter 6

Customer Needs

Learning Objectives

After reading this chapter, you should be able to:

1. Describe technology-push and market-pull innovation and explain how companies should approach market research for each.
2. Define a customer need and explain when a real customer need exists.
3. Spell out why product developers have trouble understanding what customers want.
4. Understand how product developers determine whether products are economically viable and provide better alternatives than competitors' products.
5. Explain how companies set prices for new products.
6. Define *market segmentation* and explain how companies can segment the market for new technology products and services.

7. Define *market research*, describe the different ways to collect market research data, and figure out when each approach to market research should be used.
8. Explain the advantages and disadvantages of using focus groups, survey methods, ethnography, lead user techniques, and iterative methods to identify target markets and customer needs, and identify the market characteristics favorable to each.
9. Identify when new and established firms have advantages at understanding customer needs and explain why they have advantages at understanding these needs under different circumstances.

Lead User Method: A Vignette[1]

The Medical-Surgical Division at 3M, a multinational corporation that serves a wide variety of technology markets, was looking to develop new products that would control infection in operating rooms without using antibiotics. The growth of antibiotic-resistant bacteria suggested that products in this segment would offer significant growth potential.

Despite believing strongly in customers' need for products that would control infection without using antibiotics, the managers at 3M did not know what features to incorporate in products to meet that need. The company's representatives had tried a variety of traditional market research techniques to gather information from customers, but these efforts were not very effective. The hospital staff that the 3M market researchers talked to kept suggesting that the company make incremental changes to its existing products, like lengthening their surgical drapes. As a result, the senior management of the Medical-Surgical Division did not feel that traditional market research was providing it with the right information. So the Medical-Surgical Division turned to the *lead user method*. (We will discuss this method in greater detail later in the chapter.)

The lead user method is a process for soliciting information from the likely first users of a product. First users are customers who say that they have a need for something that doesn't yet exist and are experimenting with ways to create a solution to their needs. The lead user method is based on the principal that the best way to develop a new product is to talk to the people who are struggling to find a solution to their needs. If you can understand the needs of lead users, and come up with a solution that meets those needs, then you can develop products that will meet the needs of other users of the product.

Many companies swear by the lead user method. Research has shown that it generates faster concept development than more traditional approaches to market research and allows companies to see customer needs for truly new products, rather than just market extensions.

So how does the lead user method work? First, the company forms a lead user team, which networks to find lead users in a target market. Once it has identified lead users, the team sets up a workshop in which it seeks to understand their needs. For example, when the Medical-Surgical Division at 3M was trying to understand the needs for the antibiotic-free infection control product, it brought together a MASH unit in Bosnia (which needed products that would work quickly under battlefield conditions) and a veterinary surgeon (whose patients are covered with fur and rarely bathe). Third, the lead user team uses the feedback generated in the workshop to develop new products. In the case of the 3M Medical-Surgical Division, the lead user workshop led the company to develop new draping and antimicrobial products, including a completely new product platform for infection-control devices.[2]

INTRODUCTION

Efforts to introduce new products often fail because those products don't satisfy customer needs.[3] This is particularly true for new products and services based on new technologies, which are often invented in the pursuit of scientific advance, and so are developed in the absence of information about customer needs.[4]

This chapter helps you to overcome this problem by offering the tools and techniques that you need to identify customer needs and the features that will satisfy them. The first section compares technology-push and market-pull innovation and explains how companies need to adopt different approaches to understand markets for each. The second section explains why and how companies identify real needs for new products and services, discussing how to identify new products that meet customer needs in a significantly better way than existing products, and satisfy key stakeholders. The third section explains how to price new products correctly. The fourth section describes market segmentation and explains how companies can segment the market for new technology products and services. The fifth section defines *market research* and identifies the different ways that companies conduct it, outlining the advantages and disadvantages of different approaches under different conditions.

TECHNOLOGY PUSH VERSUS MARKET PULL

Some innovation is driven by **customer needs,** or descriptions of the benefits that customers want in a product or service.[5] For instance, customers might indicate that they have a desire to wear clothes with embedded musical devices, and companies might respond by coming up with a line of clothing with built-in music players. When companies ask customers about their needs and then develop products and services to meet those needs, innovation is called **market pull.**

(Please note that needs are not the same thing as *wants,* which are a customer's beliefs about what products or services will fulfill his or her needs. Customer needs are also different from **product attributes** because a need is a description of the benefits that customers want in a product or service; whereas an attribute is the way that the product or service gives customers what they need.)[6]

Sometimes, people invent new technology in the pursuit of technological advance itself, rather than in response to market needs. Only after that technology is developed do people find the unrecognized "need" for it.[7] Remember the discussion

FIGURE 6.1
The First Laser

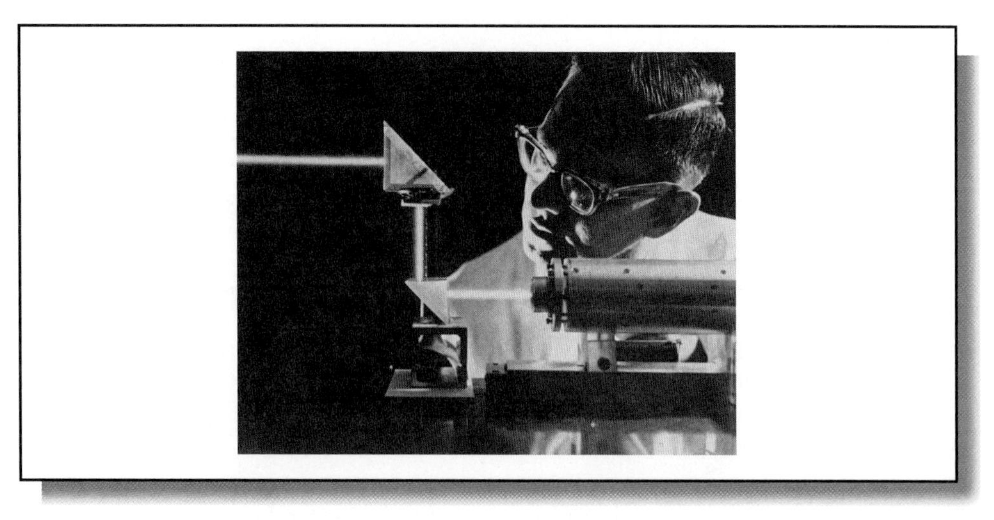

At the time that the laser was invented, no one knew what the technology could be used for; lasers are now used in CD and DVD players, bar code scanners in grocery stores, eye surgery, to cut and weld metal, to carry telephone and television signals, and to cut fabric, among a myriad of other things.
Source: Corbis/Bettmann.

of the laser in Chapter 4? Even though it represented an important technological advance, none of the people looking at the technology at the time had any idea what to do with it. Only later did people realize that you could use lasers to meet a myriad of customer needs (see Figure 6.1). Because this type of innovation occurs before anyone has figured out a market need that it can meet, it is called **technology push.**[8]

Entrepreneurs and managers need to approach market-pull and technology-push innovations differently. For technology-push innovations, recognizing market needs is not the first step in creating a new product because no market exists at the time that the innovation is developed. Instead, the innovator first needs to develop the technology. Once the technology has been developed, the innovator then needs to find or create a market by figuring out what benefits it provides and what problems it solves.

Some technology-push inventions do not require the creation of a new market, but rather, the identification of an existing market for which the technology is useful. Because the inventors of the technology have come up with the innovation without a particular customer problem in mind, they need to identify a problem that the technology can solve after it has been created. For instance, 3M developed a thin plastic film with microlouvers that worked like a tiny Venetian blind. The company thought it could be used for museum lighting, automatic teller machines, window treatments, and ski goggles, and conducted a variety of focus groups to try to identify the best application for it, without success. It was only when the executive in charge of the project noticed a secretary at 3M blocking the view of her work station from passersby that the application for computer screens was identified.[9]

Technology-push innovations have several characteristics in common that tend to differentiate them from market-pull innovations: First, they are developed without guidance from customer needs by technologists who are trying to advance knowledge or for their own purposes. Second, they are often based on radical technological changes that open up new markets, rather than on incremental technological changes within existing markets. Third, they often require a long period of

FIGURE 6.2
Identifying a Real Need

Sometimes people develop products or services for which there is no customer need; if no need ever materializes, those products will be unsuccessful.

Source: United Features Syndicate Inc. September 12, 1998.

development during which they are adapted to fit market needs. Fourth, they often face slow customer adoption, given an absence of a problem in the marketplace that the technology can solve, the magnitude of the change that customers experience to adopt the technology, and the need to iterate across market segments to find one that is interested in adopting the product.[10]

Key Points

- Developing a new product in response to customer needs is called market-pull innovation.
- When companies first develop a technology and then identify market needs, the process is called technology push.
- For some push technologies, innovators need to create a market because no market need yet exists for the technology.
- For other push technologies, innovators need to find an existing market that has a need that the technology can solve.

UNDERSTANDING CUSTOMER NEEDS

Unless a company is engaged in technology-push innovation, understanding customer needs is a central part of developing new technology products or services. Why? Because customers don't purchase product attributes—they purchase the satisfaction of their needs.[11] And it is very difficult to satisfy customer needs unless you understand what those needs are.[12]

Unfortunately, many entrepreneurs and managers do not understand their customers' needs. Take, for example, the case of Thinking Machines Corporation, a company founded to use massively parallel processing to create supercomputers. Although customers had many needs that Thinking Machines's computing technology could satisfy, the company's founders never recognized those needs, concentrating, instead, on developing the fastest computer in existence. Because customers did not need faster computers than the alternatives that existed, Thinking Machines Corporation failed.[13]

How to Identify Customer Needs

So how do you know if customers need a product or service that your company is thinking of developing? By answering two basic questions: First, do customers have a problem that no existing product or service solves? Second, is your solution to the problem significantly better than existing alternatives?

Customers need your product or service if it solves their currently unsolved problems. For example, customers need a noninvasive home test that detects hypertension because there is currently no way for people to test themselves for hypertension and learn that they should get treatment before the disease causes serious damage.

Customers also need your product or service if it offers a better solution to their problems than existing alternatives. For instance, a real need exists for Plumpy'nut, a peanut-based spread to treat malnutrition in children. In many developing countries, children suffer from malnutrition. Unfortunately, most nutritional supplements need to be mixed with water; and clean water is not available in many of the places where children are malnourished. Plumpy'nut offers a better solution than other alternatives to the problem of malnutrition in children because, as a spread, it does not need to be mixed with water.[14]

The "better" solutions to customer problems that a new product provides do not need to be as profound as that offered by Plumpy'nut. They can be as mundane as reduced complexity, lower risk, greater convenience, enhanced productivity, greater happiness, and higher satisfaction.[15] For example, online music downloads benefit customers because they eliminate the need to purchase a whole CD to listen to one song, the requirement that a CD contain music from a single artist, the need to go to a store to buy music, and the limited selection available in a physical outlet.[16]

Unfortunately, many new technology products do not offer "better" solutions to customer problems. Take, for example, HP's TouchSmart PC, an $1,800 computer for the kitchen. The computer allows people to write messages to each other, synch calendars, look up recipes, and order groceries online. However, most customers don't need an $1,800 computer to write notes to the rest of their family when they can use paper and pencils instead. Moreover, synching calendars isn't very useful in a product that is not linked to office calendars. Furthermore, few people want to pull out a keyboard in a place with limited counter space where they may not be able to sit down, especially when the alternative is to carry their laptops into the kitchen.[17]

So how do you know if customers have unsolved problems or problems that could be solved in a better way? The answer is to look to potential customers for clues. The best clue is a customer complaint, which indicates that a potential customer is unhappy with the status quo.[18] Take, for example, the case of underwriters at several insurance companies who complain that the software that they use to check the driving records of new clients is hard to use and inaccurate. The fact that several underwriters complain that they face the same problem indicates that there is a need for a better solution.

Another clue to the presence of a customer problem is the expression of an unfulfilled wish. An unfulfilled wish indicates that a customer would do something differently only if there were a way to do it. A good example of an unfulfilled wish is the number of people who indicate that they would like to vacation in outer space. The fact that people say that they would like to vacation in space when there is no way to do so at the moment provides evidence of an unmet need for the company that comes up with a way to provide "space vacations."

A third way that you can know if a real need exists is to put yourself in the customers' shoes and see if you have a need for the product you are thinking of developing. Many companies do just this. For instance, Procter & Gamble has its executives in charge of disposable diapers meet in a room that looks like a child's nursery and wear glasses that blur their vision to help them think about diapers from the point of view of children. Similarly, Meganesuper Co., a Japanese maker of eyeglasses, requires all of its employees to wear glasses to work. This requirement has helped the company to come up with innovations that meet customer needs, such as changes to nose pads that minimize slipping.[19]

The Difficulty of Identifying Customer Needs

While the process of identifying customer needs is straightforward, companies often fail to assess these needs accurately for six reasons:[20]

1. Product developers often fail to gather information about customer needs because they overestimate what they already know.[21] For example, Nokia has had problems selling its cellular telephones recently because they are much fatter than the phones of its rivals. Nokia does not offer slim mobile phones because its engineers believed that customers would want the television and video features it packed into its phones more than they cared about slimness, which turns out not to be the case.[22]
2. Not all customer needs can be discovered by talking to customers. Customers only express needs that they can think of and often don't have the imagination to think of beneficial things that don't yet exist. For instance, before e-mail existed, most people did not know that they would benefit from a system that allowed them to send computer messages to each other while walking around. Now, of course, this is such a necessity to some people that they are referred to as CrackBerry addicts because they cannot live without their BlackBerries.
3. Developers often do not know which customers to ask about their needs. Often, companies do not know the right target markets for the products that they are developing. Because customers are myopic, they cannot provide accurate information about the preferences of customers in other market segments. So if a company does not know the right segments to target, it is unlikely to gather useful information about customer needs.
4. Product developers tend to be strong believers in the value of the products that they are developing. Often, developers project their beliefs on customers, which leads them to incorrectly assess customer needs. For example, the developers of hybrid vehicles tend to care more about saving the environment than the average customer of automobiles and overweight the importance of that attribute to customers.
5. Companies have structures or processes that make it difficult for them to collect information from customers. For instance, Sun Microsystems recently reorganized its sales force so that its sales people sell multiple products. This restructuring has allowed Sun's salespeople to gather more information about what customers want in server technology rather than predetermining what they need and trying to sell that to them.[23]
6. Needs change over time. As technology advances, it creates needs for new products or products with new features. For example, when the first gigabyte hard drive was introduced in 1991 no one could figure out why anyone would need that amount of storage because the only things people were storing were spreadsheets and word processing files. Now, however, terabyte drives may not offer enough memory, given all of the video and still images that people are trying to store.[24]

Significantly Better Benefits Than Existing Products

New products must offer benefits to customers that are significantly greater than those offered by existing alternatives. Prospect theory, an important theory of behavioral decision making, explains why. When people have to choose between two alternatives, they do not perceive both alternatives in the same way. Rather, they pay attention to reference points, such as which alternative is seen as replacing the other. The attributes of the product which came first are seen as things that will be "lost" if they choose the other product, while the attributes of the product which came second are seen as things that will be "gained" if they choose it.

If people perceived the same value in products when they are viewed as both a gain and a loss, then perceived value to customers would be indicated by the 45 degree line in Figure 6.3. However, in reality, people perceive greater loss when outcomes are framed as negative (deviation below the 45 degree line) and greater gains when outcomes are seen positive (the deviation above the 45 degree line).

Prospect theory research has shown that people value losses at two to three times the level of gains. This means that customers judge the attributes of new products relative to the attributes of current products and require new products to provide benefits two to three times as large as the benefits that they received from old products before they are willing switch.

Take, for example, the efforts of an online grocery delivery service. While this service gives customers the convenience of online shopping, it also requires customers to give up selecting produce or going to the store in search of ideas for dinner. According to prospect theory, customers will not adopt online grocery delivery

FIGURE 6.3
Prospect Theory Diagram

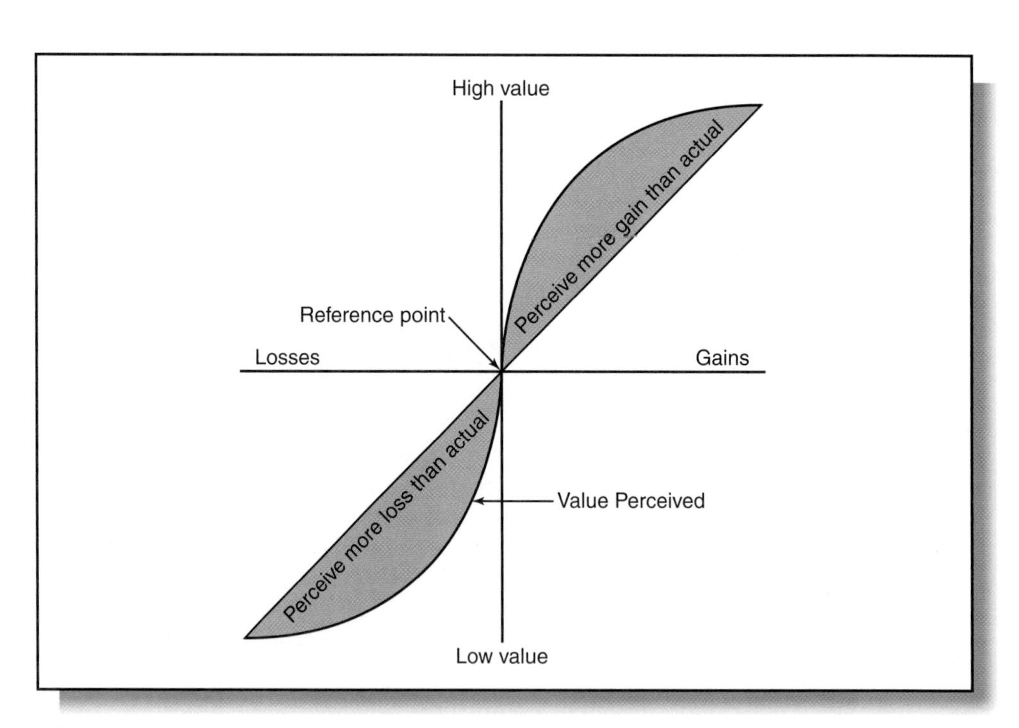

When an outcome is framed as a loss of something that people previously had, they perceive more loss than they actually experience, but when an outcome is framed as a gain of something that they never had, people perceive more gain than they actually experience.

GETTING DOWN TO BUSINESS
Going Over Like a Wet Tissue

Sometimes, managers or entrepreneurs think that they have come up with a new product or service that will meet customer needs when, really, they have not. Take, for example, the case of Kimberly-Clark and wet tissues. In 2001, Kimberly-Clark introduced a premoistened toilet tissue on a roll called "Cottonelle Fresh Rollwipes." The company believed that the product, which clips on to a toilet-paper holder, would dramatically increase the sales of toilet paper. Its managers developed elaborate projections of the sales of toilet paper, based on the assumption that consumers would use both dry and wet toilet tissues.[25]

Kimberly-Clark made a major investment in the development and launch of the product, which required over $100 million of R&D and manufacturing investment, and took $40 million of marketing investment to introduce. Confident in the importance of the new product, the company's press release at the launch of the product called Cottonelle Fresh Rollwipes "the most significant category innovation since toilet paper first appeared in roll form in 1890."[26]

While Kimberly-Clark's management expected to generate $150 million in first year sales from the product, they did not achieve anywhere near that result.[27] The company's management was surprised. The company's market research showed that 60 percent of adult consumers have used some sort of moistened cleaning method. Moreover, the market for disposable wiping products was already over $4 billion.[28]

So what went wrong? Kimberly-Clark didn't pay enough attention to the complexity of their customers' needs. While many people might benefit from moistened toilet paper, most people don't want to draw attention to that need or their use of moistened toilet paper. The product, which looked conspicuous in a bathroom, made users uncomfortable. Kimberly-Clark didn't realize that what customers needed was not just moistened toilet paper, but secret moistened toilet paper. Moreover, customers didn't want to pay more for moistened toilet paper than they would pay to wet their own paper. As a result, they were unwilling to pay the premium that Kimberly-Clark charged for the premoistened paper.[29] In short, Kimberly-Clark didn't do a very good job of assessing customer needs for the product.

services unless they perceive the convenience of online shopping to be at least twice as large as the benefits of selecting produce or going to the store in search of ideas for dinner.[30]

Meeting the Needs of Many Stakeholders

You also need to consider the needs of other stakeholders who influence your customers' buying decisions when developing a new product or service. As Figure 6.4 shows, even the decision to purchase a very simple product, like the vial stopper that goes in a syringe, can be affected by a large number of stakeholders.

In some cases, other stakeholders have such an influence on customer buying decisions that sales cannot be made if their needs are not met. Take, for example, 3M's development of a virus-proof surgical gown. While 3M's customers, the hospitals, believed that this product was great because it offered a tremendous improvement in patient safety over standard surgical gowns, this product was not widely adopted. The virus-proof gown was 10 percent to 15 percent more expensive than other surgical gowns, and managed health-care organizations, which are an important stakeholder in medical product purchases, would not pay the extra price.[31]

FIGURE 6.4 Many Stakeholders Influence a Customer's Buying Decision

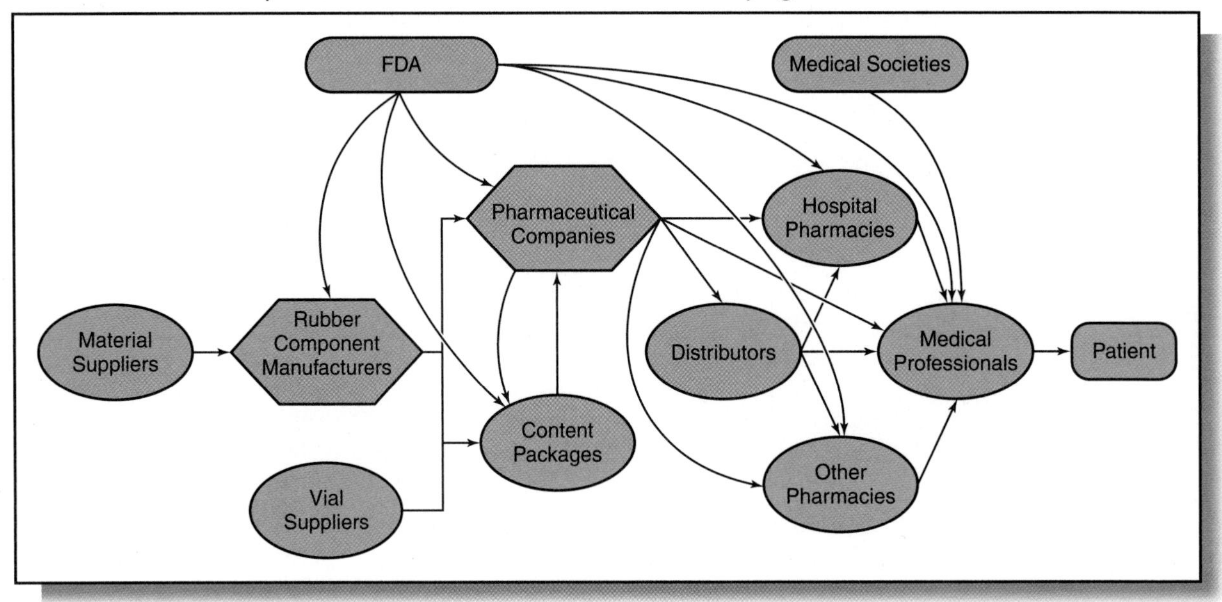

This example shows that a large number of companies are stakeholders whose preferences affect the decision to purchase the stopper that goes in a syringe.
Source: Elliot Ross.

Solutions That Work

While it might sound obvious that developing new products that work is important to having a successful technology strategy, many companies lose out to competitors because they cannot develop products that work well. Take, for example, the case of the social networking site, Friendster.

Started in 2002, Friendster was *Time* Magazine's invention of the year in 2003 and quickly became the dominant social networking Web site. However, the company suffered a dramatic fall in 2004 because engineering problems made its Web site three times as slow as the Web sites of its major competitors, MySpace and Facebook.[32] As a result, Friendster's customers defected to its competitors.

Developing Profitable Solutions

Creating a new product that meets customer needs and provides benefits to them of at least twice the benefits of existing products is still not enough to succeed at new product development. You also need to develop the new product in a way that generates a profit.

Unfortunately, many companies cannot figure out how to develop their new products profitably. Take space tourism as an example. No one has yet figured out how to get people into space at less than the cost that only a few people can afford. As a result, no one has established a space tourism business. To do so would only mean creating a business that loses money.

Developing a profitable product is difficult for two reasons. First, you have to figure out a level of sales at which you can make money. If all sales lose money, then scaling up to produce and sell more units will only create greater losses. However, in most businesses, the profitability of transactions is not the same across all levels of sales. Businesses that involve economies of scale, have increasing returns, or have large setup costs (we will talk more about these things in later chapters) invariably lose money on initial sales but find that the transactions undertaken at higher volume are profitable.

Entrepreneurs and managers often think that if they sell more units they will make money at some point. But that doesn't always happen. Take, for example, many of the Internet start-ups of the 1990s. The entrepreneurs that founded many of these companies never could figure out a sales level at which they could make money. So getting larger just caused them to lose gobs of money.

Second, you need to accurately estimate future costs and prices; incorrect estimates will cause you to fail to make money from developing new products, or fail to develop new products when you could have made money doing so. The problem, of course, is that it is difficult to figure out what costs will be in the future. While the costs of making and selling many technology products decline as companies get better at making them, predicting this pattern with any kind of accuracy is very difficult. For example, Johnson & Johnson, an early leader in disposable diapers, was driven out of that business because it failed to predict that Procter & Gamble would dramatically reduce the cost of the product.

Similarly, it is difficult to predict what prices for products will be in the future. For example, several oil companies have developed ways to turn extra-heavy crude from tar sands into crude oil. However, the cost of this process is approximately $25 per barrel, which makes it worth doing when oil prices are high, but did not make sense when the technology was first developed, and oil sold for $12 per barrel.[33]

Key Points

- A customer need is a description of the benefits that customers want in a product or service; it is different from a product attribute, which is a description of how that need gets satisfied.
- To succeed at new product introduction, you need to understand customer needs; many companies have failed because they did not understand their customers' needs.
- A product meets a customer need if it solves an unsolved problem or provides a better solution than existing alternatives to problems that have already been solved.
- Customer complaints and unfulfilled wishes are clues to unmet customer needs.
- Companies often have trouble identifying customer needs because product developers overestimate their understanding of customer needs, incorrectly project their beliefs about the value of product attributes on customers, do not know which target markets to talk to, have structures and routines that inhibit information gathering, try to gather information only by asking customers for it, and fail to recognize how needs change over time.
- Because customers weight losses more highly than gains, new products must provide benefits two to three times better than existing products, or customers will not adopt them.
- Companies need to consider the interests of all stakeholders that influence the customers' buying decisions.

PRICING PRODUCTS CORRECTLY

Whether your innovation is technology push or market pull, correct pricing of your product is an important element of a successful technology strategy. The price of your product affects the willingness of potential customers to adopt it, your ability to take customers away from your competitors,[34] and your profit margin.

As a technology entrepreneur or manager, you have two basic approaches to setting the price of your new products: **price skimming,** in which you set a high price to earn the highest possible profit, and **penetration pricing,** in which you set a low price to get the highest possible market share.[35] The choice between the two approaches depends on the balance between the benefits and costs of charging a high price. Setting a high price will deter potential customers from adopting the product and will risk losing customers to your competitors. But setting a low price will undermine your profitability.

Moreover, a penetration pricing strategy can be very profitable if you use it to gain customers who can be charged a higher price later. However, doing so is risky because you might lose money indefinitely if you can't raise the price of your product. The effectiveness of a penetration strategy depends a lot on the kind of business you are in. As we will see in Chapter 13, for businesses in which the value of the product to customers increases with the number of customers already using the product, a penetration strategy can be very effective. For instance, PayPal actually paid its customers to use its online payment system initially because it knew that if customers adopted its service, it would be hard for competitors to dislodge it, and it could raise prices later.

Setting a Price

In addition to the topics identified in the previous section, and, of course, the demand for your product or service, you should consider four things when you set your pricing strategy: the timing of your market entry, the nature of the market into which you are selling, how products are paid for in your industry, and your cost structure.

First, you need to understand how the timing of your company's entry into the market will affect what customers will pay for your new product. Not only is there a competitive aspect to this question—entering after a competitor has launched their product means that your competitor's price will influence the price for your product—but also you need to think about the relationship between the price of a product and customer adoption patterns. Some customers will pay more to be the first people to have a product.

However, early market entry does not always mean that you should price high and skim the market. If your product is based on increasing returns (more about this in Chapter 13), you might want to do just the opposite. Your company will benefit greatly from broad initial adoption because the profitability of selling your product will increase with the volume sold. Therefore, setting a low price to encourage adoption is often a more profitable approach in the long run if your product is based on increasing returns.

Second, you need to consider the market in which you are selling the product or service. The range of prices commonly charged in a market is a basic factor affecting prices. Even totally new products are limited in their price range because customers will substitute related products if the price of the new product is not within the expected range. For example, suppose you developed a levitating car that can fly over traffic jams in rush hour traffic. Even though there are no other levitating cars on the market, the price that you set for the car is going to be limited by the price of helicopters, which would be a substitute for the levitating car. If you try to price your levitating car much higher than the price of helicopters, customers will shift to the substitute product.

Another aspect of your market that influences the price that you can charge for your new products is the structure of the sales channels in the industry. Products often travel through intermediaries before going to the end customers, and those intermediaries want to make a profit on their activities. Consequently, the price that you set needs to consider the profit margins that these intermediaries expect to earn because the price that your end customer is willing to pay is limited, and includes these profit margins.

Third, you need to consider how products are paid for in your industry. In some industries, products are paid for in cash, while in others they are paid for on credit. In some markets, companies routinely discount their prices; whereas, in others, prices are largely fixed. You should consider the effects of hidden costs, discounts, and credit when you set prices for your new product because these factors will affect the actual revenues that you receive from your customers.

Credit is a good example of this. If you are going to provide credit to your customers, you need to factor that credit into your calculation of the price for your products. What you will earn is actually the list price that you charge less the cost of the credit that you provide.

Fourth, you need to consider your cost structure, including the cost of the materials and labor needed to make your product, as well as your development and marketing costs. For instance, if you are going to provide a great deal of after-sales service, have a lot of sales people, or do a lot of advertising, then you need to price in a way that allows you to pay for these marketing costs. In short, whatever price you set has to allow you to make a profit over the long term. And to figure out if you are going to make a profit, you need to know your costs.

One very important aspect of your costs that you need to factor into your pricing is the relationship between fixed and variable costs. As Figure 6.5 shows, fixed costs, things like your rent, do not vary with the quantity of your product that you make; while variable costs, things like the cost of a assembling a product, are a function of the number of units that you produce. Because variable costs depend on how much of your product you make, and the volume of sales is difficult to estimate, you might find it difficult to estimate your company's per unit portion of fixed costs. Consequently, your company's pricing will be inaccurate. If you believe that your company will produce more than it actually does, then your per unit price will be too low for your costs and your company will lose money.

The relationship between fixed costs and variable costs is particularly important in those high-technology businesses that sell products based on information, like digital music. The fixed cost of producing information goods is very high, while the variable cost is almost zero. Therefore, the per unit cost depends almost exclusively on the number of units sold, which makes it is very difficult to forecast costs per unit in these industries.

Versioning, fixed fee pricing, and bundling are very effective ways to set prices for information goods. Versioning means offering different versions of a product of different quality at different prices to different sets of customers based on their needs. Because competition will drive costs down to variable costs, but fixed costs are high, reducing competition is critical for making profitable the sale of products

FIGURE 6.5
Fixed and Variable Costs

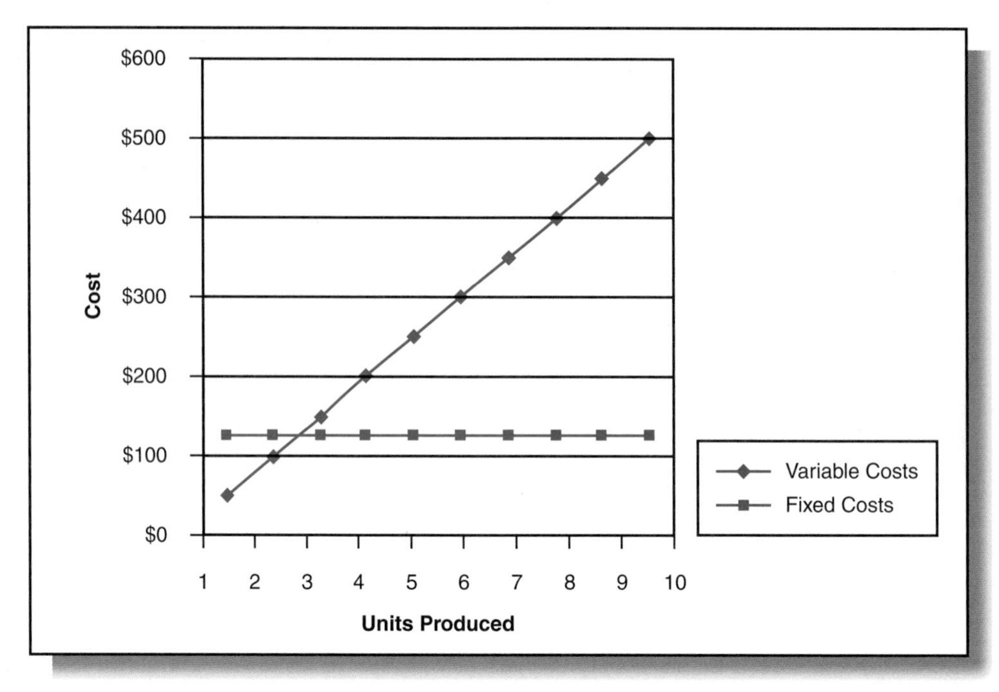

This figure shows the relationship between the number of units produced and fixed and variable costs; variable costs increase with the number of units produced, while fixed costs stay constant.

based on information. By offering different versions of the product to different customers, companies can segment the market and reduce the competition that drives prices down.[36]

Providing unlimited use for a fixed fee is also valuable because the variable cost of providing the product is close to zero. Consequently it costs little to serve customers who exceed the average per unit cost under the fixed-fee arrangement, while allowing extra profit to be captured from those customers who do not reach the average per unit cost.

Bundling is the process of offering components to customers for purchase as a group, usually for less than the price of components separately. For instance, software products are often sold in bundles, as is the case when a database spreadsheet and word processor are sold as part of a single package for less than the sum of both of those components alone.[37]

Bundling is a good pricing strategy, particularly when a product has a high fixed cost and a low variable cost, because it increases the price that a company can charge on some components, helps it expand into other markets, gives it leverage over suppliers, and deters entry by the providers of the unbundled parts of the product. Consequently, often you can make more money selling one component at a low margin to encourage customers to purchase the other components at a high margin than you can by maximizing the price of each individual component.[38]

The effectiveness of bundling depends a great deal on the nature of the industry in which you operate. In addition to the fact that bundling works better in industries that sell a product based on information, bundling also tends to be more effective when economies of scale exist in production and economies of scope exist in distribution because the scale and scope economies generate cost savings, permitting higher profit margins, even when the bundled components are sold at a lower price.

Bundling has important implications if you are thinking of entering a new market. When new technologies are introduced, companies in related industries are often better positioned than incumbent firms to take advantage of them because their assets allow them to create more valuable product bundles than incumbent firms can offer. Take, for example, the entry of large cable companies, such as Cablevision Systems Corporation, into the telephone business. These firms are better positioned to offer VOIP phone service than the telephone companies because their assets in cable, Internet, and wireless services permit them to economically bundle telephone, cable television, Internet, and wireless services, while the phone companies cannot bundle cable television with these other products.[39]

While bundling is a useful pricing strategy, you need to be careful how you use it. Your company can violate antitrust laws if it has monopoly power in a market and you bundle your products because bundling can be used to exert market power to limit consumer choice. For example, Microsoft faced antitrust action when it bundled its Web browser with its Microsoft Office suite because the company possessed market power in the office software market.

Finally, to determine the right price for a product, you need to understand how the relationship between its different components influences their respective sales. One common relationship between components occurs when one component is purchased once (for example, a cable box) and another component is purchased repeatedly (for example, monthly cable service). For these types of products, you probably want to price the nonrepeat component low to encourage customers to make an initial purchase into your system. If you can get customers to make the initial purchase, then they will begin to make the repeat purchases that are profitable for you. For instance, with cable service, you might want to charge a low price for an initial setup of the box and make your money on the monthly service fee.

Even if you factor all of these things shown in Figure 6.6 into the price that you set, you have to be ready to dramatically change your pricing structure after you enter the market if your pricing is incorrect. Even giant companies, like Microsoft,

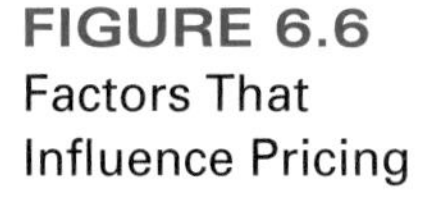

FIGURE 6.6
Factors That Influence Pricing

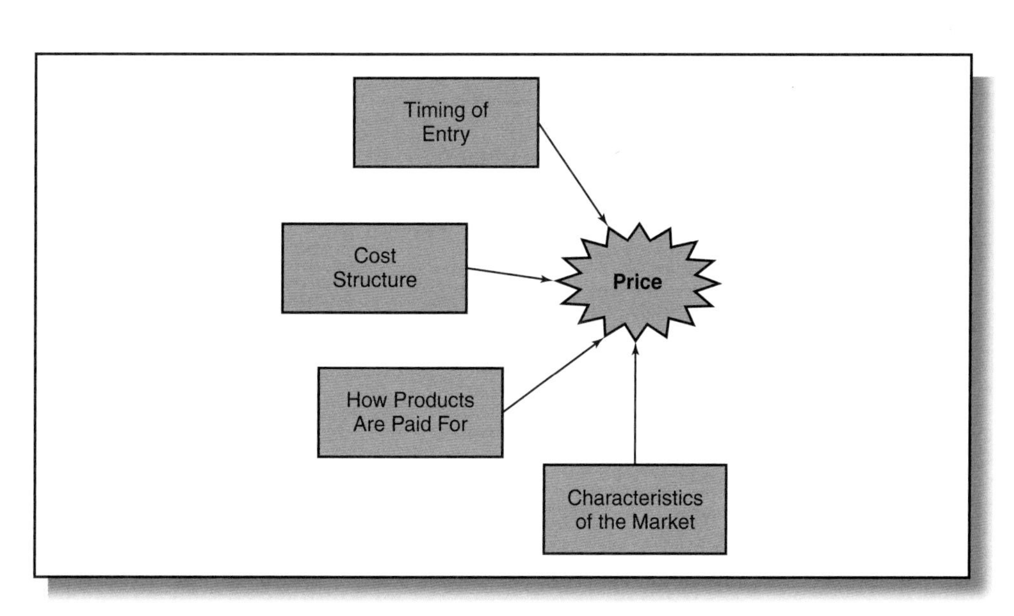

Four categories of factors influence the prices charged for technology products.

have had to make huge changes in the prices of their products after introduction when the information that they gathered from the market revealed that their pricing structure was wrong. For example, when Microsoft introduced the Xbox, the company quickly found out that it had priced the product too high and had to cut the price to $100 to get consumers to buy it.[40]

Key Points

- To launch new products successfully, companies must decide whether to use a skimming or penetration strategy and must set the right price, considering all relevant factors.
- Effective pricing depends on many factors, including how the timing of your company's entry into the market will affect what customers will pay for your new product, your cost structure, the market in which you are selling the product or service, and how products are paid for in your industry.
- One very important aspect of your costs that you need to factor into your pricing is the relationship between fixed and variable costs, particularly in those high-technology businesses that sell products based on information.
- The high fixed and low variable cost of information goods makes versioning, fixed fee, and bundling good pricing strategies in information-based industries.
- Bundling is a good pricing strategy because it increases the price that a company can charge on some components, helps it expand into other markets, gives it leverage over suppliers, and deters entry of the providers of the unbundled parts of the product.
- Even if you factor all relevant information into the price that you set, you have to be ready to dramatically change your pricing structure after you introduce a new product if the price that you set turns out to be incorrect.

MARKET SEGMENTATION

Companies generally face one of three types of markets: homogenous markets, in which all of the customers have the same needs; diffuse markets, in which few customers have the same needs; and clustered markets, in which customers fall into a small number of groups on the basis of their needs. When markets are composed of customers with clustered demand, **market segmentation,** the process of dividing a market into groups which have common needs, is often an effective approach to identifying those needs.

Market segmentation is particularly helpful when:

1. The technology underlying the product or service is only effective in solving some customer problems and not others. For example, a heart drug named BiDil works on African Americans, but does not work on people of other races, suggesting the value of segmentation in marketing this drug.[41]
2. The importance of the solution that the company can provide varies across groups of customers, with some having a much greater need to buy the product than others. For example, Match.com, an Internet dating site, targeted older, less Internet savvy customers when it found that the ease of use of its social networking Web site made it more attractive than competitor sites to the older segment of the on-line dating market.[42]

3. The organization lacks the ability to serve all of the market at once, perhaps because it is a new company that cannot scale up quickly. For instance, a start-up software company many not have the resources to develop a version of its product for both the consumer and industrial segments of the market, and so will benefit from the opportunity to segment the market into these two groups and go after only one initially.
4. Competitors respond differently to efforts to sell to different groups of customers. For instance, Canon succeeded in getting inroads into the photocopier market by targeting small and medium-sized businesses. Xerox, which was focused on large corporations, did not respond quickly to this attack because it did not care that much about smaller customers.[43]

In an ideal world, companies would always segment the market by dividing customers into categories based on their needs. However, many needs, such as a desire for high quality, are unobservable. Therefore, to divide the customers into groups that have common needs, companies use "profilers," which are measurable dimensions, like geographic location, industry, demographics (e.g., age, gender, education), or psychographics (e.g., lifestyle, personality), that are correlated with customer needs.

To segment the market, companies identify key profilers and gather information about customer needs. They then divide the market by those profilers and identify the minimum number of profilers necessary to specify the groups of customers that have similar needs within the group, and different needs between the groups.

Market segmentation is particularly important to new technology firms. New firms will be more successful if they can avoid drawing the attention, and direct competition, of large, established companies, which they can do if they avoid the latter's primary customers.[44] In addition, they tend to benefit relatively little from economies of scale, given their small size. Segmented markets provide an opportunity for them to enter with small-scale production and focus on underserved niches.

Take, for example, the efforts by Nucor, a minimill, to enter the steel industry. Because Nucor initially targeted the segment of the steel market where the profit margins were the slimmest and where only the marginal customers of the integrated steel mills were found, the major steel makers accommodated its entry. Had Nucor gone after the high margin mainstream of the market first, the major steel makers would likely have retaliated and possibly driven it out of business.

Key Points

- When markets are composed of customers with clustered demand, market segmentation, the process of dividing a market into groups that have common needs, is often an effective approach to identifying those needs.
- Market segmentation is particularly helpful when the technology underlying the product or service is only effective in solving some customer problems, the importance of the solution that the company can provide varies across groups of customers, the organization lacks the ability to serve all of the market at once, and competitors respond differently to efforts to sell to different groups of customers.
- When needs are unobservable, companies use "profilers," which are measurable dimensions, like demographics, that are correlated with customer needs.
- New firms benefit from market segmentation because they can avoid drawing the attention, and direct competition, of large, established companies if they focus on a niche, and because they tend to benefit relatively little from economies of scale.

MARKET RESEARCH

Companies often engage in **market research,** the process of gathering and analyzing information about customer needs, preferences for products, and purchasing decisions. Although market research has important limitations, it is also valuable to companies in five ways:[45]

- First, it identifies potential customers and the market segments that are most interested in a product or service.
- Second, it specifies the product features that customers prefer, and which product developers need to include in new products and services.
- Third, it identifies the factors that influence customer purchasing decisions.
- Fourth, it provides information about the attractiveness of a market, its size, and its growth.
- Fifth, it indicates ways that companies can influence sales, such as cutting prices.

Market Research Techniques

Market research can take a variety of different forms. For example, it can involve the collection of primary data, or it can rely on the analysis of secondary data. It can be qualitative, or it can be quantitative. As a technology strategist, you need to be aware of the advantages and disadvantages of the different forms of market research so that you can adopt the right approach when your company develops a new product.

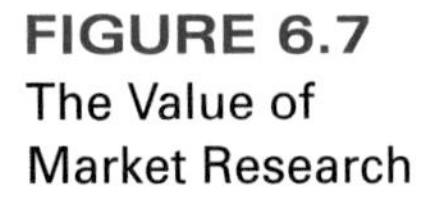

FIGURE 6.7
The Value of Market Research

"Forget about softness. Tests show we could triple sales if we add pepper."

Market research helps you to formulate a technology strategy.
Source: http://www.cartoonstock.com.

Newness of the Market

The effectiveness of many market research techniques rests on three crucial assumptions. First, potential customers understand their own needs and preferences. While this is a pretty fair assumption for products and services that already exist, such as personal computers or automobiles, it may not be true for new-to-the-world products, which potential customers can't visualize.[46] Take, for example, Internet shopping. When the Internet was first created, people really did not know how it could be used to meet their shopping needs. As a result, efforts to survey people about their preferences for it were very ineffective. People did not yet understand why they might want to shop online, or even what the concept of online shopping meant.

The second assumption is that companies know who the right customers are for their new products. However, when products are based on new-to-the-world technologies, companies often do not know the right potential customers to talk to because these technologies turn out to be most appropriate for markets that are different from those that the developers intended.[47] Take the laser, for example. When this technology was first invented, it was not powerful enough to be used for the weapons that the military hoped to use it for. It turned out that the earliest applications of lasers were for surgery. Thus, the developers of the laser initially thought to talk to the wrong set of customers—the U.S. military—about the use for the technology.[48]

The third assumption is that customers can understand the product concept before the new product has been developed. However, this is often not the case for new-to-the-world products. Take, for instance, the photocopying machine. When the first photocopier was introduced by Haloid Corporation, the precursor to Xerox, people simply could not easily reproduce existing documents in libraries, universities, offices, and so on. This made it difficult for the developers to explain the product concept to potential customers before they could demonstrate a prototype.

Research has shown that using traditional market research techniques to gather information about new-to-the-world products, whether those products are video cassette recorders, xerox machines, e-mail, personal computers, or optical fibers, almost always creates problems. When companies cannot be sure who the right potential customers are, potential customers do not understand their own needs and preferences, and potential users cannot grasp product concepts in advance of their development, then traditional market research techniques just lead to the collection of bad data. For instance, when Corning first developed optical fibers, it went to AT&T, the dominant phone company at the time, and asked about its interest in fiber optic cables. AT&T's response was that there was no need for this technology in long-distance phone lines, despite the fact that MCI developed its entire phone network on the basis of fiber optics shortly thereafter.

As a technology entrepreneur or manager, you need to figure out if your products and services are new to the world, like the Apple Newton (the first personal digital assistant),[49] or merely a new variation of an existing product, like the latest Intel microprocessor.[50] If your products are new to the world, then you need to use market research techniques that are very different from the standard market research techniques that you might have learned about in a market research course. As Table 6.1 shows, when conducting market research on new-to-the-world products, you need to rely on a more inductive approach, and you need to get deeply involved with potential customers to understand the context in which they might use your products.

TABLE 6.1 Techniques for New-to-the-World and Less Novel Products
This table shows the different approaches to conducting market research for less novel and new-to-the-world products; the latter demand more inductive approaches than the former.

	Less Novel	New-to-the-World
Philosophy of Market Research	Deductive analysis	Inductive reasoning
Techniques for Gathering Customer Information	Focus groups, surveys	Lead user method, Delphi technique, ethnography
Examples	New computers, new car models	First Internet auction houses, initial photocopiers

Source: Based on information contained in Barton, D. *Commercializing Technology: Imaginative Understanding of User Needs.* Harvard Business School Note 9-694-102.

When companies do not know who the right potential customers are for their new products, and when they cannot ask customers about their needs and preferences, they often need to engage in an iterative approach to market research, developing an initial version of the product and testing it out on the market to gather the information necessary to further refine it.[51] This approach might involve the creation of prototypes with features that would be included in future versions of the product if customers were to respond positively.[52] It might also involve **beta testing,** or the release of an early version of a product to see how customers react to it before it reaches commercial production.[53]

By taking what they learn from testing a product with customers, companies can modify their products to meet customer needs. This allows them to launch new-to-the-world products with features that customers want, even though potential customers are unable to communicate their needs or preferences. For example, GE identified the features for its first CT scanner by introducing a crude version on which the company solicited feedback. The feedback enabled GE to improve imaging resolution and speed, and design a product that met customer needs.[54]

However, the use of an iterative approach to market research is not without risks. When a company introduces an early, and, necessarily flawed, version of its product to obtain customer feedback, it risks its reputation. If the initial version is too flawed and cannot be improved to meet customer needs, then the customers' view of the company will suffer.

Because taking an iterative approach to market research has costs and benefits, you need to balance the risk of coming out with a failed product with the cost of obtaining inaccurate customer feedback. When risk of new product failure is low, but the cost of obtaining inaccurate feedback is high, then you should take an iterative approach to market research. However, when cost of obtaining inaccurate customer feedback is low, but risk of a failed product is high, then you are better off not taking that approach.[55]

Ethnography. When your potential customers don't understand their own needs and preferences, and so cannot articulate them, you need to identify customer needs by using market research techniques that involve observation, such as **ethnography**—the description of a group and its activities based on observation and participation.[56] By observing potential customers, you can identify needs that customers may not even know they have, let alone are able to articulate. For instance, a product designer might observe a surgeon at work and see the value of a new surgical tool that the surgeon was unaware that he or she needed.[57]

Even when customers understand their needs, ethnography often helps companies to get a detailed understanding of how customers might use new technology products. For example, when SIRIUS Satellite Radio wanted to figure out how to make a device that would allow the company to gain market share against its larger competitor, XM Satellite Radio, it had a team of ethnographers follow 45 people over a one-month period to see how they listened to music, watched television, and used other electronic devices. The information gathered by the team of ethnographers helped Sirius come up with the S50, a small music playing device that met customer needs much better than the comparable product from XM Radio.[58]

The use of ethnography is so important to market research in some industries that some companies even have ethnographers on staff whose job is to use the technique to come up with new product ideas. For instance, one Intel ethnographer came up with the idea for the Community PC, which allows millions of farmers in developing countries to access the World Wide Web, after spending two years traveling through the developing world; while another thought of the Centrino wireless technology after observing that Alaskan fishermen needed wireless transmission of information to adhere to government limits on the amount of different species they were allowed to fish.[59]

Ethnography is also useful when companies want to enter an industry that is new to them. Take, for example, the case of GE, which wanted to enter the plastics fiber business. Because GE couldn't just go to potential customers and ask them how to take the market away from the established players, it persuaded several customers to let it visit and tape record conversations between their executives. From this effort, GE learned that its senior executives were wrong in their initial assumptions about what customers wanted. Customers, it turned out, did not care as much about price as about the opportunity to collaborate with suppliers to produce exactly the kind of materials that they needed. So that's what GE focused on providing.[60]

Lead User Method. The **lead user method** is a market research technique, developed by Professor Eric Von Hippel of the Sloan School of Management at MIT, which is particularly useful for gathering information about customer needs for new-to-the-world products.[61] Standard market research is often limited by the imagination of customers, who do not give much thought to possible new products or services that they might need someday. Consequently, when asked by market researchers about their needs and preferences, most customers request incremental extensions to existing products or services.

Lead users are often a better source of information than representative customers because they are actively seeking new solutions to their problems. As a result, they are a good source of information about the needs of customers for features in new-to-the-world products.[62] Moreover, lead users are very interested in becoming involved in beta tests and are very likely to be early adopters of any product that solves their problems.[63]

The lead user method differs from standard market research methods because information is solicited only from people who are so unhappy with existing products that they are trying to develop products to meet their own needs. In addition, lead users can help to design new products, rather than just give feedback on them, because they are involved in product development.[64]

Lead user research is conducted as follows: A company forms a team of between four and six people from marketing and engineering to manage the process. In the first phase, the team identifies the products and markets that they would like to pursue and engages in activities to learn about them, such as reading trade journals and talking to industry experts. In the second phase, the team selects a specific user trend to focus on and develops a plan for interviewing lead users. In the third phase, the team interviews lead users and gathers information to help understand customer needs and to identify a potential product concept. In the fourth phase, the team conducts a workshop with 10 to 15 lead users to fill in information that could not be gathered in the prior phases.[65]

Many companies swear by the lead user method and use it religiously to develop new product concepts. For example, 3M has developed a policy of working with lead users to develop new products and has invested significant resources in the lead user method. As a result, this method has been used to identify a variety of product concepts, including those for mountain bikes, Gatorade, and sports bras.

Focus Groups. When new products are not new to the world, companies usually know who the right customers to talk to are, potential customers generally understand their own needs, and product concepts are typically easily explained. As a result, companies can use traditional market research techniques to gather information. For instance, they often use **focus groups,** or discussion sessions of between eight and twelve individuals, who meet for two to three hours to discuss their needs, preferences for different product features, or interest in new products and services. Focus groups are usually led by a trained moderator whose goal is create an atmosphere that motivates the participants to engage in discussion.

Focus groups have several advantages as market research tools. First, they have high response rates, reducing the nonresponse bias present with other market research tools, like surveys. People tend to prefer to participate in focus groups than to respond to surveys because focus groups involve the social interaction of a discussion. Moreover, focus group participants are more likely to answer all of the questions asked of them because of the effect of being in a group setting.

Second, the focus group format permits market researchers to ask more open-ended questions than is the case with many other methods of collecting market research data, like surveys. The use of open-ended questions allows market researchers to gain a greater understanding of customer needs and preferences than is present with closed-ended questions.[66]

On the other hand, focus groups have several important limitations. First, they are relatively expensive for the amount of information they provide. Research has shown that two one-on-one interviews elicit approximately the same amount of information as a single focus group, but cost far less to conduct.[67]

Second, the data obtained from focus group respondents are rarely independent of the views of other focus group members. Not only do a small number of people usually dominate a focus group discussion, leading to an overrepresentation of their views in the information obtained, but also the comments made by one participant influence the subsequent comments made by other participants. As a result, many comments made in focus groups do not represent the participants' true preferences, but, rather, their interpretations of the preferences of the other participants.

Third, focus groups are hard to control, with the participants taking the discussion in directions that market researchers often do not want to go. As a result, focus group moderators are often confronted with the choice of gathering large amounts of irrelevant information or adversely affecting the free flow of information by intervening in the discussion. Unfortunately, both alternatives reduce the value of focus groups as a market research tool.

Fourth, it is often difficult to assemble a focus group that is representative of a larger population. Because focus group participation takes time and effort, many people are unwilling to participate in them. This means that focus groups are often selected samples, making it difficult for market researchers to generalize from focus group data.[68]

Survey Research. Companies also use **survey research,** which consists of efforts to ask a sample of people questions about something of interest, such as their preferences for a product or service. For example, you might conduct a survey to find out how product features influence computer purchasing decisions.

Companies need to survey a **sample,** or portion of the population, when they can't gather market research information from the entire population. For example, you might want to survey a sample of car buyers about their preferences for product features because the entire population of automobile users is over a billion people.

To gather accurate information through the use of surveys, you need to develop questions that people are willing to answer and phrase them in a way that is clear to the respondents. You also need to gather information from a large enough number of people to differentiate statistically between the answers of the different respondents. Furthermore, you need to identify a **sampling frame,** or targeted group of respondents, that represents the population about which you seek to gather information. If the subjects in the sampling frame are not **representative** of the population from which it is drawn (that is, they don't have the same distribution of characteristics), you won't be able to generalize, or draw inference, to the broader population.

Groups that differ from the overall population on characteristics that affect customer preferences or purchasing decisions make poor sampling frames for market research. For example, students standing in line at a concert might be a convenient sample to ask about interest in iPods, but they would not be representative of the population of all students. They are very likely to differ from the overall population in terms of their interest in music, particularly live musical events, both of which might be related to their interest in iPods.

New and Established Businesses

Established companies have an advantage over new companies when the market and customer preferences for the innovation are well known. Why? Because these innovations are ones for which the capabilities of established firms tend to be the most beneficial. Large, established firms often develop expertise in market research based on large sample data collected from surveys and focus groups. New firms cannot easily compete with established firms at gathering and analyzing these data because they have limited resources. Therefore, when the market and customer preferences for the innovation are known, established companies have an advantage over new businesses.

TABLE 6.2 Advantages and Disadvantages of Different Approaches to Market Research
This table shows that different approaches to market research have advantages and disadvantages that you need to consider when selecting an approach.

Approach to Market Research	Advantages	Disadvantages
Iterative Approaches	Gets useful information from customers in a new market	Hurts reputation if product fails
Ethnography	Gets information when customers cannot articulate their needs	Takes a lot of time and requires special skills
Lead User Method	Gets information from customers very interested in adopting the new product	Focuses on the needs of a very select part of the market
Focus Groups	Provides open-ended information from a group with high response rates	Expensive, requires expertise to control, and cannot provide independent data from a representative sample
Survey Research	Gets information relatively cheaply from a representative sample	People may be unable or unwilling to answer and might not act in the ways that they say they will

However, when the demand for the innovation is uncertain (because the market is new or customer preferences are unknown), the large sample market research that established companies do well is not very effective. Instead, what works well are small sample efforts to interact closely with lead customers. (If you do not realize why, you should go back and reread the earlier sections of this chapter.) These situations favor new firms because entrepreneurs can often make better and faster decisions on the basis of smaller amounts of information than established firms, which have to adhere to established decision-making rules and norms. Therefore, new firms are often better than established firms at gathering information about customer needs in truly new markets.

Key Points

- Market research involves the collection and analysis of information about customer needs, preferences for products, and purchasing decisions.
- It consists of both primary research (collecting data from respondents) and secondary research (analyzing data collected by others), and both qualitative and quantitative analysis.
- Appropriate market research techniques for new-to-the-world products are based on induction and intuition; while appropriate market research techniques for less novel products are based on deduction and analysis.
- When products are new to the world, companies are often better off using an iterative approach to market research, even though such an approach risks their reputations if they can't develop those products successfully.
- Ethnography is a technique that involves observing customers to learn their needs and preferences, and is useful when customers cannot articulate those needs and preferences.
- The lead user method is a way to develop truly new products or services by gathering information from people who are so unhappy with existing products that they are trying to come up with new solutions.
- Focus groups are sessions in which eight to twelve people come together to discuss their needs or preferences for new products or services; they are valuable

for gathering fine-grained information from customers about existing product categories but are not very useful for drawing inference to a larger population, particularly for new-to-the-world products.

- Survey research asks respondents questions about their needs and preferences; its accuracy depends on the existence of the product category, the selection of a sampling frame that is representative of the population of interest, a large enough sample size to draw inference to the population with confidence, and an effective method of data collection.

Discussion Questions

1. When should you be concerned with customer needs in the process of product development, and when should you not focus on these needs? Why?
2. Why do many companies fail to develop products that meet customer needs?
3. Why do companies fail with products that *do* meet customer needs?
4. Why don't customers adopt new products that are just as good as competitors' products?
5. How should companies price their new products? What factors should influence their pricing decisions?
6. How should companies segment the market? What factors are most important to consider in segmenting a market?
7. What market research techniques do you think are most effective and least effective? Are there conditions under which some market research techniques work better than others? If so, what are those conditions, and why do they influence the effectiveness of market research techniques?
8. What are the advantages and disadvantages of the lead user research methodology?
9. Can companies be too market-oriented? Why or why not? What are the advantages and disadvantages of being market-oriented?

Key Terms

Beta Testing: The release of early versions of a product to see how customers react to the product before it reaches commercial production.

Bundling: The process of offering components to customers for purchase as a group, usually for less than the price of the components separately.

Customer Need: A description of the benefits that customers want in a product or service.

Ethnography: The description of a group and its activities based on observation and participation.

Focus Group: A meeting of people under the direction of a trained moderator to discuss needs or preferences for products.

Lead User Method: A market research technique that gathers data from customers who would likely be the first to adopt the product or service.

Market Pull: The situation in which companies ask customers about their needs and then develop products and services to meet those needs.

Market Research: The collection and analysis of information about customer needs, preferences for products, and purchasing decisions.

Market Segmentation: The process of dividing a market into groups that have common within group needs and different between group needs.

Penetration Pricing: A pricing strategy in which a company sets a low price to get the highest possible market share.

Price Skimming: A pricing strategy in which companies set a price to earn the highest possible profit.

Product Attribute: The way that a customer need is satisfied by a product or service.

Representative: Having the same distribution of characteristics as the population from which a sample is drawn.

Sample: A portion of the population contacted to gather market research data.

Sampling Frame: A targeted group of respondents, which represents the population to which one seeks to generalize.

Survey Research: Research that asks a sample of respondents questions about their needs or preferences.

Technology Push: The situation that occurs when companies develop a technology and then find a need in the marketplace that the technology can satisfy.

Putting Ideas into Practice

1. **Is There a Real Need?** Identify three technology products that were introduced in the past couple of years. Identify the customer need that made the introduction of those products worthwhile. Then think about how badly customers needed each of these products. Which one did customers need the most? The second most? The least? Explain why you rank these products the way that you do.

   Now think about the attributes of these products. How well does each of them meet customers' needs? Rank order the three products from meets-needs-the-most to meets-needs-the-least. Explain why you rank the three products in this order?

   Is there a relationship between the ranking that you gave the products on the magnitude of customer needs and how well the products meet customer needs? Why or why not?
2. **The Role of Stakeholders** Identify a technology product or service with which you are familiar. Make a list of all of the stakeholders that influence the buying decision of the end user. Explain why each of these stakeholders affects the buying decision. Make a chart (similar to the one in Figure 6.4) that shows how the different stakeholders influence the end customer's buying decision.
3. **Approach to Market Research** Think of three unfulfilled customer needs that a technology product or service could fill. For each of these needs, describe in one paragraph a product or service that would meet that need. Now assume that you need to gather information from customers about their needs, the attributes that the products would need to meet those needs, and how well the product you described would do this. For each of the three cases, identify the market research technique that you believe would be the best for gathering this information. Explain why you would recommend that approach.

Notes

1. Adapted from Henderson, C. Finding, examining lead users push 3M to leading edge of innovation, www.refresher.com/!leadusers.
2. Anonymous. 2001. The innovation engine. *3M Stemwinder*, March 20–April 9.
3. Cooper, R. 1979. The dimensions of industrial new product success and failure. *Journal of Marketing*, 43: 93–103.
4. Burgelman, R., C. Christiansen, and S. Wheelwright. 2004. *Strategic Management of Technology and Innovation*. New York: McGraw-Hill/Irwin.
5. Urban, G., and J. Hauser. 1993. *Design and Marketing of New Products*. Upper Saddle River, NJ: Prentice Hall.
6. Bayus, B. Forthcoming. Understanding customer needs. *Blackwell Handbook of Technology and Innovation Management*. Oxford: Blackwell.
7. Mokyr, J. 1990. *The Lever of Riches*. New York: Oxford University Press.
8. Ettlie, J. 2000. *Managing Technological Innovation*. New York: John Wiley.
9. Bartlett, C., and A. Mohammed. 1999. 3M optical systems: Managing corporate entrepreneurship. *Harvard Business School Case*, Number 9–395–017.
10. Markides, C., and P. Geroski. 2005. *Fast Second: How Smart Companies Bypass Radical Innovation to Enter and Dominate New Markets*. San Francisco: Jossey-Bass.
11. Wind, J., and V. Mahajan. 1997. Issues and opportunities in new product development: An introduction to the special issue. *Journal of Marketing Research*, 34(1): 1–12.
12. Zirger, B., and M. Maidique. 1990. A model of new product development: An empirical test. *Management Science*, 36: 867–883.
13. Mullins, J. 2003. *The New Business Road Test*. London: Financial Times Prentice Hall.
14. Thurow, R. 2005. In battling hunger, a new advance: peanut-butter paste. *Wall Street Journal*, April 13: A1, A14.
15. Kim, W., and R. Mauborgne. 2000. Knowing a winning business idea when you see one. *Harvard Business Review*, September-October: 129–137.
16. Moon, Y. 2006. Online music distribution in a post-Napster world. *Harvard Business School Teaching Note*, Number 5–506–058.
17. Boehret, K. 2007. A new push to put a PC in the kitchen. *Wall Street Journal*, January 24: D4.
18. Cooper, R., S. Edgett, and E. Kleinschmidt. 2002. Optimizing the state-gate process: What best-practice companies do—I. *Research Technology Management*, 45(5): 21–27.
19. Dvorak, P. 2007. Seeing through buyers' eyes. *Wall Street Journal*, January 29: B4.
20. Stein, E., and M. Iansiti. 1995. Understanding user needs. *Harvard Business School Note*, Number 9–695–051.
21. Gourville, J. 2005. Why developers don't understand why consumers don't buy. *Harvard Business School Note*, Number 9–504–068.
22. Bryan-Low, C. 2006. Nokia joins slimming trend. *Wall Street Journal*, November 28: B2.

23. Lawton, C. 2006. Sun Microsystems remakes its sales force. *Wall Street Journal*, December 26: B3.
24. Gomes, L. 2007. As disk drives reach new milestone, flash gains new currency. *Wall Street Journal*, January 17: B1.
25. Nelson, E. The tissue that tanked, http://www.wsjclassroomedition.com/archive/02sep/MKTG.htm.
26. Ibid.
27. Bayus, Understanding customer needs.
28. Barblova, I. Uncertain fortunes for wet tissue, http://www.paperloop.cpm/db_area/archive/tw_mag/2003/0307/exit_issues.html.
29. Ibid.
30. Gourville, Why developers don't understand why consumers don't buy.
31. Thomke, S. 2002. Innovation at 3M (A), *Harvard Business School Case*, Number 9–699–012.
32. Piskorski, M., and C. Knoop. 2006. Friendster (A). *Harvard Business School Case*, Number 9–707–409.
33. Gold, R. As oil prices surge, oil giants turn sludge into gold. *Wall Street Journal*, March 27: A1, A15.
34. Jedidi, K., and J. Zhang. 2002. Augmenting conjoint analysis to estimate consumer reservation price. *Management Science*, 48(10): 1350–1368.
35. Schilling, N. 2005. *Strategic Management of Technological Innovation*. New York: McGraw-Hill.
36. Shapiro, C., and H. Varian. 1998. Versioning the smart way to sell information. *Harvard Business Review*, November–December.
37. Bakos, Y., and E. Brynjolfsson. 1999. Bundling information goods: pricing, profits and efficiency, *Management Science*, 45 (12): 1613–1630.
38. Ibid.
39. Brown, K., and A. Latour. 2004. AT&T will offer Internet phone calls in selected markets. *Wall Street Journal*, March 30: B1, B2.
40. Lohr, S. 2007. Preaching from the Ballmer pulpit. *New York Times*, January 28: Section 3, 1, 8–9.
41. Saul, Stephanie. 2005. FDA approves heart drug for African-Americans. *New York Times*, June 24, http://www.nytimes.com/2005/06/24/health/24drugs.html?ex=1172984400&en=51190b4fb32616a3&ei=5070.
42. Silver, S. 2007. How Match.com found love among boomers. *Wall Street Journal*, January 27–28: A1, A7.
43. Markides and Geroski, *Fast Second*.
44. Romanelli, E. 1989. Environments and strategies of organization start-up: Effects on early survival. *Administrative Science Quarterly*, 34: 369–387.
45. Cooper, R., and E. Kleinschmidt. 1987. New products: What separates winners from losers? *Journal of Product Innovation Management*, 4: 169–184.
46. Barton, D. 1994. Commercializing technology: Imaginative understanding of user needs. *Harvard Business School Note*, Number 9–694–102.
47. Friar, J., and R. Balachandra. 1999. Spotting the customer for emerging technologies. *Research Technology Management*, 42(4): 37–43.
48. http://www.lasers.org.uk/laser_welding/briefhistory.htm
49. Barton, Commercializing technology.
50. Sathe, V. 2003. *Corporate Entrepreneurship*. Cambridge, UK: Cambridge University Press.
51. Afuah, A. 2003. *Innovation Management*. New York: Oxford University Press.
52. Brown, S., and K. Eisenhardt. 1997. The art of continuous change: Linking complexity theory and time-paced evolution in relentlessly shifting organizations. *Administrative Science Quarterly*, 42(1): 1–34.
53. Schilling, *Strategic Management of Technological Innovation*.
54. Lynn, G., J. Morone, and A. Paulson. 1996. Marketing and discontinuous innovation: The probe and learn process. *California Management Review*, 38(3): 8–27.
55. Krubasik, E. 1988. Customize your product development. *Harvard Business Review*, November–December, 4–9.
56. Bayus, Understanding customer needs.
57. Leonard, D., and J. Rayport. 1997. Spark innovation through empathic design. *Harvard Business Review*, November–December: 102–113.
58. The Science of Desire. *Business Week*, June 5, 2006, http://images.businessweek.com/ss/06/05/ethnography/index_01.htm.
59. Ibid.
60. Ibid.
61. Sathe, *Corporate Entrepreneurship*.
62. Brown and Eisenhardt, The art of continuous change.
63. Day, G. 2000. Assessing future markets for new technologies. In G. Day and P. Schoemaker (eds.) *Wharton on Managing Emerging Technologies*. New York: John Wiley.
64. Lilien, G., P. Morrison, K. Searls, M. Sonnack, and E. von Hippel. 2002. Performance assessment of the lead user idea-generation process for new product development. *Management Science*, 48(8): 1042–1059.
65. Thomke, S., and A. Nimgoade. 1998. Note on lead user research. *Harvard Business School Note*, Number 9-699-014.
66. Allen, K. 2003. *Bringing New Technology to Market*. Upper Saddle River, NJ: Prentice Hall.
67. Griffin, A., and J. Hauser. 1993. The voice of the customer. *Marketing Science*, 12(1): 1–27.
68. Mullins, J. 2003. *New Business Road Test*. London: Financial Times Prentice Hall.

Chapter 7

Product Development

Learning Objectives

After reading this chapter, you should be able to:

1. Explain how to identify and prioritize customer features for new products.
2. Produce a perceptual map, conduct a concept test, do a conjoint analysis, and use the Kano method.
3. Spell out the pros and cons of reducing product cycle time, and describe how companies do it.
4. Explain how to use prototyping, house of quality, design for manufacturing, and Web-based tools to improve the product development process.
5. Describe the stage gate process and explain how it helps to make decisions about innovation.
6. Define *concurrent development* and explain the advantages and disadvantages of engaging in it.
7. Define *modularity*, and explain the advantages and disadvantages of developing modular products.

Product Development: A Vignette

In 1993, Apple Computer introduced the first personal digital assistant (PDA), the Apple Newton, after spending $200 million on the project.[1] Although the Newton had a number of innovative features, including spreadsheet software and printing capability, it was a failure (see Figure 7.1). The product cost as much as a desktop computer, ran very hot, and was heavy.[2] Moreover, customers did not want many of the features of the Newton, including a hard drive and infrared long distance networking capability.

In contrast, Jeff Hawkins created the Palm Pilot, a handheld organizer with far fewer features than the Newton: a calendar, address book, to-do list, memo pad, and computer connectivity.[3] The Palm Pilot was small enough to be placed in the pocket of a shirt, was very light, and cost considerably less than the Newton. The product was a huge success, and, within a few years, Palm took control of more than two-thirds of the market.[4]

Why was the Palm Pilot a success when the Apple Newton was a failure? Part of the answer lies in the development of supporting technologies. The Newton predated the Palm by several years, and many of the supporting technologies that have made PDAs so popular had not yet been developed at the time the Newton was introduced. But part of the answer also lies in how the two companies conducted product development. Hawkins discovered that customers did not want handheld computers to replace their desktop computers; they wanted handheld computers to replace their calendars. So, unlike Apple Computer, Hawkins created a product that met customer needs without including unnecessary features that undermined the appeal of the product, and raised its costs.[5]

To understand how he did that, and how you can too, read on. Effective product development is the subject of this chapter.

FIGURE 7.1 The Problems with the Apple Newton

The Apple Newton was not very successful, in part because it used novel handwriting recognition software that didn't work very well.

Source: http://images.ucomics.com/comics/db/1993/db930827.gif

INTRODUCTION

This chapter discusses **product development**, or the creation of new products by companies, and its role in technology strategy. Although product development involves a variety of activities, from making improvements to existing products, to developing new product platforms, to creating new-to-the-world products, all product development efforts have one thing in common: They influence the way in which firms are organized and compete. Therefore, you need to think about product development in the context of your company's technology strategy.

Take, for example, the efforts of two companies to develop microchips for cellular telephones. One competitor, Texas Instruments, has concentrated on meeting a wide range of design specifications; whereas another competitor, Silicon Labs, has focused on keeping costs down. These different strategies have led the two companies to develop very different products. Texas Instruments has developed a product that places transmission activities on one chip and the power function on another chip; while Silicon Labs, has developed a chip that puts all of these functions together in one place.[6]

This chapter is organized as follows: The first section discusses why reducing product development time is an important issue for companies and describes how to do it. The second section looks at concurrent product development and discusses the advantages and disadvantages of using it. The third section examines modular product development, explaining why companies use it, and what its limitations are. The fourth section describes how to identify the right features for your products and discusses several tools that help you to do that: perceptual mapping, concept testing, conjoint analysis, and the Kano method. The fifth section examines several important product development management tools, including prototyping, the house of quality, design for manufacturing, and Web-based search tools, and explains how you can use them to improve your product development efforts.

PRODUCT CYCLE TIME

Product cycle time is a measure of how long it takes companies to develop new products or services. While research shows that the length of the product cycle depends on the type of product (with cycle time being longer for more complex and more novel products)[7] and several industry characteristics, companies in a wide variety of industries are reducing product cycle time across a wide variety of new product types.[8] Therefore, as a technology manager or entrepreneur, you need to understand how to reduce product cycle time. But, before we turn to explaining how you can do this, you first need to understand why reducing product cycle time is such an important aspect of technology strategy today.

Reduced product cycle time helps companies in four ways:

- First, it increases the amount of time that your products can be on the market before becoming obsolete, which lengthens the period over which the cost of product development can be amortized.[9]
- Second, it speeds the development of product adaptations, which gives you the option to launch more products or to develop products that are more tailored to customer needs.

- Third, it allows you to delay product development efforts until you learn more about markets, competitor actions, and complementary technologies, which reduces your need to rely on long-term forecasts, and increases the likelihood that market conditions and customer needs will be the same when the product is introduced as when product development was initiated.[10]
- Fourth, it gives your company the potential to develop the advantages of being the first to market.[11]

However, reducing product cycle time will not help your company if you do it by cutting out important steps in the product development process, such as initial product design or market research (that will only cause you to produce undesirable products quickly).[12] Rather, you need to reduce product cycle time while developing products that have the same features as those introduced more slowly.[13]

Research has shown that several policies reduce product development time without reducing the quality of product attributes, including strong senior management support for rapid product development, an incentive structure designed to reward fast product development, focused project teams, powerful project leaders, experienced product development teams, autonomous product development efforts, and strong external linkages.[14] Therefore, you should consider adopting these policies if reduced product cycle time is important to your competitive strategy.

Key Points

- Companies want to reduce product cycle time to generate revenue sooner, respond faster to customer needs, have more time to gather information, and have a greater chance of creating a first mover advantage.
- Product development time can be reduced by having strong senior management support for rapid product development, an incentive structure designed to reward fast product development, focused project teams, powerful project leaders, experienced product development teams, autonomous product development efforts, and strong external linkages.[15]

CONCURRENT DEVELOPMENT

As Figure 7.2 illustrates, you can reduce cycle time by using **concurrent development,**[16] which is the process of undertaking some product development steps simultaneously, rather than sequentially.[17] For example, when you make the body of a car, a giant stamping machine called a die presses the metal into a shape of a part of the car, such as the door. In sequential design, automobile companies design the die first, then obtain the steel needed to make the parts, and finally begin cutting it. In concurrent development, the automakers get the necessary blocks of steel and start cutting them at the same time that the die is first designed.[18]

Concurrent development helps you to reduce cycle time because it does not require you to complete each step in the development process before you start subsequent steps.[19] It also provides you with an early warning about potential problems at a later stage in development, while corrections can still be made upstream.[20] For example, you might find that you cannot manufacture products at a desired quality level with a particular design. Because manufacturing engineers are working on manufacturing the product at the same time that product designers are finishing

FIGURE 7.2 Concurrent and Sequential Development

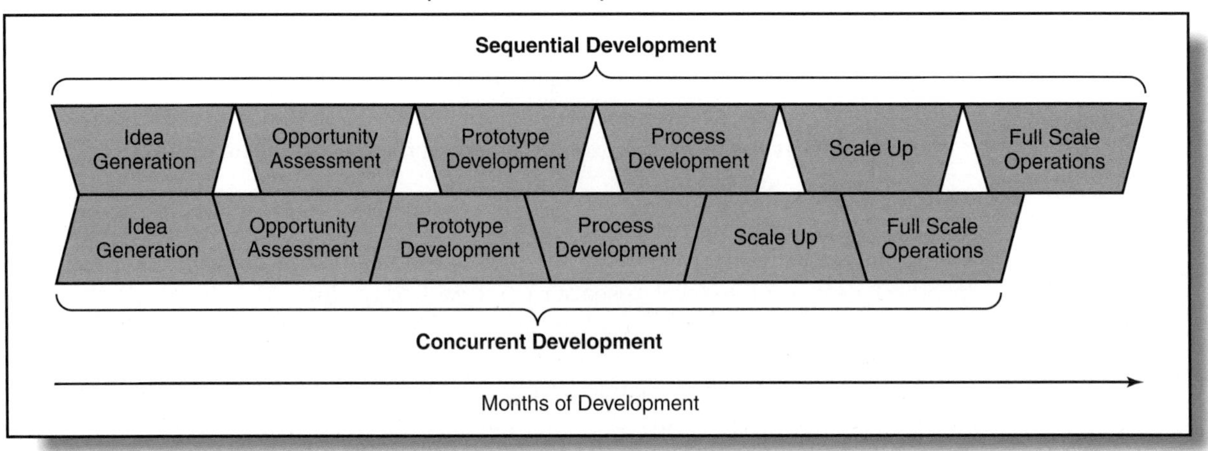

This figure compares concurrent and sequential product development, and shows that concurrent development reduces product cycle time.

their designs, with concurrent engineering you can address manufacturability problems before the product design process has been completed.[21]

Concurrent development is difficult to implement because it imposes additional requirements on organizations. In particular, the overlap in the activities of manufacturing and design personnel demands a high level of communication between them.[22] Moreover, concurrent development requires the exchange of information between these groups to take place in real time, or its benefits will be lost.[23] Consequently, to undertake concurrent development, you need to establish communication mechanisms that permit integrated problem solving across parts of your organization.

Concurrent development is also risky. By undertaking steps of the development process before completing the prior stages of the process, you risk making product development decisions without adequate information. If your lack of information leads to poor decisions that require you to redo part of the development process,[24] you might have to spend much of your budget on something that would have been avoided with sequential development.[25]

Key Points

- Concurrent development, or the process of undertaking some product development steps simultaneously, helps companies to reduce product cycle time.
- However, it is risky and requires high levels of communication between parts of the organization.

Modularity and Product Platforms

Modularity is the degree to which a complex product can be built from smaller **components** (units from which something is made) that can be created independently, but function together.[26] For example, cell phones are modular because they are composed of a number of components, such as the microchip and the screen.

When products are modular, the functionality between the components is established through the creation of design rules and parameters that cover the product architecture, interfaces between the modules, and standards for testing conformity to design rules.[27]

The use of modularity will help you to accelerate your company's pace of product development because each part of the organization responsible for the development of a module can conduct its design experiments concurrently with those of other units.[28] As a result, you can test your designs much faster than you can if you don't take a modular approach.

Modularity also offers several other benefits. It makes it possible to undertake very complex product development efforts. Take the case of Microsoft Windows. Over the years, Windows software has become too complex for designers to make major changes to it and still ensure that the software will work. Therefore, Microsoft has created different versions of Windows by linking together thousands of tiny pieces of software code written by different engineers. As Windows has grown, this approach has caused the software to act in unpredictable ways, making it harder and harder to add new features. Moreover, it made conducting daily tests of the interface of different pieces of the code impossible, requiring manual searches of thousands of lines of code to find bugs in the system.

To overcome this problem, Microsoft shifted to a modular approach to software development for its Windows Vista release. With a modular approach, different development teams were responsible for producing different parts of the software program, like graphics and spreadsheets, according to common specifications. The modules were then linked together according to a planned design.

Modularity also permits you to customize your products to customer needs by allowing you to design in many alternative combinations of components.[29] This breadth of alternatives, in turn, allows you to target new market segments and to respond more quickly to changing customer demands.[30]

Moreover, modularity makes product development easier to do. Because the smaller subsystems that compose a module are easier to develop than a whole nonmodular product, modularity reduces the expertise necessary for successful product development.

Furthermore, modularity helps you to lower your cost of product development. Because you can use components that have worked in the past, your design costs are lower when you take a modular approach. And you can develop the components of your product without costly coordination between the different teams developing them because changes in one component will not affect the functionality of another.[31] Perhaps most importantly, modularity allows you to purchase off-the-shelf components, which are often cheaper than custom-made ones.

For example, take the cost savings that Teradyne, a scientific equipment manufacturer, achieved by adopting a modular approach to the development of a new semiconductor testing device. The device needed to have a spreadsheet feature for engineers to use when conducting tests on semiconductors. By using a modular approach to the development of the device, the company was able to purchase a Windows-based spreadsheet component for a few hundred dollars, rather than develop, on its own, a customized Unix-based spreadsheet at many times the cost. The latter approach, however, would have been necessary if the company had taken a nonmodular approach to product development.[32]

However, modularity has several drawbacks. First, as the personal computer business illustrates, you often have to compete on cost when products are modular because specialization by different component manufacturers drives down the cost of the overall finished product. Thus, when products are modular, the winner in competitive battles is often companies, like Dell Computer, that are the most efficient, not the most innovative.

Second, modularity does not work well unless you can define technical standards that ensure that different components interface correctly.[33] In industries in which components are produced by different companies, this means that companies have to converge on a common set of technical standards for modularity to work.[34] And even when technical standards exist, the linkages between the components of modular products sold by different companies are often worse than the linkages between the components of modular products sold by a single company because their design and production involves less information exchange between the developers of the different components. For instance, Apple's iPod device, iTunes software, and iTunes music store interface much more smoothly than the alternative that Microsoft and its collaborators offer because Microsoft's products are composed of modules made by several different companies.[35]

Third, the architecture of modular products is very inflexible because changing the architecture of a modular product requires convergence of the firms in an industry on a new technical standard. Therefore, modularity does not make sense when product architectures need to be fluid and change from year to year.[36]

So, should you adopt a modular approach to the design of your product? The answer depends. Research has shown that modularity makes the most sense when a product is composed of components that are very different from each other; when

FIGURE 7.3
A Computer Is a Modular Product

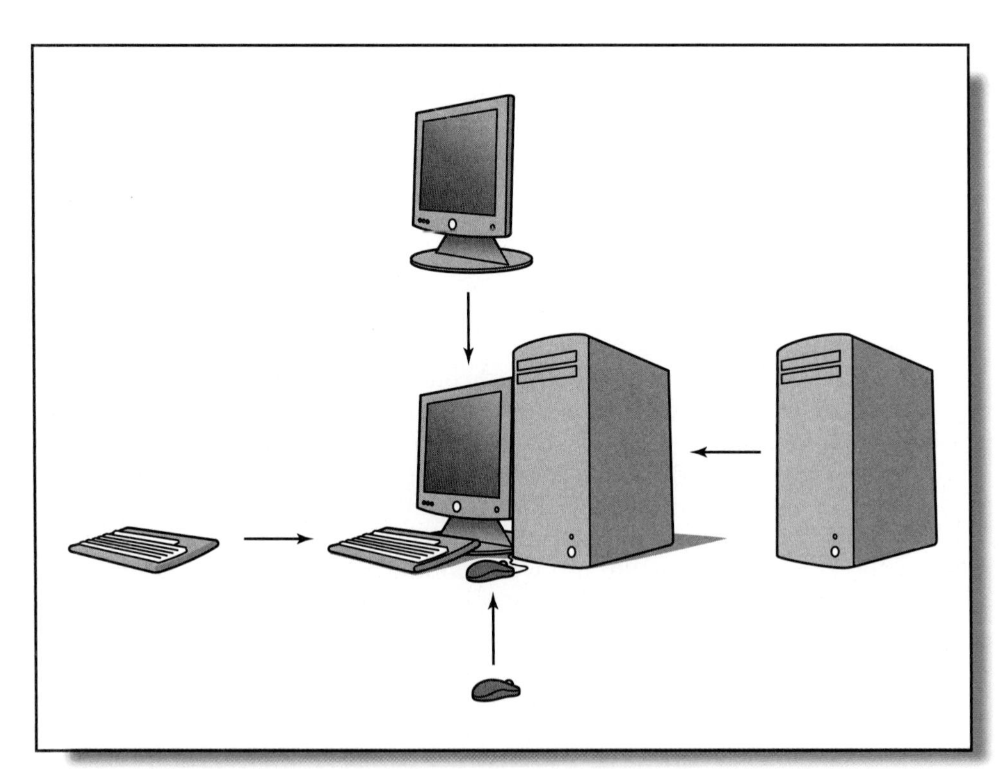

A computer is a modular product because it is composed of a set of smaller components—the CPU, mouse, monitor, and keyboard—that are combined to produce a larger system.

customer demand is heterogeneous; when technical standards have been established; when the pace of technological change is high; when the firms producing different components have very different capabilities; and when the components do not have to be highly integrated.[37]

Product Platforms

Many technology companies use a **platform**, or a common technological base to which different features are added, to create a family of products, each targeted at different customers. For instance, Intel has produced several microprocessor platforms, including the 8088, the 486, and the Pentium, off of which it has created many different products.[38] (Table 7.1 shows Intel's product platforms and derivative products during the 1990s and indicates how the derivative products varied on processor speed.)

Product platforms can be built on a variety of different types of common technologies, including physical architecture, as is the case with automobiles; process technology, as is the case with microprocessors; or a common set of components, as is the case with cameras.[39] They can be created by changing existing products to conform to common components, as Lutron did with its lighting control products.[40] However, they are typically designed in advance to support derivative products, as Dell does with its computers.

Using a product platform can benefit your company in a wide variety of ways. First, it allows you to offer greater variety to customers without straining organizational efficiencies. If you create different derivative products off of a common platform, you won't have to conduct as much new product development or process redesign as you will if you create multiple products without the benefit of a platform.[41] You can also better coordinate resource use between products, plan for successive generations of products, and manage the timing of market entry.[42] Using a product platform may even lead you to invest more in designing your operations, which could lead to superior product architecture and better linkage between components.[43]

Second, using a product platform will reduce your costs. As long as the incremental costs of making derivative products are low relative to the cost of creating the platform in the first place, companies can benefit by developing a platform strategy.[44] Moreover, you can use a platform strategy to reduce manufacturing, materials, and procurement costs because products built off the same platform typically use common machinery, parts, and other inputs,[45] and change equipment setup less frequently.[46] As a result, if you use a product platform, you will incur lower costs from changes to the setup and achieve better scale economies in purchasing inputs and raw materials.[47]

TABLE 7.1 Intel Product Platforms

This table shows the products that make up three different Intel product platforms: the 486, the Pentium, and the Celeron.

486	PENTIUM	CELERON
SX 16–25 MHz	Pentium 60–66 MHz	266 MHz
SL 20–33 MHz	Pentium II 233–300 MHz	300 MHz
DX2 20–66 MHz	Pentium III 350–600 MHz	533 MHz

Source: Adapted from Mohr, J., S. Sengupta, and S. Slater. 2005. *Marketing of High Technology Products and Innovations* (2nd edition). Upper Saddle River, NJ: Prentice Hall.

Third, using a platform strategy helps you to segment the market on the basis of price and performance because it helps you to develop different product variants for different groups of customers. This allows you to capture smaller market niches, as well as to hold on to customers as they develop interest in different product features. For instance, many companies develop a low-cost product variant with the goal of attracting low-end users. Then they offer product variants that provide greater performance on a variety of product dimensions, which gives them the opportunity to offer new products to customers as they become interested in the performance benefits that come from those product dimensions.[48] As a result, a platform strategy allows companies to keep competitors from taking their customers by filling in gaps between their product offerings.[49]

Fourth, using product platforms shortens product cycle time. If you don't use a product platform, you will need to conduct additional R&D before you can introduce a new variant of your product. However, if you develop a product platform, you won't have to conduct unique R&D for each new product that you introduce, but, instead, will be able to use the same development knowledge across all of the products derived from the platform. As a result, you'll be able to get your products to market more quickly if you use a platform strategy.[50]

However, using a platform strategy has several costs. First, products created off of a product platform are less distinctive than products created independently.[51] Consequently, your company will not be well positioned to serve customers that have a strong preference for distinctiveness if it uses a platform strategy.

Second, to use a platform strategy, you will need better cross-functional coordination in your organization. Because the designs of the different derivative products built off of a common platform need to be coordinated, developing products off of a platform is more complex and costly than the development of stand-alone products.[52]

Third, with a platform strategy, lower-level products have to be "over designed" so that they can share systems and components with higher-level products.[53] This "over design" makes low-end products from a platform more costly than low-end independent products, and this greater cost results in either higher prices and lower sales, or reduced profit margins, on those items.

Given the pros and cons of product platforms, when should you use them? In general, product platforms are more useful if your company operates in an industry that is not very dynamic but is very capital intensive, such as aerospace or automobile manufacture.[54] In addition, platforms are more useful when you can achieve scale economies at the platform level, when the likelihood of overdesigning lower-end products is low, and when your markets are neither extremely homogenous nor very heterogeneous. When markets are homogenous, creating one standard product is often better than using a product platform, while focusing on the most attractive niche is often the best approach when markets are very heterogeneous.[55]

The disadvantages of product platforms also lead many companies to combine modularity with product platforms to balance efficiency with flexibility when they develop new products. A standardized product platform with modular components provides the efficiency advantages of standardization and scale economies at the platform level, with the variation at the product level necessary to meet customer needs. For example, combining a standard platform with a modular product allowed Sony to create 75 different models of the Walkman just by offering different combinations of standardized common modularized components, such as color, size, music format, and so on.[56]

Key Points

- Modularity speeds product cycle time, facilitates customization, makes product development easier, and reduces its cost.
- However, modularity creates cost-based competition, makes product architecture inflexible, and requires technical standards to be developed.
- Modularity makes the most sense when a product is composed of components that are very different from each other, when customer demand is heterogeneous, when technical standards have been established, when the pace of technological change is high, when the firms producing different components have very different capabilities, and when the components do not have to be highly integrated.
- Companies use platforms, or a common technological base for a family of products, to offer more variety to customers without straining organizational efficiencies, to facilitate market segmentation, and to reduce product cycle time; however, the use of platforms reduces product distinctiveness, increases demands for organizational coordination, and leads to the "over design" of low-end products.

IDENTIFYING THE RIGHT PRODUCT FEATURES

You need to develop products with features—or characteristics of the product —that are better at meeting customer needs than the features of competitors' products, or your competitors' products will be more appealing to potential customers than your products.[57] Take, for example, the problem faced by Philips with its home entertainment system. This product failed to sell because it was complicated, costly, and difficult to operate in a market where competing products were simple, inexpensive, and easy to operate.[58]

Unfortunately, you can't ensure that your product will appeal to customers by including every possible feature that customers might want. If customers don't want those features, their inclusion will impose unnecessary costs, which will require you to raise the price of your product and make it less appealing to customers. Moreover, as the opening vignette to the chapter indicates, adding unnecessary features can complicate your product, also making it less appealing.

As Figure 7.4 indicates somewhat humorously, to be successful, you need to identify and include just the features that will make your product attractive to customers. This is not easy because customers will often express preferences for hundreds of different features if you ask them what they want. Because these preferences will not be of equal importance and will not be expressed by equal numbers of people, you will need to assess their importance and prioritize them.[59] The next section describes several ways to identify and prioritize customer needs and to select product features that satisfy those needs.

Conjoint Analysis

Conjoint analysis is a statistical tool that allows you to assess the relative importance to customers of different product features (including price) and to identify the best combination of features to meet their needs, even when customers are not aware of the value that they ascribe to those features.[60] Conjoint analysis is particularly useful when you need to make trade-offs between different product attributes, and you want to see the relative contribution of those attributes to customers' preferences.[61]

FIGURE 7.4
Getting the Features Right

"We have a calendar based on the book, stationery based on the book, an audiotape of the book, and a videotape of the movie based on the book, but we don't have the book."

Not all companies develop new products with the right set of features to meet the needs of customers.
Source: The Cartoon Bank.

You can even see the relative contribution when different factors interact or when preferences are not linear, as is the case with the relationship between two brands.

For example, suppose you want to know how important each of the following features of a laptop computer is to potential customers: speed, warranty length, price, screen size, and weight.[62] Potential customers might not be able to tell you this directly because they don't think about individual features of a computer. But with conjoint analysis, you don't need customers to know how they feel about each of these features. Customer preferences can be inferred from choices that they make between products with different combinations of these features.

In a typical conjoint analysis, a **factorial design** is used. A factorial design presents all different combinations of product features to potential customers. For example, a new computer might have two dimensions of interest to customers: high and low price and laptop versus desktop. A two-by-two factorial design would present all four computer combinations: high-price laptop, low-price laptop, high-price desktop, and low-price desktop.

In a conjoint survey, potential customers are asked questions about their preferences for different versions of the product. (For example, they might be asked how likely they would be to purchase each version of the product on a scale of 0 to 100.) Once the data on these preferences are collected, regression analysis is used to identify the contribution of each product feature to the overall choices. That is, the regression weights provide a quantitative measure of the importance of each feature to customers.[63]

For example, suppose, you work for Michelin, and you want to know how much your brand name affects the likelihood that customers would buy your mountain bike tires rather than the mountain bike tires made by competitors. (The information in this paragraph is taken from the Sawtooth Software mountain bike simulation.) You know that mountain bikes have five important features: brand name, tread type, tread wear, weight, and tire type. You can collect data from respondents, asking them about their preference for products that represent different combinations of these five attributes. For example, you might ask, "on a scale of 0 to 100, how likely are you to buy a Tioga mountain bike with very firm, off-road treads that last 2,000 miles, and weigh 2 pounds?" Then you might ask, "On a scale of 0 to 100, how likely are you to buy a Michelin mountain bike with very firm, off-road treads that last 2,000 miles, and weigh 2 pounds?" Then you might continue, "On a scale of 0 to 100, how likely are you to buy a Tioga mountain bike with very firm, off-road treads that last 3,000 miles, and weigh 2 pounds?" (Of course, in reality, you need more than just three questions. Because you have five attributes of the product, each of which has three possible categories, a full factorial design would have 125, or (5^3), combinations, requiring you to have a conjoint survey with 125 questions in it.)

Figure 7.5 shows the output from a conjoint analysis on this product produced by a company named Sawtooth Software. Sawtooth has already conducted the conjoint survey and has inputted the data from it. This allows us to conduct the analysis and see what kind of results we get.

We can use the "market simulator" feature in the software to create a scenario in which all of the products are the same on all dimensions, except that one is made by Michelin, another by Tioga, and a third by Electra. We can then have the computer run a regression analysis on the market research data by selecting "purchase likelihood." The software then runs a conjoint regression analysis and produces the output shown in Figure 7.5 for the purchase likelihood and standard error based on the preferences given by respondents to the conjoint survey. (See the tire simulator at Sawtooth Software for the actual analysis, which can be downloaded at www.sawtoothsoftware.com/downloads.shtml).

The analysis shows that customers prefer the Tioga brand the most, followed by Michelin and then Electra. However, the differences are not huge. The Tioga brand only makes customers about 2 percent more likely to buy a mountain bike tire than the Michelin brand. Moreover, you cannot even be sure, with 95 percent confidence, that there is any difference between the brands because of the size of the standard error.

FIGURE 7.5
Conjoint Analysis Input and Output

Input

	Brand	Type	Tread	Weight	Wear
Product 1	Michelin	1	1	1	1
Product 2	Tioga	1	1	1	1
Product 3	Electra	1	1	1	1

Output

	Purchase Likelihood	Standard Error
Product 1	35.61	3.68
Product 2	37.77	3.50
Product 3	31.78	3.64

The purchase likelihood estimates in this conjoint analysis output show that customers are more likely to purchase a mountain bike tire produced by Tioga than an identical tire produced by Michelin or Electra.

Source: This output was generated from the Sawtooth Software tire simulator, which is available as a demo from http://www.sawtoothsoftware.com/downloads.shtml.

While conjoint analysis is a very useful tool, it is subject to several important limitations. First, it can only be conducted if a product or service can be specified as a collection of independent attributes. (Otherwise, you can't disentangle the relative contribution of different product features to potential customers' preferences.) Therefore, conjoint analysis is more useful for functional products, like computers, where different product features are observable, than image products, like perfume, where they are not.

Second, the respondents must be familiar enough with the product category to express their preferences. If your product is so new that potential customers do not understand it, conjoint analysis will yield poor results. This means that conjoint analysis is not a very effective technique for developing new products in new markets, and works much better for developing new products in existing markets.

Third, you need to know what product attributes are salient to potential customers. Conjoint analysis only compares the relative importance of product attributes, and cannot be used to identify attributes to consider in the first place. Moreover, you can only present so many combinations of attributes to potential customers before fatigue sets in, so there is a limit to the number of attributes that you can include in a conjoint analysis.[64]

Kano Method

When you develop new products, you also need to consider the functional form of the relationship between product features and the satisfaction of customer needs. Conjoint analysis assumes that this relationship is linear—the more of the feature the greater the satisfaction of customer needs—but it need not be. One feature might be absolutely necessary to satisfy customer needs, while another might be nice to have but not essential. Another might delight customers if it were present but not affect their satisfaction if it were absent. Finally, a feature might be something to which customers are completely indifferent.

Take, for example, a hand sanitizer. It is absolutely necessary for the product to kill bacteria. If it fails to get rid of the E. coli, it is of no use to anyone. But dead is dead, so including a feature that makes bacteria "more dead than dead" will not satisfy customers any more than a feature that just kills bacteria. On the other hand, it isn't absolutely necessary for a hand sanitizer to smell nice. People might prefer it if it smelled good, but if it smelled like chemicals, it would still satisfy their needs. The mechanism that is used to get the chemicals into their container won't affect customer satisfaction; customers are indifferent to that. In the end, cost might be the only feature that has a linear relationship with customer satisfaction. That is, the less the product cost, the more customers would prefer it.

The **Kano method** is a technique for evaluating customer preferences when those preferences are different if a feature is present than if the feature is absent.[65] The Kano method is similar to a conjoint analysis in that it gathers data from potential customers about their preferences for different product attributes. However, it differs from conjoint analysis because data is gathered from potential customers about how both the presence and absence of a product feature affects their satisfaction. As a result, the Kano method can be used to differentiate product attributes into four categories: *must haves, delighters, linear satisfiers,* and *indifferents.*[66]

As Figure 7.6 shows, must haves are those attributes that do not provide any additional satisfaction to customers as performance becomes more functional, but whose absence makes customers dissatisfied. In contrast, delighters are those attributes

FIGURE 7.6
The Kano Method

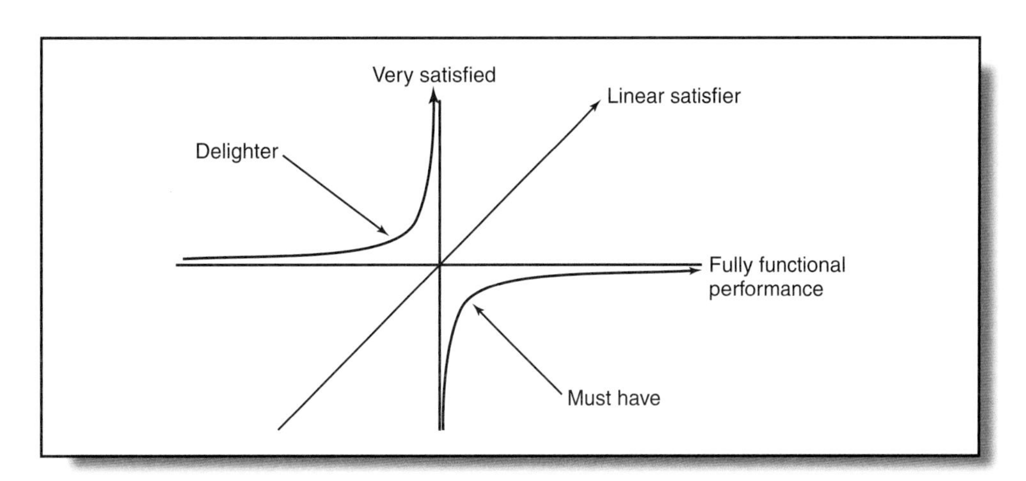

This figure is a visual representation of three of the four Kano categories: For "linear satisfiers" there is a linear relationship between customer satisfaction and functional performance; for "must haves" more performance does not increase satisfaction, but less performance reduces satisfaction; and for "delighters" less performance does not reduce satisfaction, but more performance increases it.

whose lack of functionality does not make customers dissatisfied, but whose presence will add satisfaction. Indifferents are those attributes whose functionality has no effect on satisfaction either positively or negatively. Linear satisfiers are those attributes whose lack of functionality decreases satisfaction and presence of functionality increases satisfaction.

Using the Kano method will help you to identify "must have" and "delighter" features, which are often missed in conjoint analysis. Take, for example, the case of the received message function on an e-mail program. A typical market research survey might ask "how would you feel if your e-mail program verified whether a message had been received?" Suppose that the average respondent answered this question with a score of 5 on a 5-point scale in which 1 equals "not interested" and 5 equals "very interested." You might think that you need to put the feature in your product because respondents like it so much.

However, now suppose you administered a Kano survey that also asked the respondents "how would you feel if your e-mail program did not have a feature that verified whether a message had been received?" and the average response was a 3 on the 5-point scale. The combination of the 5 points on the first question and the 3 points on the second question would indicate that the product was a "delighter." Customers would still be interested in the product if it lacked the feature but would be very interested if the feature were present.

Similarly, suppose your market research survey just asked customers "how would you feel if your e-mail program verified whether a message had been received?" and the average respondent indicated a score of 3. You might think that the feature doesn't matter much to customers (since it doesn't make them like the product) and decide not to include the feature. This could be a wrong interpretation. If you also administered a Kano survey that asked the respondents "how would you feel if your e-mail program did not have a feature that verified whether a message had been received?" and the average response was a 1 on the 5-point scale, you would find out that the feature was a "must have" and would need to be included in your product to satisfy your customers.

Perceptual Mapping

To know if your new product meets customer needs in a significantly better way than existing alternatives, you need to compare the features of your product with those of your competitors. This can be done by collecting information from potential customers about their perceptions of your competitors' products as well as your own.

This information can be presented in a **perceptual map**, or a visual display of the perceptions of customers about the different features of competing products. On a perceptual map, competing products are displayed on the different dimensions that potential customers have identified as important to them. By looking at the lines for different products on the map, one can tell whether one product is perceived as better than another at meeting customers' needs, and what features would need to be changed for your new product to be perceived as better than those of your competitors.

For example, Figure 7.7 shows a perceptual map in which customers clearly perceive product 1 as better than its competitors on the dimensions that customers have identified as important to them: price, size, speed, cost, and functionality. This perceptual map indicates that products 2, 3, and 4 will not be able to attract customers away from product 1, given the current product attributes. Based on this information, companies developing products 2, 3, or 4 should redesign their products or decide not to enter the market.

Concept Testing

Concept testing is a procedure in which customers are presented with a new product and are observed for, or asked for, their reaction.[67] Concept testing is valuable because companies would like to know if customers are positively disposed to their product concepts before they go through the time and expense of introducing their products to the market. For example, a maker of outdoor grills might conduct a concept test to determine if a new grill concept that emphasizes design over function would be appealing to customers.

Concept tests come in two varieties. A **positioning concept test** presents the product concept along with mock advertisements; whereas a **core idea concept test**

FIGURE 7.7
An Example of a Perceptual Map

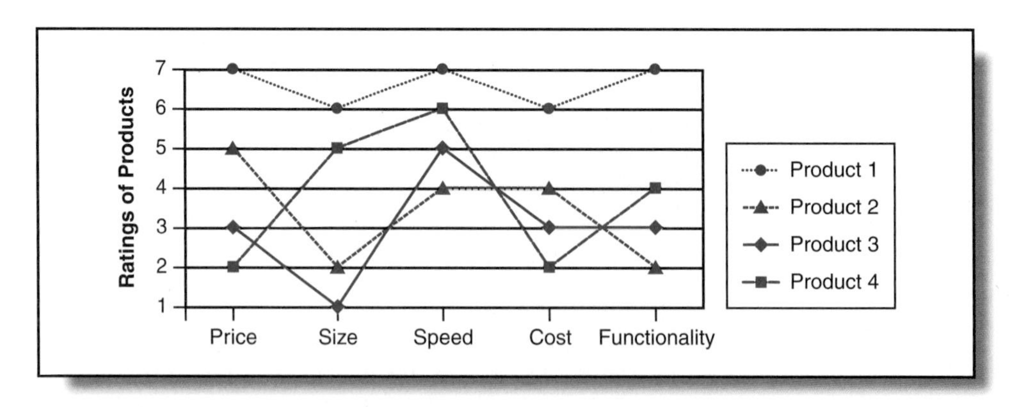

By presenting the results of market research visually, in the form of a perceptual map, market researchers can tell which, of a set of competing products and services, are perceived as better by customers; in this case, Product 1 is clearly preferred by customers on all of the relevant dimensions.

presents the idea without a marketing message. While the former approach offers a better prediction of customer reactions, it confounds the concept with the marketing message. In contrast, the latter approach avoids the confounding of positioning, but it doesn't predict customer reactions very well.[68]

Concept tests have a very important limitation for you to think about. If potential customers rate your product concept poorly, the concept test tells you not to develop the new product. But it doesn't tell you anything about how to fix the product concept to make it acceptable to customers.[69]

Key Points

- To satisfy a customer need, you need to come up with a product that has features that are better at meeting that need than those of competing products; this is not easy to do because simply including every feature that customers might be interested in having will make your product too expensive and complicated to be appealing.
- Conjoint analysis is a statistical tool that allows you to assess the relative importance of different product features to customers by looking at their preferences for products with different combinations of features.
- The Kano method examines the relationship between the presence of a product feature and customer satisfaction when that relationship may not be linear.
- A perceptual map is a visual display of the perceptions of customers about the different features of competing products; it facilitates the identification of the features that make one product better than another at satisfying customer needs.
- Concept testing is a procedure in which customers are presented with a product idea and are asked for their reaction; it provides information about whether a company should move forward with the introduction of a new product.

Product Development Tools

Many new products never reach the market,[70] and more than half of all new products introduced to the market generate negative economic returns.[71] This low success rate is the result of the complexity of developing a new product. As was discussed in the previous chapter, to have a successful new product, you must develop something that is technically feasible, meet the needs of customers in a much better way than existing alternatives, develop and produce the product in a profitable manner, satisfy the demands of relevant stakeholders, and price the new product correctly.

Regardless of whether you are running a start-up or an established company, you can increase your company's success at new product development by using product development tools. (However, limitations on time and money mean that entrepreneurs need to balance the value of using these tools against the financial and time constraints that they face.) Web-based tools help you to gather information about customer needs that you might otherwise miss. The house of quality helps you to ensure that your new product has the features that your market research indicated that customers value. Prototyping and stage gates help you to ensure that your new products don't fail late in the development process when the cost of failure is very high. Finally, design for manufacturing helps you to ensure that you can actually manufacture any new products that you design.

Web-Based Tools

Companies are beginning to use Web searches to figure out what products should be offered to customers and what attributes those products should have. For instance, National Instruments Corporation, which makes computer hardware and software for engineers, modified its products to have USB interfaces when it learned that engineers conducting Web searches for hardware tended to include "USB" in their search queries. The company also made a product for testing auto quality that was compatible with a particular data protocol based on the results of a Web search. Especially for entrepreneurs, who often have limited budgets for gathering information about desired product attributes, looking at Web search patterns provides an inexpensive way to access a large amount of data about customer preferences.[72]

House of Quality

To ensure that your new products have features that meet the needs of your customers, you need to create **specifications**, or rules about how you will get desired product features. For example, a customer might have a need to "record her child's birth." This need might lead a product developer to create a digital camera with the feature of 24 hours of battery life. This feature might lead to a specification of a Li-ion rechargeable battery.[73]

The **house of quality** is a product development tool that helps companies to specify product attributes that satisfy customer needs.[74] Specifically, it is a visual matrix that compares weighted product attributes with weighted customer needs, providing a basis for an improved conversation between marketing and design personnel about those things.[75] Thus, the tool helps you to avoid developing new products with attributes that have no purpose in meeting customer needs, and to develop all of the product attributes that are necessary to satisfy customer needs.[76]

For example, Figure 7.8 shows a house of quality for a software product. The figure shows that customer needs for an "on time" and "high quality" software product correspond to certain software design and subcontractor management techniques. On the left side of the house of quality is the list of customer needs. Along the top of the house of quality are the attributes of the product. The intersections of the rows and columns show the fit between the needs of customers and the attributes of the product, which are weighted by the company's assessment of how well they do on the fit between those two dimensions. The triangle on the left side indicates the intersection between different customer needs and the company's evaluation of the importance of those intersections; while the triangle on the top marks the intersection between different product attributes and the company's assessment of the importance of those intersections. The line titled "direction of improvement" marks the company's assessment of whether the company's product development efforts are getting better or worse at matching customer attributes and product demands.

Although the house of quality works better for less complex projects than for more complex ones,[77] it is valuable in a wide variety of product development settings for several reasons. First, it helps you to make trade-offs between product attributes in an informed manner.[78] Second, the house of quality helps make your designs coherent by showing you the complementary changes that you need to make to satisfy customers or to create a competitive product.[79] Third, it allows the voice of the customer to be carried through to manufacturing because the outputs from one house of quality can be used as inputs into another. For instance, the weight of a car

FIGURE 7.8 A House of Quality

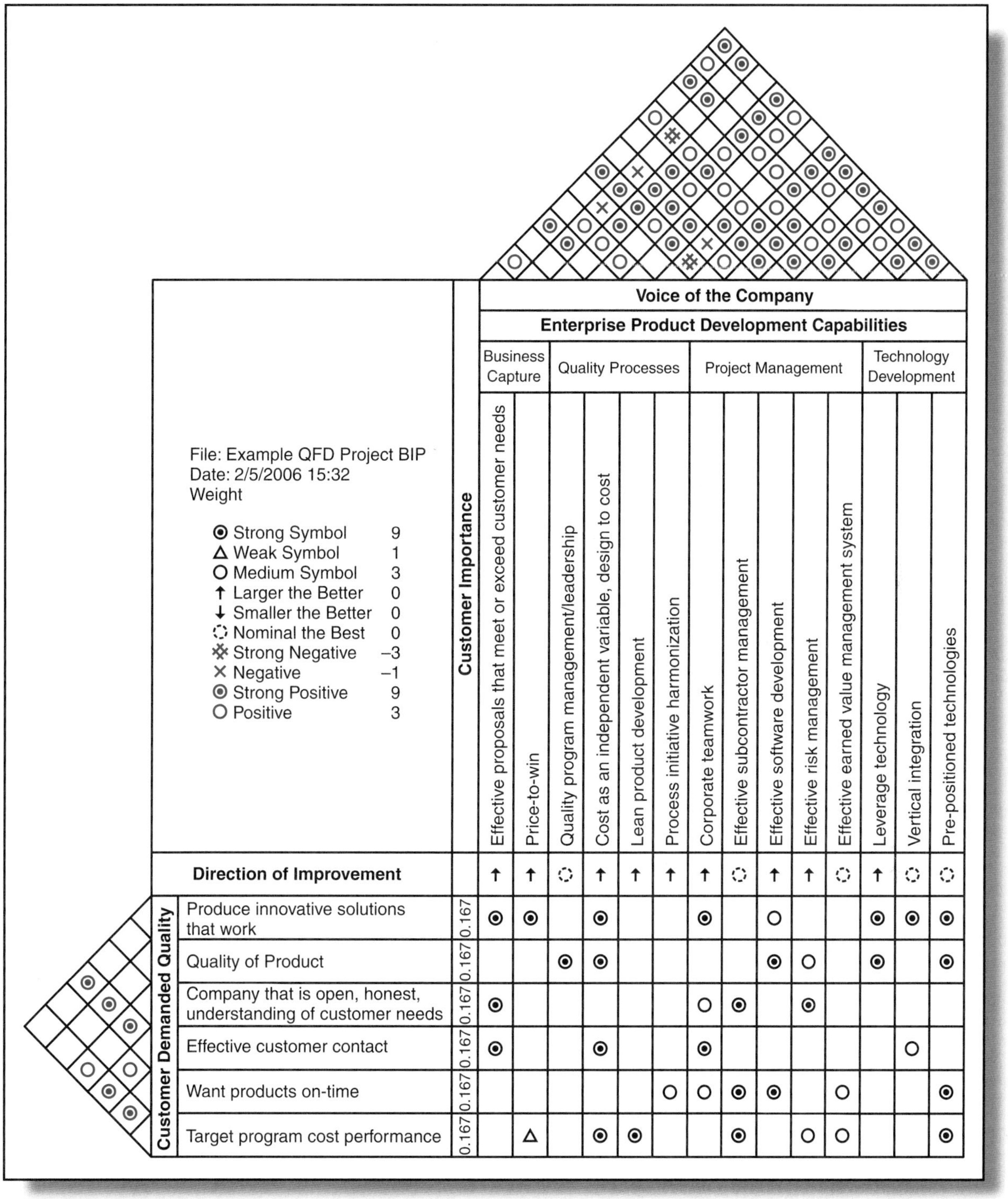

File: Example QFD Project BIP
Date: 2/5/2006 15:32
Weight

- ◉ Strong Symbol 9
- △ Weak Symbol 1
- ○ Medium Symbol 3
- ↑ Larger the Better 0
- ↓ Smaller the Better 0
- ◌ Nominal the Best 0
- ✖ Strong Negative −3
- × Negative −1
- ◉ Strong Positive 9
- ○ Positive 3

		Voice of the Company													
		Enterprise Product Development Capabilities													
		Business Capture		Quality Processes				Project Management					Technology Development		
Customer Demanded Quality	Customer Importance	Effective proposals that meet or exceed customer needs	Price-to-win	Quality program management/leadership	Cost as an independent variable, design to cost	Lean product development	Process initiative harmonization	Corporate teamwork	Effective subcontractor management	Effective software development	Effective risk management	Effective earned value management system	Leverage technology	Vertical integration	Pre-positioned technologies
Direction of Improvement		↑	↑	◌	↑	↑	↑	↑	◌	↑	↑	◌	↑	◌	◌
Produce innovative solutions that work	0.167	◉	◉		◉			◉		○			◉	◉	◉
Quality of Product	0.167			◉	◉					◉	○		◉		◉
Company that is open, honest, understanding of customer needs	0.167	◉						○	◉		◉				
Effective customer contact	0.167	◉			◉			◉						○	
Want products on-time	0.167						○	○	◉	◉		○			◉
Target program cost performance	0.167		△		◉	◉			◉		○	○			◉

This figure shows a house of quality—a tool to match product attributes with customer needs—for a piece of enterprise software. *Source:* http://en.wikipedia.org/wiki/QFD.

door needed to satisfy customers about crash safety might be used as an input into parts deployment, where the weight of the door is used to determine the properties of hinge parts that are sourced.[80]

Prototyping

Effective product development involves minimizing the losses that come from failed efforts to develop products that work or meet customer needs. Because failure is an inevitable part of doing something new, you cannot prevent it from occurring. But you can change the timing of when it happens.

As Figure 7.9 shows, costs increase as companies move through the stages of product development. Therefore, if you can increase the proportion of failures that occur early—say, during the design phase—you can reduce the costs of failure.[81] Moreover, you can increase the value of learning from failure by increasing the amount of learning that occurs at the earliest stages of the product development process.

One of the main ways that you can encourage early rather than late failure is through prototyping. **Prototypes** are approximations of actual products, including drawings, physical objects, and computer simulations.[82] Although prototyping adds costs to the product development process, and cannot completely remove the risk of design errors,[83] it provides two important benefits. First, prototyping allows companies to consider design alternatives before the actual product is developed, which permits design errors to be detected at an earlier stage of development, when the cost of fixing them is lower.[84] Second, prototyping helps firms to gather accurate information from their customers about their product preferences by providing a focal point for a discussion of the product's design.[85]

Today, many prototypes take the form of computer simulations. In the typical example, product developers use computer-aided design to create prototypes of products with different attributes, which can be tested through the use of computer simulations (see Figure 7.10).[86]

Computer-based prototyping offers four advantages over the development of physical prototypes: lower cost and faster development,[87] better information about trade-offs that comes from using the information processing capabilities of computers to test more precise alternatives,[88] the creation of full-scale versions that do not have to be changed when scaling up actual products to real-world levels, and the reuse of models from other products, which reduces mistakes, and accelerates the pace of the product development process.[89] For instance, Dassault Aviation used computer-based prototyping to design its Falcon 7X business aircraft. As a result, it didn't need to create a mock-up or test plane and cut its product development time in half to seven months.[90]

FIGURE 7.9 The Product Development Process

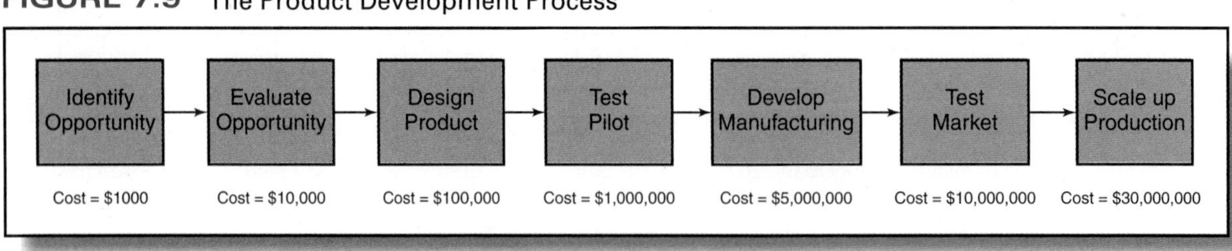

Cost increases as you move to later stages of the product development process.

FIGURE 7.10
Computer-Aided Design

This figure shows an example of a computer-aided design for a solid model assembly created in NX, a commercial CAD/CAM software package.

Source: http://en.wikipedia.org/wiki/Image:Cad_crank.jpg.

GETTING DOWN TO BUSINESS

Prototyping at IDEO[91]

IDEO is a consulting firm that helps clients in a variety of industries to design their products. Among the products that they have helped to design are the Palm V, Steelcase's Leap Chair, and Apple's first computer mouse.[92]

When designing products, IDEO's staff creates prototypes out of cardboard, clay, LEGOs, and foam so that their clients, IDEO personnel, and end users can all have a common point of reference when discussing products. Typically, an IDEO designer will develop an initial prototype of key product components to use as a "straw man" for all participants to criticize so that he or she can come up with a better design. Then, IDEO designers may make several prototypes, one after another, to refine the product design, discussing each one with the clients and end users. The company views each prototype as an "experiment" that results in a "failure" on some dimension. Through these failures the designers learn what features to put into the products.[93]

IDEO's management believes in making rough prototypes quickly. They think that if the designers spend too much time making detailed prototypes, they will become enamored of the prototypes and will not change them in response to feedback. They also believe in never going to client meetings without a prototype to ensure that in all discussions the designers and clients focus on the same things.[94]

Stage Gates

As Figure 7.11 shows, a **stage gate** is a decision tree that permits evaluation of the option to continue a project at particular milestones.[95] When using a stage gate process, you make "go" or "no-go" decisions at different milestones, called stage gates. These decisions are informed by answers to key questions about the technology, the market, and expected financial returns,[96] such as whether the estimated internal rate of return on the project at that point in time is greater than the cost of

FIGURE 7.11 The Stage Gate Process

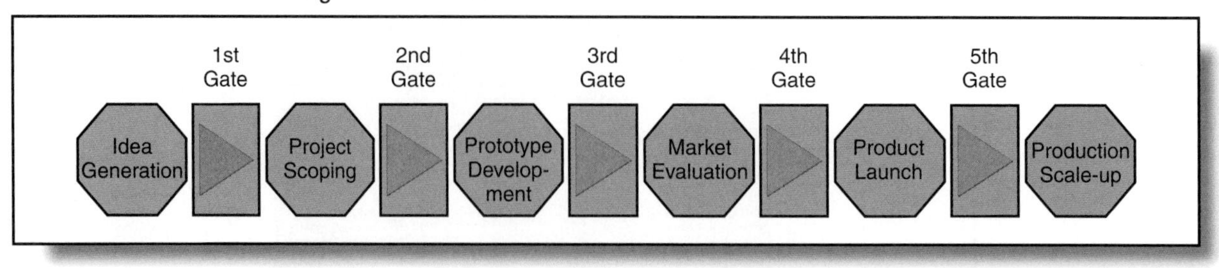

The stage gate process, developed by Robert Cooper, is a conceptual map for bringing a new product idea to market; when projects meet milestones, called stage gates, they are allowed to move to the next stage of development.

Source: Adapted from http://www.prod-dev.com/stage-gate.shtml.

capital.[97] The typical stages include idea screening, building a business case, design and development, product verification and validation, and product launch.[98]

Many companies use stage gates to manage the product development process. For example, Activision, a maker of video games, uses a stage gate process to decide whether or not to put game ideas into production. The first stage of the process is called "concept review" and is designed to figure out if the game's concept makes technical sense and can be sold. If the concept review is positive, the idea passes to the second stage, called "assessment." In the assessment stage, the evaluators determine whether the design for the game makes technical sense and can be sold. If the game design and positioning are viewed positively, then the idea passes to the "prototype" stage. At this stage, the game's look and feel are examined, and the promotion and launch are considered. If the idea passes through this milestone, it goes to the "first playable" stage. At this stage, evaluators determine if the game development process is adhering to budget limitations, deadlines, and quality. If the milestones at the first playable stage are met, the game is sent to the "alpha" stage, at which point evaluators determine whether the game plays as well as it was intended.[99]

While the stage gate process is more effective for innovation efforts that emerge from research than ones that do not,[100] it is, in general, a very useful tool. Because it restricts investment at the next stage until the previous stage has been completed, a stage gate analysis helps you to avoid undertaking activities for which insufficient information about upstream stages has been gathered. This is particularly important because each progressive stage in the innovation process takes more time and costs more money than the previous stage. As a result, using stage gates will help you to ensure that your innovation efforts achieve key milestones before you spend large amounts of money on them.[101]

Design for Manufacturing

Design for manufacturing is a set of rules that structure the new product design process to make products easier to manufacture, thereby reducing costs and improving quality.[102] For instance, one design for manufacturing rule is to reduce the number of components in a product by combining different features and simplifying assembly.[103] If you use fewer components, your customers can benefit by purchasing fewer parts, having higher manufacturing yields (a result of fewer part failures), and by spending less time testing components.[104] Another design for manufacturing rule is to design products for easy fabrication, which reduces labor costs and delays in production.

A third rule is to design products to simplify tooling and minimize setup, which speeds production, and makes it less expensive to change production lines.[105] A fourth rule is to design products so that the costs of the parts themselves are reduced.[106]

The use of design for manufacturing is helpful to companies seeking to develop new high-technology products by increasing the fit between product attributes and customer needs, as well as by shortening product development time.[107] For example, when NCR used design for manufacturing to develop an electronic cash register, it reduced both the time to produce its product and the cost of that production by over 50 percent.[108]

Key Points

- Web search tools provide a cheap and effective way to gather some types of information useful for product development.
- To ensure that new products have features that meet the needs of customers, many companies use the house of quality, a matrix that compares product attributes with customer needs.
- Prototyping helps companies to reduce the cost of failure at new product development by making it occur earlier in the development process.
- Stage gates are a decision-making tool that restricts investment in the more costly stages of the innovation process until key milestones have been met.
- Companies use design for manufacturing, a set of rules that structure the new product design process to make products easier to manufacture.

Discussion Questions

1. Should companies seek to reduce product cycle time? Why or why not?
2. What are the pros and cons of concurrent product development? Do the pros outweigh the cons? Why or why not?
3. Explain modularity. What are the advantages and disadvantages of developing products in a modular way? When should you develop modular products? When should you not develop modular products?
4. Why is managing a technology platform different than managing a single innovation? When is platform management most important? Why?
5. Compare perceptual maps, conjoint analysis, and the Kano method. What are the advantages and disadvantages of using each of them to figure out what attributes your product or service should have to meet customer needs?
6. What are the pros and cons of using Web-based tools to support product development? Do the pros outweigh the cons? Why or why not?
7. Compare prototyping and the house of quality. How does each of these tools help you to develop a product that customers want? What are the drawbacks of each of these tools?
8. What are stage gates? How do they help you to manage product development?
9. What are some important "design for manufacturing" rules? Why are these rules useful?

Key Terms

Components: The units from which something is made.

Concept Testing: A procedure in which customers are presented with a product idea and are observed for, or asked for, their reaction.

Concurrent Development: The process of undertaking some product development steps simultaneously, rather than sequentially.

Conjoint Analysis: A statistical tool that allows you to assess the relative importance of different product features and to identify the best combination of features to meet the needs of customers.

Core Idea Concept Test: A concept test that involves the presentation of the idea without a marketing message.

Design for Manufacturing: A set of design rules that structure the new product development process.

Factorial Design: A research design that presents all different combinations of product features to potential customers.

House of Quality: A matrix that compares customer requirements and product attributes, which is designed to improve communication between marketing and product design personnel.

Kano Method: A technique for evaluating customer preferences when those preferences are different if a feature is present than if it is absent.

Modularity: The degree to which a complex product can be built from smaller subsystems that can be created independently but function together.

Perceptual Map: A visual display of how people perceive the different features of competing products.

Platform: A common technological base off of which a family of products can be created to meet a common market application.

Positioning Concept Test: A concept test that involves the presentation of the product along with mock advertisements.

Product Cycle Time: A measure of how long it takes to develop new products.

Product Development: The creation of new products.

Prototypes: Any kind of approximation of actual products across attributes of the product.

Specifications: The rules about how you would design a product with certain features.

Stage Gate: A decision tree in which a project can be evaluated for the option to continue at particular milestones.

Putting Ideas into Practice

1. **Web-Based Tools** Many businesses are now using Internet search tools to identify attributes for the products that they sell. The purpose of this exercise is to familiarize you with the use of these tools and allow you to evaluate them. Go to the following Web sites: http://inventory.overture.com/d/searchinventory/suggestion, http://adlab.microsoft.com/ForecastV2/KeywordtrendsWeb.aspx, http://trends.google.com, and www.keyworddiscovery.com/search.html. Type key words into each of the sites (for example, type in "hybrid vehicle" and "fuel cell" vehicle). Now look at the results of those searches. How do the results help you to conduct product development? What are some pitfalls with using Web searches to gather information for product development?
2. **Conjoint Analysis** To do this conjoint analysis, you need to use a demo of some software. Go to www.sawtoothsoftware.com/downloads.shtml#ssiweb and download Sawtooth Software's SMRT v4.7 demonstration software. Once you have the software installed, open the demo and go to "file" to select "Tires1.smt." This demo contains data from a survey of 30 customers about their preferences for different mountain bike tires. Customers have been asked about brand (Michelin, Electra, and Tioga), tire type (very firm puncture proof solid foam core, somewhat firm puncture proof solid foam core, less firm puncture proof solid foam core, and standard inner tube tire), tread type (off-road tread, on-road tread, and off/on road tread), tire weight (2, 3, and 4 pounds), and tread wear (tire lasts for 1,000, 2,000, and 3,000 miles).

   You should first look at the questionnaire that has been presented to customers so that you can get a sense of the information that has been collected. In the Sawtooth software package, this means that you need to select the "Tires1.smt" simulation, and then go to "compose," and pull down on "run questionnaire." Answer the questions as they come up to get a feeling for the market research information that has been gathered. Once you are done with that, figure out what combination of attributes will create a tire with the highest purchase likelihood. (Hint: You need to go to market simulator in the Sawtooth software. You need to create different scenarios that hold all other features constant and vary one feature to figure out which value is highest. Then, you need to create a scenario with that combination of features and select "purchase likelihood" under "simulation method.")
3. **Design for Manufacturing** This exercise examines design for manufacturing. Your company is considering developing a new cordless, rechargeable, battery-powered drill. Before the product designers come up with a prototype of the product, you have been tasked with establishing several guidelines for them to follow to make sure that the product that

they design is easy to manufacture. Please specify five design-for-manufacturing rules for the product. (Hint: Think about rules that reduce the number of parts, facilitate the assembly of the product, make testing of the product easier, reduce the difficulty of servicing the product, use common parts and materials, and take a modular approach.)

1.
2.
3.
4.
5.

Notes

1. Corts, K., and D. Freier. 2003. The rise and fall (?) of Palm Computing in handheld operating systems. *Harvard Business School Case*, Number 9–703–519.
2. Markides, C., and P. Geroski. 2005. *Fast Second: How Smart Companies Bypass Radical Innovation to Enter and Dominate New Markets*. San Francisco: Jossey-Bass.
3. Teague, P. 2000. Father of an industry. *Design News*, March 6: 108.
4. Markides and Geroski, *Fast Second: How Smart Companies Bypass Radical Innovation to Enter and Dominate New Markets*.
5. Teague, Father of an industry.
6. Ramstad, E. 2005. Cellphone squeeze play: Key parts on one chip. *Wall Street Journal*, November 17: B3, B5.
7. Griffin, A. 1997. The effect of project and process characteristics on product development cycle time. *Journal of Marketing Research*, 34: 24–35.
8. Schilling, M., and C. Hill. 1998. Managing the new product development process: Strategic imperatives. *Academy of Management Executive*, 12(3): 67–81.
9. Schilling and Hill, Managing the new product development process.
10. Wind, J., and V. Mahajan. 1997. Issues and opportunities in new product development: An introduction to the special issue. *Journal of Marketing Research*, 34(1): 1–12.
11. Wheelwright, S., and K. Clark. 1992. *Revolutionizing Product Development*. New York: Free Press.
12. Wind and Mahajan, Issues and opportunities in new product development.
13. Cohen, M., J. Eliashberg, and T. Ho. 1996. New product development: The performance and time to market trade-off. *Management Science*, 42: 173–186.
14. Kessler, E., and A. Chakrabarti. 1996. Innovation speed: A conceptual model of context, antecedents, and outcomes. *Academy of Management Review*, 21(4): 1143–1191.
15. Ibid.
16. Clark, K., W. Chew, and T. Fujimoto. 1987. Product development in the world auto industry. *Brookings Papers on Economic Activity*, 3: 729–771.
17. Schilling and Hill, Managing the new product development process.
18. Poppendieck, M. 2005. Tactics and benefits of concurrent software development. *Concurrency*, 14(2): 1–19.
19. Wheelwright and Clark, *Revolutionizing Product Development*.
20. Eisenhardt, K., and B. Tabrizi. 1995. Accelerating adaptive processes: Product innovation in the global computer industry. *Administrative Science Quarterly*, 40(1): 84–110.
21. Schilling and Hill, Managing the new product development process.
22. Terwiesch, C., C. Loch, and A. De Meyer. 2002. Exchanging preliminary information in concurrent engineering: Alternative coordination strategies. *Management Science*, 13(4): 402–419.
23. Wheelwright and Clark, *Revolutionizing Product Development*.
24. Allen, K. 2003. *Bringing New Technology to Market*. Upper Saddle River, NJ: Prentice Hall.
25. Terwiesch, Loch, and De Meyer, Exchanging preliminary information in concurrent engineering.
26. Schilling, M. 2000. Toward a general modular systems theory and its application to interfirm product modularity. *Academy of Management Review*, 25(2): 312–334.
27. Baldwin, C., and K. Clark. 1997. Managing in the age of modularity. *Harvard Business Review*, September–October: 84–93.
28. Ibid.
29. Schilling, Toward a general modular systems theory and its application to interfirm product modularity.
30. Sanchez, R., and J. Mahoney. 1996. Modularity, flexibility, and knowledge management in product and organization design. *Strategic Management Journal*, 17 (Winter Special Issue): 63–76.
31. Fleming, L., and O. Sorenson. 2003. Navigating the technology landscape of innovation. *Sloan Management Review*, Winter: 15–23.
32. Bower, J. 2005. Teradyne: The Aurora project. *Harvard Business School Case*, Number 9-397-114.

33. Sanchez and Mahoney, Modularity, flexibility, and knowledge management in product and organization design.
34. Fleming and Sorenson, Navigating the technology landscape of innovation.
35. Mossberg, W. 2006. In our post-PC era, Apple's device beats the PC way. *Wall Street Journal*, May 11: B1.
36. Ernst, D. 2005. Limits to modularity: Reflections on recent developments in chip design. *Industry and Innovation*, 12(3): 303–335.
37. Schilling, Toward a general modular systems theory and its application to interfirm product modularity.
38. Robertson, D., and K. Ulrich. 1998. Planning for product platforms. *Sloan Management Review*, Summer: 19–30.
39. Christiansen, C. 1998. We've got rhythm! Medtronic Corporation's cardiac pacemaker business. *Harvard Business School Teaching Note*, Number 5-698-056.
40. Farrell, R., and T. Simpson. 2003. Product platform design to improve commonality in custom products. *Journal of Intelligent Manufacturing*, 14(6): 541–556.
41. Meyer, M., and J. Utterback. 1995. Product development cycle time and commercial success, *IEEE Transactions in Engineering Management*, 42(4): 297–304
42. Wind and Mahajan, Issues and opportunities in new product development.
43. Krishnan, V., and S. Gupta. 2001. Appropriateness and impact of platform-based product development. *Management Science*, 47(1): 52–68.
44. Mohr, J., S. Sengupta, and S. Slater. 2005. *Marketing of High Technology Products and Innovations* (2nd edition). Upper Saddle River, NJ: Prentice Hall.
45. Meyer, M. 1997. Revitalize your product lines through continuous platform renewal. *Research Technology Management*, 40(2): 17–28.
46. Robertson and Ulrich, Planning for product platforms.
47. Meyer, M., and A. DeTore. 1999. Product development for services. *Academy of Management Executive*, 13(3): 64–76.
48. Meyer, Revitalize your product lines through continuous platform renewal.
49. Mohr, Sengupta, and Slater, *Marketing of High Technology Products and Innovations*.
50. Krishnan and Gupta, Appropriateness and impact of platform-based product development.
51. Robertson and Ulrich, Planning for product platforms.
52. Gottfredson, M., and M. Booker. 2005. Finding your innovation fulcrum. *Wall Street Journal*, December 20: B2.
53. Krishnan and Gupta, Appropriateness and impact of platform-based product development.
54. MacMillan, I., and R. McGrath. 2002. Crafting R&D Project portfolios. *Research Technology Management*, 45(5): 48–59.
55. Krishnan and Gupta, Appropriateness and impact of platform-based product development.
56. Schilling, N. 2005. *Strategic Management of Technological Innovation*. New York: McGraw-Hill.
57. Cooper, R., and E. Kleinschmidt. 1987. New products: What separates winners from losers? *Journal of Product Innovation Management*, 4: 169–184.
58. Schilling and Hill, Managing the new product development process.
59. Griffin, A., and J. Hauser. 1993. The voice of the customer. *Marketing Science*, 12(1): 1–27.
60. Jedidi, K., and J. Zhang. 2002. Augmenting conjoint analysis to estimate consumer reservation price. *Management Science*, 48(10): 1350–1368.
61. Dolan, R. 1999. Analyzing consumer preferences. *Harvard Business School Case*, Number 9-599-112.
62. Ibid.
63. Ibid.
64. Ibid.
65. Stein, E., and M. Iansiti. 1995. Understanding user needs. *Harvard Business School Note*, Number 9-695-051.
66. Bayus, B. Forthcoming. Understanding customer needs. *Blackwell Handbook of Technology and Innovation Management*. Oxford: Blackwell.
67. Dolan, Analyzing consumer preferences.
68. Ibid.
69. Ibid.
70. Schilling and Hill, Managing the new product development process.
71. Page, A. 1991. PDMA's new product development practices survey: Performance and best practices. PDMA 15th Annual International Conference. Boston, MA: October 16.
72. Delaney, K. 2007. The new benefits of web-search queries. *Wall Street Journal*, February 6: B3.
73. Bayus, Understanding customer needs.
74. Hauser, J., and D. Clausing. 1988. House of quality. *Harvard Business Review*, May–June: 1–13.
75. Schilling and Hill, Managing the new product development process.
76. Griffin and Hauser, The voice of the customer.
77. Schilling and Hill, Managing the new product development process.
78. Griffin and Hauser, The voice of the customer.
79. Hauser and Clausing, House of quality.
80. Ibid.
81. Buggie, F. 2002. Set the "fuzzy front end" in concrete. *Research Technology Management*, 45(4): 11–14.
82. Ulrich, K., and S. Eppinger. 2004. *Product Design and Development*. McGraw–Hill/Irwin.

83. Thomke, S. 1998. Managing experimentation in the design of new products. *Management Science*, 44(6): 743–762.
84. Ibid.
85. Ibid.
86. Schilling, *Strategic Management of Technological Innovation*.
87. Schilling and Hill, Managing the new product development process.
88. Millson, M., S. Raj, and D. Wilemon. 1992. A survey of major approaches for accelerating new product development. *Journal of Product Innovation Management*, 9: 53–69.
89. Afuah, A. 2003. *Innovation Management*. New York: Oxford University Press.
90. Poppendieck, Tactics and benefits of concurrent software development.
91. Adapted from Thomke, S., and A. Nimgade. 2000. IDEO product development. *Harvard Business School Case*, Number 9-699-143.
92. http://en.wikipedia.org/wiki/IDEO
93. Thomke, S. 2003. IDEO product development. *Harvard Business School Teaching Note*, Number 5-602-060.
94. Thomke and Nimgade, IDEO product development.
95. Walryn, D., D. Taylor, and G. Brickhill. 2002. How to manage risk better. *Research Technology Management*, 45(5): 37–42.
96. Ibid.
97. Ibid.
98. Cooper, R., and E. Kleinschmidt. 1991. New product processes at leading industrial firms. *Industrial Marketing Management*, 20(2): 137–148.
99. MacCormack, A., and E. D'Angelo. 2005. Activision: The Kelly Slater's pro surfer project. *Harvard Business School Case*, Number 9-605-020.
100. Schilling and Hill, Managing the new product development process.
101. Schilling, *Strategic Management of Technological Innovation*.
102. Ulrich, K., D. Sartorius, S. Pearson, and M. Jakiela. 1993. Including the value of time in design-for-manufacturing decision making. *Management Science*, 39(4): 429–447.
103. Ibid.
104. Ramstad, Cellphone squeeze play.
105. Schilling and Hill, Managing the new product development process.
106. Ulrich, Sartorius, Pearson, and Jakiela, Including the value of time in design-for-manufacturing decision making.
107. Schilling, *Strategic Management of Technological Innovation*.
108. Clark, K., and S. Wheelwright. 1993. *Managing New Product and Process Development*. New York: Free Press.

Chapter 8

Patents

Learning Objectives

After reading this chapter, you should be able to:

1. Explain why firms can easily and quickly imitate most of their competitors' products and services, and why firms need to obtain intellectual property protection.
2. Explain what a *patent* is, and identify the characteristics of an invention that are necessary to obtain one.
3. Identify the different types of patents, and explain what they protect.
4. Identify the key parts of a patent, and explain what they do.
5. Outline the trends over time in the expansion of what is patentable, and discuss the pros and cons of these trends.

6. Define *patent infringement* and explain how patent owners can use the legal system to enforce their patent rights.
7. Discuss the benefits and limitations of patenting, and explain when patenting makes the most sense.

Patents: A Vignette[1]

In 2001, J.M. Smucker Co. instructed its attorneys to send a letter to a small caterer in Gaylord, Michigan, named Albie Foods Inc., demanding that Albie cease making and selling crustless peanut butter and jelly sandwiches (PB&Js). Smucker's attorney explained that Albie's PB&Js infringed on Smucker's patent for the sealed crustless sandwich patent, which the jelly maker uses to protect its Uncrustables sandwich.

Much to the chagrin of Albie's management, Smucker's does, in fact, own the patent on the sealed crustless sandwich, U.S. patent number 6,004,596, which it bought from the inventors of this sandwich, Len Kretchman and David Geske. As the abstract of the patent indicates, Smucker's has a monopoly right to "a sealed crustless sandwich for providing a convenient sandwich without an outer crust which can be stored for long periods of time without a central filling from leaking outwardly."[2]

Patents protect things that are stated in their claims. As will be explained more fully in this chapter, a claim identifies the parts of an invention that others may not imitate. Smucker's attorney pointed out to Albie that the patent for the sealed, crustless sandwich makes 10 claims, the first of which is "a sealed crustless sandwich, comprising: a first bread layer having a first perimeter surface coplanar to a contact surface; at least one filling of an edible food juxtaposed to said contact surface; a second bread layer juxtaposed to said at least one filling opposite of said first bread layer, wherein said second bread layer includes a second perimeter surface similar to said first perimeter surface; a crimped edge directly between said first perimeter surface and said second perimeter surface for sealing said at least one filling between said first bread layer and said second bread layer; wherein a crust portion of said first bread layer and said second bread layer has been removed."[3] This claim means that no one else has the right to make a sealed, crustless sandwich with at least one filling and a crimped edge. Moreover, if anyone makes, imports, or sells such a sandwich, Smucker's has the right to sue them to collect financial damages.

Albie decided to fight Smucker's, charging that it did not infringe on Smucker's patent because the patent was invalid. Because sealed, crustless sandwiches have been popular in the Midwest since the nineteenth century, Albie claimed that the technology described in the patent wasn't novel. And, if an invention can't be shown to be novel, a patent on it is invalid.

In the end, this patent dispute was never decided in the courts. As is often the case, both sides agreed to settle. But the lessons from the example are, nonetheless, instructive. Patents protect a wide variety of products. They provide the right for people to sue to collect damages if others make, import, or sell anything that uses the part of the invention protected by the patent claims without the written permission of the inventor. Consequently, as a technology entrepreneur or manager, you need to understand the role of patents in technology strategy.

INTRODUCTION

Intellectual property is of great importance to many companies. In fact, at some public companies, intellectual property now accounts for as much as 70 percent of the value of the business.[4] Because intellectual property is an important source of value in companies in high-technology industries, managing intellectual property is an important part of technology strategy.

As a technology entrepreneur or manager, you need to understand the different legal mechanisms that you can use to protect your intellectual property, and you need to develop ways to employ those mechanisms to your advantage. The next two chapters focus on the four ways that companies protect their intellectual property by legal means: through patents, trade secrets, copyrights, and trademarks. This chapter focuses on patents, while the next chapter discusses the other three mechanisms.

The first section of the chapter discusses why companies need intellectual property protection. The second section identifies what is patentable. The third and fourth sections identify the parts of a patent, and describe how to use a patent, respectively. The final section helps you to decide whether or not you should patent your inventions. By reading this chapter, you will learn how to use patents to protect your intellectual property, and how to avoid violating your competitors' patents.

WHY YOU NEED INTELLECTUAL PROPERTY PROTECTION

Before you can understand how to use patents, trade secrets, copyrights, and trademarks as strategic tools, you first need to understand why you need intellectual property protection at all. The answer lies in the components of a successful technology strategy. Introducing an innovative new product or service that meets the needs of customers is a necessary, but not sufficient, condition for success. Success also depends on protecting your product or service, or the way it is produced and sold, against imitation by competitors. Otherwise, your competitors, rather than you, will capture the profits that flow from your innovation.

Why is the ability of competitors to imitate your products and services so problematic to your profitability as an innovator? The answer is simple. You need to earn a profit on the sale of those innovative products and services to recoup the investment that you made to develop them. Initially, when you introduce a new product or service, you'll have a monopoly; no one else yet offers a product or service to meet the same market need. Your monopoly position allows you to charge high enough prices to generate the profit margins that you need to recoup your investment.

Unfortunately, any success that you have will motivate your competitors to copy what you are doing. If your competitors can come up with a product or service that meets the same customer need, or if they can undermine your advantages in producing or selling your product by copying how you do those things, then they can capture some of the profits that you are earning. To make matters worse for you, the more successful you are at the introduction of the new product or service—and the less you want to be imitated—the more motivated your competitors will be to imitate what you are doing. Your success makes it more obvious that competitors *should* imitate; and in many cases, it provides them with the information that they need to imitate your new product or service, or your method of producing and selling it.

If imitators are not stopped, they'll undermine all of your profit. To produce their copies of your initial product or service, imitators need to get access to the same resources as you are using—the employees, the capital, and the raw materials. As a result, they bid up the prices of these resources, causing your profit margins to fall. Moreover, the imitators take away some of your customers. Each customer that they woo away from you drives down your revenues, further hurting your profit margins.

Clearly, you need to stop your competitors from imitating your new products or services if you are going to be successful. Unfortunately, doing this isn't easy. Most new products are simple to copy, particularly for large, established firms. One study by Richard Levin and his colleagues showed that approximately half of the time, the average unpatented new product can be duplicated by between six and ten competitors, at less than half the cost of the original development.[5] Another study, this one by Edwin Mansfield, showed that, on average, one-third of new products can be imitated in six months or less (see Figure 8.1).[6]

Companies can figure out how to imitate your new products and services in a wide variety of ways. Many new products can be reverse engineered, with your competitor's technical staff simply taking apart your new product and figuring out how it works. Once their engineers figure out how your product works, it is often very easy for them to come up with another way to do exactly the same thing.[7]

Your competitors can easily hire your employees as a way to learn what they know. Labor markets are free in most countries, and people often leave their jobs to go work for competitors. So your competitors could figure out how to imitate your products and services by offering a higher salary to your employees to get them to jump ship. Then they can use the knowledge that your employees have developed in the course of their careers to create products and services that imitate yours.

FIGURE 8.1
The Amount of Time It Takes to Imitate New Products

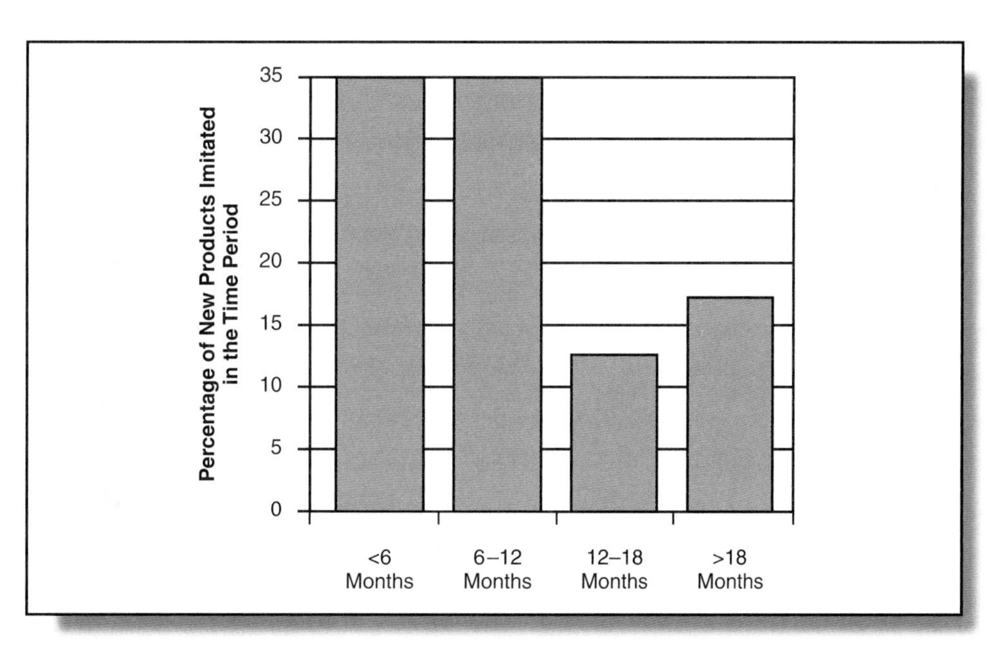

Imitation doesn't take very long; on average, more than two-thirds of the time it takes competitors less than 12 months to figure out how to imitate an innovator's new product.

Source: Based on information contained in Mansfield, E. 1985. How rapidly does industrial technology leak out? *Journal of Industrial Economics*, 34(2): 217–223.

Sometimes simply working on similar new products allows your competitors to figure out how to copy your new products or services. Most competitors are working on new products and services that are similar to each other, and just knowing that your company has figured out a way to, say, make a product smaller or add features to it, is sometimes enough to allow your competitors to come up with an imitative product or service on their own.[8]

Your competitors can often look at public documents and publications to figure out how to copy your new product or service.[9] Because engineers and scientists have strong expertise in the areas in which they work, they can often extrapolate from partial information obtained in public sources and figure out how to imitate your new products or services, just on the basis of information that you have made public.

Key Points

- Intellectual property protection helps you to deter imitation by competitors, which will undermine your profits from innovation.
- Most new products and services are easy to copy, and a large number of firms are able to duplicate those products and services at a low cost.
- Competitors can imitate your products by reverse engineering them, by hiring your employees, by working on similar projects, and by reading your publications and patent disclosures.

What Is Patentable?

A **patent** is a government-granted monopoly that precludes others from using an invention for 20 years (for utility patents) in return for the inventor's disclosure about how the invention operates. Patents are based on a fundamental trade-off. In return for showing others how an invention works, and thereby advancing the level of technical knowledge in a country, inventors receive monopoly rights to their invention for a period of time.

Patents have a complex set of effects on technological innovation. On the one hand, they provide people with an incentive to innovate. In the absence of the monopoly right provided by patents, inventors often would be unable to capture the value coming from their inventions and, therefore, be unwilling to develop or exploit them. Moreover, the disclosure that patents require makes it possible for other parties to learn from inventions and make further advances, which would not be possible if the inventors kept the inventions secret.

On the other hand, patents can deter technological innovation by making it difficult for others to reap commercial value from undertaking further developments in an area, given the inventor's property rights. For instance, some observers believe that patents on genes deter follow-on innovation because they give the patent holder too much protection, thus deterring others from developing genetic tests based on the initial inventor's discovery.

Patenting is an old, and established, form of intellectual property protection. The first uses of patents go back centuries; and patents were important mechanisms to protect basic inventions of the industrial revolution, such as the steam engine. The patent system is so important that, in the United States, it is enshrined in Article I of the Constitution, and has existed since the birth of the nation.

Despite being around a long time, patenting appears to be increasing in importance. Since 1983, the number of patents granted by the United States Patent and Trademark Office (USPTO) has increased by approximately 5.7 percent annually.[10] Currently, approximately 350,000 patent applications are made to the USPTO each year, and approximately 200,000 patents are awarded. And inventors now spend in excess of $5 billion per year to obtain U.S. patents to protect their inventions.[11]

What Can Be Patented?

As Figure 8.2 shows, many brilliant business concepts cannot be protected with patents because only the mechanisms for exploiting ideas can be patented, not the ideas themselves.[12] For example, you can't patent the idea of a fast-food restaurant drive-through window. All you can do is patent a mechanism for exploiting the idea, such as the window itself.

Because they are not embodied in physical form, most services are difficult to patent.[13] So you can't patent courteous service, even if it provides your business with a competitive edge. The best that you can do is to patent the process by which that service is created, as would be the case if you developed a robotic employee that could be programmed to be courteous all of the time, even if it were having a bad day.

You also cannot patent laws of nature or any substances that appear naturally,[14] such as chemical elements, because the government thinks of nature, not the person discovering them, as the inventor. The best that you can do to protect a discovery of a natural substance is to patent the mechanisms for obtaining it, such as the process of leaching iron from rock.[15]

FIGURE 8.2
Not Everything Is Patentable

Many great inventions are not patentable.
Source: http://www.cartoonstock.com.

You *can* get a utility patent, which is given for new or improved products and processes, for one of four things: a process (such as a chemical reaction), a machine (such as a laser), an article of manufacture (such as a diskette), or a composition of matter (such as a genetically altered bacterium).[16]

FIGURE 8.3
A United States Patent

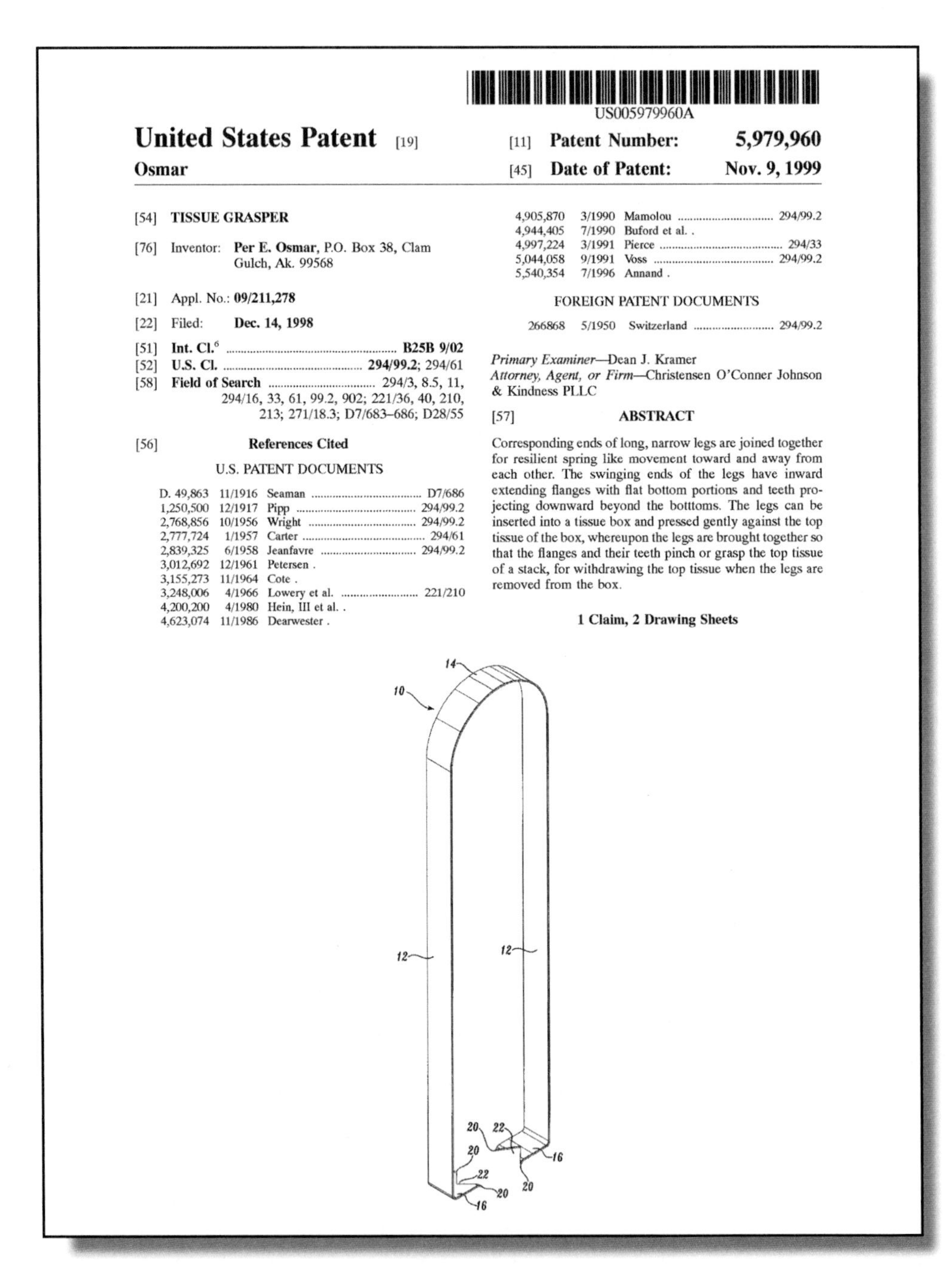
US005979960A

United States Patent [19]
Osmar

[11] **Patent Number:** **5,979,960**
[45] **Date of Patent:** **Nov. 9, 1999**

[54] **TISSUE GRASPER**

[76] Inventor: **Per E. Osmar**, P.O. Box 38, Clam Gulch, Ak. 99568

[21] Appl. No.: **09/211,278**

[22] Filed: **Dec. 14, 1998**

[51] **Int. Cl.**^{6} **B25B 9/02**
[52] **U.S. Cl.** **294/99.2**; 294/61
[58] **Field of Search** 294/3, 8.5, 11, 294/16, 33, 61, 99.2, 902; 221/36, 40, 210, 213; 271/18.3; D7/683–686; D28/55

[56] **References Cited**

U.S. PATENT DOCUMENTS

D. 49,863	11/1916	Seaman	D7/686
1,250,500	12/1917	Pipp	294/99.2
2,768,856	10/1956	Wright	294/99.2
2,777,724	1/1957	Carter	294/61
2,839,325	6/1958	Jeanfavre	294/99.2
3,012,692	12/1961	Petersen .	
3,155,273	11/1964	Cote .	
3,248,006	4/1966	Lowery et al.	221/210
4,200,200	4/1980	Hein, III et al. .	
4,623,074	11/1986	Dearwester .	
4,905,870	3/1990	Mamolou	294/99.2
4,944,405	7/1990	Buford et al. .	
4,997,224	3/1991	Pierce	294/33
5,044,058	9/1991	Voss	294/99.2
5,540,354	7/1996	Annand .	

FOREIGN PATENT DOCUMENTS

266868	5/1950	Switzerland	294/99.2

Primary Examiner—Dean J. Kramer
Attorney, Agent, or Firm—Christensen O'Conner Johnson & Kindness PLLC

[57] **ABSTRACT**

Corresponding ends of long, narrow legs are joined together for resilient spring like movement toward and away from each other. The swinging ends of the legs have inward extending flanges with flat bottom portions and teeth projecting downward beyond the botttoms. The legs can be inserted into a tissue box and pressed gently against the top tissue of the box, whereupon the legs are brought together so that the flanges and their teeth pinch or grasp the top tissue of a stack, for withdrawing the top tissue when the legs are removed from the box.

1 Claim, 2 Drawing Sheets

The figure shows the first page of a U.S. patent for a device to grasp tissues from a box.
Source: http://www.freepatentsonline.com/5979960.pdf?s_id=919c52aeb0632d7cfd9d86498edd018c.

Novel, Nonobvious, and Useful

Patents are only granted for inventions that the patent office determines are novel, nonobvious, and useful. The USPTO defines an invention as "novel" if it has not been previously invented and is not a trivial improvement on an existing invention—an improvement that comes from exchanging one material for another, changing the size or shape of a device, increasing its portability, or exchanging one element for an equivalent other.[17]

The patent office deems an invention to be "obvious" if it is a clear next step in technological development to a person who is an expert in the field (for example, an electrical engineer would be considered an expert on electrical circuits) or if the elements of the invention all were present in existing patents.[18] For example, J.M. Smucker Co. was denied a patent on its method of applying filling to its Uncrustables sandwich product because the patent examiner assigned to the case believed that the concept of applying peanut butter on one slice of bread and applying jelly on another would be "obvious" to anyone trained in the art of making a sandwich.[19]

For the USPTO to view an invention as "useful," it has to work, have a use, and be functional.[20] So you can't patent something like a piece of music, which isn't functional. However, being "useful" doesn't mean that an invention has to have commercial value; in fact, most patented inventions generate no financial returns. Take, for instance, U.S. patent 5,023,850, for a dog watch that moves at seven times the rate of a normal watch.[21] While this device works, is functional, and has a purpose, it has no commercial value. Perhaps the number of dogs who can tell time and have disposable income is too small for that.

First to Invent

As Figure 8.4 indicates somewhat humorously, the U.S. patent system differs from the patent systems in all other countries (except the Philippines) because the United States awards patents to the first party to invent something, not to the first inventor

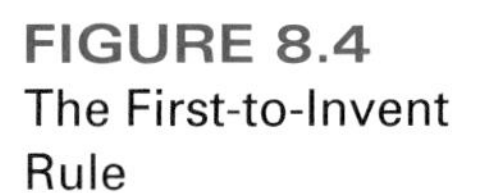
FIGURE 8.4
The First-to-Invent Rule

In the United States, if two inventors come up with the same invention, the patent is awarded to the inventor who can prove that he or she invented it first.

Source: Estate of Charles Addams. © Tee and Charles Addams Foundation.

to file for a patent. The importance of the first-to-invent rule can be seen in the case of the rotational wheel interface in Apple Computer's iPod. Microsoft was the first company to file for a patent on this technology, which is currently the subject of a patent dispute between Apple and Microsoft. For Apple Computer to prevail in this dispute, it has to file a declaration to the USPTO, which states that it invented the device before Microsoft, and the USPTO has to determine, as a result of an investigation of the declaration and Apple's records, that Apple's invention predated Microsoft's patent application.[22]

Nondisclosure

In the United States, patents are only awarded for inventions that have not been offered for sale and have not been publicly disclosed, either in an open forum or in print, more than one year earlier.[23] While experimental testing of your invention is not considered public use, you can't advertise the invention, issue a press release about it, present it in a seminar, or even offer a description of it at a trade show.[24] In fact, in some cases, even showing an invention to your friends might constitute a public disclosure. So, to keep your invention secret, you need to have anyone who looks at the invention before you file for a patent sign a nondisclosure agreement.[25]

GETTING DOWN TO BUSINESS
Patenting a Snowman Accessory Kit

Do you know how to build a snowman? Are you so good at it that you think your approach can be patented? Believe it or not, someone has already done it. In 1993, Robert Kenyon of Scranton, Pennsylvania, received U.S. patent number 5,380,237 for a "Snowman Accessory Kit." According to the abstract of this patent, the invention is of an accessory kit "having a head, an upper torso body and a lower torso body all made of packed snow, which consists of a set of decorative articles. A structure is for inserting each of the decorative articles into the packed snow of the head, the upper torso body and the lower torso body, to enhance the appearance of the snowman. Components are for retaining each of the inserting structures within the packed snow of the head, the upper torso body and the lower torso body of the snowman, so as to prevent the decorative articles from falling off of the snowman."[26]

Given the claims of this patent, no one else has the right to make or sell "an accessory kit for a snowman having a head, an upper torso body and a lower torso body, all made of packed snow, which comprises: a set of decorative articles having a circumference and a longitudinal axis; means for inserting each of said decorative articles into the packed snow of the head, the upper torso body and the lower torso body, to enhance the appearance of the snowman; and retaining means on each of said articles having a plurality of appendages, each of said appendages being circumferentially spaced from every other appendage and each of said appendages further being longitudinally spaced from every other appendage whereby when an article is inserted into said snowman and turned, each of said appendages follows the same insertion path in a corkscrew-like manner, thus retaining said article in said snowman."[27] This means that if you go outside and build a snowman by taking a carrot and putting it in the head of the snowman for its nose and inserting some sticks clockwise into the torso for its arms, then you have violated Robert Kenyon's patent. He can sue you and collect damages. To make a snowman in this way legally, you need to get a license from Kenyon.

But, before you think Keynon has a blockbuster, pioneering, snowman patent, you need to check out the prior art that he had to cite to get his patent. You see, Kenyon doesn't have the rights to a kit to make the hat for the snowman, which was patented by Brian Schwuchow of Hobart, Indiana or a snowman feature and accessory system, which was patented by Phillis Lorenzo of White Marsh, Maryland.[28]

FIGURE 8.5 Growth in Number of Utility Patents Issued

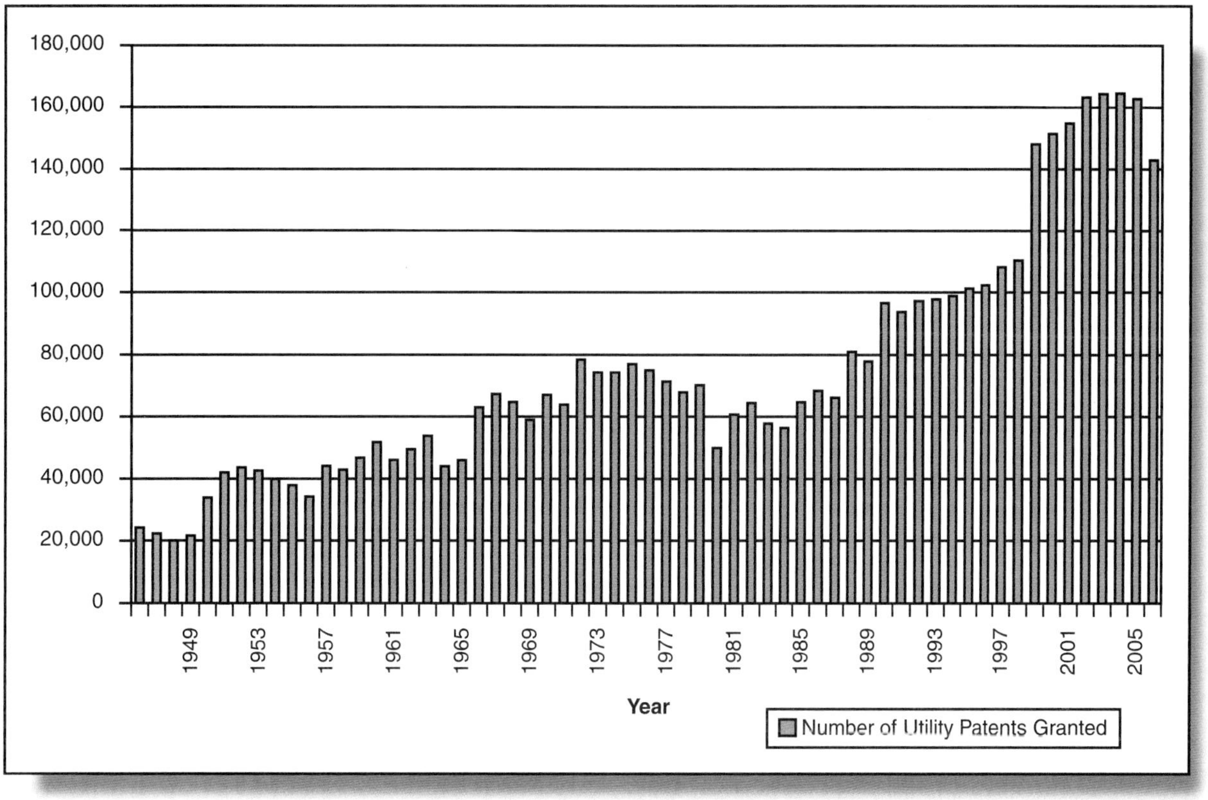

Since the early 1980s there has been a dramatic increase in the number of utility patents issued in the United States every year.

Source: Created from data downloaded from the U.S. Patent and Trademark Office.

Expansion of What Is Patentable

Over time, the U.S. government has steadily expanded the types of things that can be protected by a patent, leading to a rise in the number of patents issued (see Figure 8.5). Since 1980, patents can be obtained on genetically engineered organisms, like mice, genetically altered substances, like yeast, and human genetic sequences.[29]

In addition, while mathematical formulas are considered to be natural phenomena and cannot be patented,[30] those formulas that are applied to a structure or process, as occurs in software, have been patentable since 1981.[31] As a result, patents now protect a wide range of computer software, from training tools to investment and insurance systems to e-commerce payment mechanisms.[32] Furthermore, since 1998, business methods like Amazon.com's "One-Click" system, which allows repeat purchasers to make Internet purchases without reentering information, can be protected by patents.[33]

The expansion of what is patentable has raised a number of important issues that you need to consider as you formulate your company's technology strategy:

1. The increase in the volume of patent applications has created backlogs in the patent office, making the process of getting a patent less efficient than it used to be.

2. The growth in the patent "thicket" has resulted in a lot of cumulative and overlapping patents, increasing the number of patent disputes, as well as the rate at which firms license their patents to each other (cross-licensing).
3. The expansion into business method patents, which tend to be broader and more obvious than other patents, has raised questions about the degree to which innovation is being hindered by property rights.
4. The expansion into genetically engineered organisms has raised questions about whether patents block follow-on research and are making it difficult to come up with new medical and pharmaceutical innovations.

Design and Plant Patents

There are two types of patents other than utility patents, which we have been discussing: design patents and plant patents. Design patents are given for the appearance of products.[34] For example, U.S. patent number D339456 protects an ornamental design for a shoe sole. Design patents differ from utility patents because they have only one claim and protect a piece of intellectual property for only 14 years.[35]

Plant patents are given only for engineered plants that are reproduced asexually.[36] An example is Tropicana's patent for the varieties of oranges used in its Pure Premium orange juice. Like utility patents, plant patents protect a piece of intellectual property for 20 years and have multiple claims.

Key Points

- A patent is a government-granted monopoly that precludes others from using an invention for a specified period of time in return for the inventor's disclosure about how the invention works.
- Three types of patents can be obtained: utility patents, which are given for new or improved products or production processes; design patents, which protect the appearance of a product; and plant patents, which protect plants that are reproduced asexually.
- Patents cannot be obtained on natural substances or ideas; only a working process, machine, manufacture, or composition of matter that is novel, nonobvious, and useful, has not been disclosed either in an open forum or in print more than one year earlier, and has not been offered for sale, can be patented.
- In recent years, the variety of things that can be patented has increased and now includes genetically engineered organisms, computer software, and business methods.

The Parts of a Patent

Patents have two key parts: the specification and the claims. The **specification** is a description of how the invention works, and may include accompanying illustrations. The specification is what you must trade off in return for the monopoly right that you receive. Its purpose is to allow those skilled in the relevant technical area to reproduce your invention.[37]

The other major part of a patent is the set of **claims**, or the statements that identify a particular feature or combination of features that are protected by the patent.

The claims are what indicate whether another patent infringes on your patent (that is, violates your monopoly right).

As an inventor, you want to obtain a patent with broad scope claims (as long as you can enforce them) because a patent prevents imitation of only those things specified in the claims. The broader the scope of the claims, the harder it is for other firms to make changes to your invention and work around your claims. For example, Dry-Dock Systems Inc. has a patent for floating dry docks made of connected plastic cubes. But instead of having a patent that covers just the cubes or their method of assembly, Jet Dock Systems Inc. has a patent on the process of drive-on docking. As a result, the company's patent bars competitors that use a different method of assembly of a floating dry dock from imitating its product.[38]

While broad scope claims are valuable for the reasons just described, you also need to ensure that your patent will be held valid if challenged, and that you will be able to prove infringement. The broader the scope of your claims, and the better it protects your innovation, the more likely other people will be to challenge it, in the hopes that they can get rid of your monopoly right and do what you are doing.

Unfortunately (for the patent holder), patents with very broad claims are often difficult to enforce. The less specific the claims, the less easily judges and juries can interpret the words in them.[39] In addition, patents with very broad claims that cover entire methods of approaching problems, rather than just specific products, are more likely to be deemed invalid. For instance, a federal appeals court recently invalidated a University of Rochester patent on Cox-2 inhibitors, saying that the patent's claims were too general for the patent to be valid, causing the University of Rochester to lose its lawsuit against Pfizer Inc. for infringement of the university's patent by its drug, Celebrex.

So how do you know how strong your patent's claims are? You need to look at the patent to see if part of the claims could be changed or dropped and still yield the same level of protection. For example, if a patent claims the process for using a particular adhesive for attaching two pieces of metal, but you could easily attach the two pieces of metal with another adhesive, the patent has weak claims. All a competitor has to do to get around your adhesive patent is to substitute a different adhesive for the one that you have claimed.

Sometimes companies cannot get one or two broad claims, and, therefore, try to protect their inventions with a large number of claims. As Figure 8.6 shows, sometimes the number of claims that inventors seek on their patents is quite large indeed.

Pioneering patents—basic patents in a technical area on which a wide range of inventions build—are a special case of patents with strong claims. Control of these patents is important because they can be used to generate royalty payments from a large number of users. For example, the holders of pioneering patents in genetic engineering earned hundreds of millions of dollars in royalties from a variety of companies using genetic engineering techniques to make drugs. Similarly, NEC Corp. has made a great deal of money licensing its pioneering patents on carbon nanotubes, which are being used for fuel cell batteries in notebook computers, transistors, wide-screen televisions, and sensors.[40]

Pioneering patents and patents with broad claims are especially important to you if you are starting a company. New firms often lack other forms of competitive advantage when they are first established. Strong patents allow you to create the value chain for your new business and establish additional competitive advantages before your new product or service is imitated by other firms.

FIGURE 8.6 Number of Claims

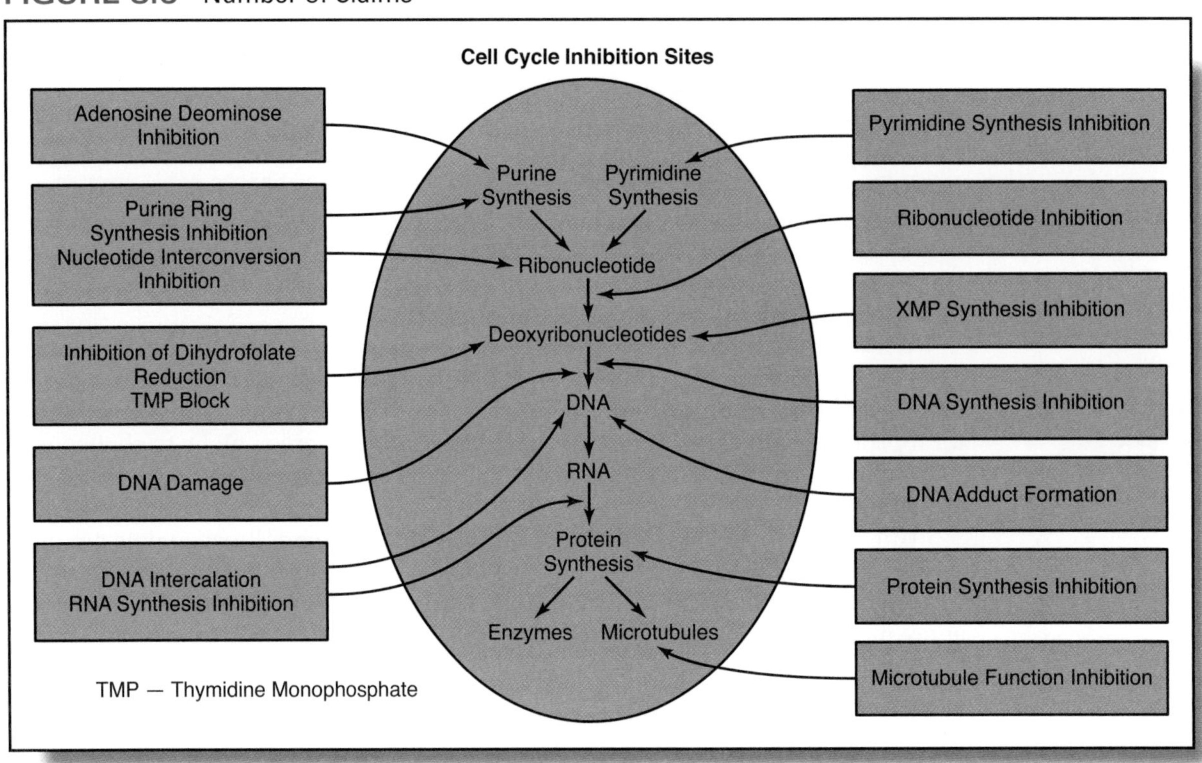

The patent application for the electrical device and anti-scarring agent shown here was originally submitted to the USPTO with 13,305 claims; however, the patent was issued with only about 50 of those claims.

Source: http://www.gelsing.ca/blog/?p=106.

The more pioneering your patent, and the broader its scope, the more competitor firms you can deter from imitating your new product or service. For instance, Friendster recently obtained a patent on the method of searching for people on the Web as a function of their social ties. Because this patent is so broad, Friendster can use it to stop Facebook and MySpace from engaging in similar Web-based social networking activities, or at least to obtain licensing fees from them. Consequently, the patent will help the start-up, whose performance has lagged as a result of competition from these two companies.[41]

Defining the Claims

The claims that you are allowed to make are limited by what previous inventors, whose patents have already been granted, have claimed. The initial arbiter of what claims you can get is the patent examiner. To help patent examiners determine what claims should be granted, as an inventor, you have a duty to provide citations to previous patents whose technical art you build upon in creating your inventions. These citations limit your property right to only those things not claimed in previously cited patents.

To disallow claims, patent examiners must provide legal reasons for their actions. If an examiner denies a claim, an inventor can file a response to the examiner, and have that response evaluated. If the decision remains negative after this revision, then an inventor can appeal to the Board of Patent Appeals for a conference with senior patent examiners. This appeal can result in either a reversal of the examiner or a continued rejection of the patent claim. If an inventor fails to obtain positive decisions at the level of the Board of Patent Appeals, he or she can ask the Board of Examiners to consider the issue.[42]

Although patent examiners are very good at determining what claims should be granted, they sometimes grant patents that are too broad or that overlap, particularly in new technical fields where examiners lack expertise. For example, many observers have criticized the USPTO for initially allowing very broad claims on genetic engineering and Internet business method patents.[43] These broad claims may have increased the cost of developing new products by creating a complex morass of conflicting claims.

It is important to note how human the patenting process is. Patents are not an objective exclusion device created by machines. Rather, they are the result of interpretations by human beings and a complex negotiating process. As a result, some issued patents make sense and others do not. Moreover, sometimes inventors who deserve patents are denied them.

Who Can Apply?

Only inventors can apply for, and be awarded, patents. If you employ, or contract for, work done by others, and would like to own the patents on any resulting inventions, you need your employees or contractors to agree in writing to assign those patents to you. Otherwise, U.S. law assumes that employees and independent contractors own the rights to the inventions that they make. To facilitate this assignment process, many companies specify it in their employment agreements.[44]

The limitation on who can be awarded a patent raises an important issue for start-up firms. Inventors who found companies based on their inventions own the patents awarded for those inventions unless they assign them to their companies. To protect themselves, investors in start-up companies typically require inventor-founders to assign those patents to their companies. That way, if the ventures don't do well, the investors have access to the businesses' intellectual property assets and can sell them to recoup some of their losses. Of course, this means that inventors may lose control of their own inventions if their start-ups, which use these inventions, are unsuccessful.

Key Points

- Patents have two key parts: the specification, which describes how the invention works, and the claims, which identify the features protected by the patent.
- Inventors want the broadest scope claims that can be enforced.
- To get a patent, inventors craft claims that do not infringe on those of previous patents.
- Only inventors can apply for patents, which they do by applying to the USPTO.

Using a Patent

As a technology strategist, you also need to understand how to use patents effectively. This section discusses two important issues: the use of multiple patents to protect technology and the use of the legal system to enforce patents.

Picket Fences and Brackets

Although you might try to obtain a broad patent with strong claims to protect your invention, such patents are not always possible. As a result, you might want to apply for more than one patent to protect your new product or service. For example, if you run a biotechnology company that has invented a new drug, you might try to patent both the molecule itself and the process for producing it.

If you cannot get a single broad patent to protect your invention, then building a **picket fence** of patents around the core invention can be a way to protect your new product or service. For example, when Gillette developed the Sensor razor, it obtained 22 different patents to protect the product, including patents on the blades, the handle design, and the packaging container.[45] Because your competitors' ability to imitate your invention without infringing on your patents depends on coming up with a way to do the same thing as your invention in a way not indicated in your patent claims, then obtaining a variety of patents can deter imitation by closing off alternative paths.

Because companies build picket fences of packet protection around their inventions, other companies engage in **bracketing**—the process of keeping an inventor from using his or her invention by patenting around it—to counteract their efforts. For example, if your competitor has obtained a patent on a filament for a new, higher intensity light, you can keep that company from using its filament patent by patenting ancillary inventions, such as a new bulb, new housing, new connections, and a new shade, thereby bracketing the filament patent with your own patents on the rest of the light.[46]

Patent Litigation

Because patents only provide you with the right to sue to collect damages from others who infringe on your patents, successful strategies for exploiting patents invariably involve legal action. Therefore, intellectual property litigation is, and has always been, an integral part of technology strategy.

But, before we can discuss how you can enforce your patents, you first have to know what patent infringement is. **Infringement** occurs when someone to whom you have not licensed your invention makes, uses, sells, or imports something covered by the claims of your patent.[47] Your invention does not have to be duplicated in its entirety for your patent to be infringed. If any part of what is claimed is duplicated, infringement has occurred. Moreover, infringement can occur inadvertently as well as deliberately and can occur if one party induces another to build something that is covered by your patent.[48]

So how do you know if your patent has been infringed? Infringement occurs if another invention does substantially the same thing, in the same way, with the same result, as a patented invention. For example, a new bicycle wheel would infringe on an existing bicycle wheel patent even if the second bicycle wheel was not exactly the same as the first but performed the same function in a similar manner to the patented wheel.[49] But a patent on the engine of a fuel cell powered car would not infringe a patent on a car powered by an internal combustion engine. Even though the two technologies both make the car go, the two kinds of engines do so in a very different way, one by turning hydrogen into water and the other by burning gasoline.

Still unsure about what constitutes infringement? Take a look at Figure 8.7. Using the example of a bucket, the figure shows you what questions to ask to figure out if another invention infringes on your patent.

If you suspect that your patent has been infringed, you should take legal action immediately. If you wait too long to enforce your patent rights, then the courts will presume that you know about the infringement and do not care to take action.[50]

After filing a patent infringement lawsuit, you can ask the court for an injunction to stop the alleged infringer from engaging in the actions that violate your patent. An injunction alone may be enough for you to achieve your goal of stopping the infringement. For instance, when Amazon.com sued Barnes and Noble for violation of its patent on its One-Click purchase method, Amazon.com obtained an injunction against Barnes and Noble, which had the effect of shutting down Barnes and Noble's Express Lane purchasing system, and achieved Amazon.com's objective.[51]

If you initiate a patent infringement lawsuit, the defendants are likely to fight back by seeking to prove that your patent is invalid. (If your patent is invalid, then no one can infringe it.)[52] Patents will be declared invalid if the invention is deemed obvious to people trained in the relevant technical art, or if the patent holder is shown to have publicly disclosed or sold the invention more than one year before filing the patent application.[53] For instance, eBay Inc. recently won an infringement lawsuit

FIGURE 8.7
Does It Infringe?

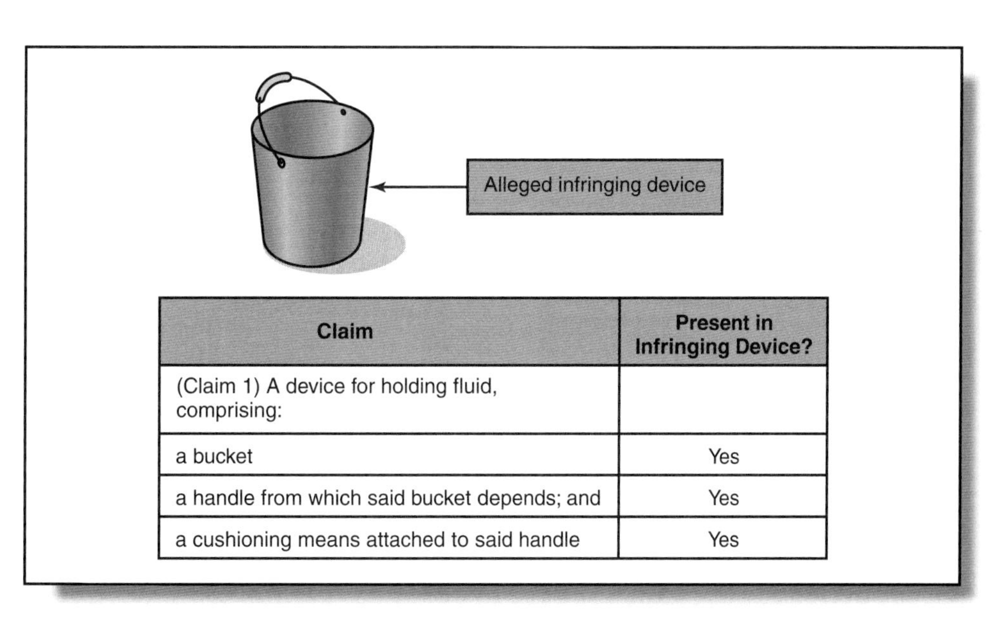

Claim	Present in Infringing Device?
(Claim 1) A device for holding fluid, comprising:	
a bucket	Yes
a handle from which said bucket depends; and	Yes
a cushioning means attached to said handle	Yes

To determine if a product infringes a patent (in this case a bucket), you look to see if what is claimed in the patent is present in the device of the alleged infringer.

Source: Adapted from http://www.smithhopen.com/ip_litigation.asp.

against it by MercExchange LLC by showing that MercExchange's e-commerce business process patents were obvious technical improvements to people trained in the art.[54]

The United States Supreme Court recently made it easier to challenge the validity of someone else's patent. A licensee can now challenge validity while still paying royalties to the patent holder, which protects the licensee against the risk of a countersuit for patent infringement. Because patent infringement can result in large damages, the previous requirement that the licensee stop paying royalties to challenge patent validity kept many small, start-up companies from challenging the validity of established companies' patents. They were simply afraid of a countersuit that would cause them to go bankrupt if they lost.[55]

If you win a patent litigation lawsuit, you can obtain monetary damages and/or a permanent injunction that prohibits the infringer from using the patented technology. The amount of damages that you can obtain depends largely on the intent of the infringer. If the infringement was not deliberate, then the penalty is very lenient, usually a reasonable royalty that might be based on what the infringer would have paid you had they licensed the invention in the first place, or the amount that you lost because of the infringement.[56] Moreover, if the infringement is unintentional, the courts may require you to license the technology to the infringer in return for royalties.

It is a very different story if the infringement is willful. Willful infringement occurs if the infringer deliberately copied your patented idea, tried to conceal its effort, or acted in bad faith.[57] Then the courts impose triple damages.[58] Therefore, the compensation that you can be awarded if you win a patent litigation lawsuit can be extremely large. For example, the University of California and Eolas Technologies Inc. were awarded $565 million from Microsoft for the infringement of a software patent.[59]

The largest patent infringement penalty ever awarded went to Polaroid. In that case, Kodak had to pay $990.5 million for infringement of Polaroid's instant camera technology.[60] Moreover, Kodak had to close down a $1.5 billion manufacturing operation, lay off 700 workers, and spend $500 million to buy back cameras it had sold using the infringing technology.[61]

While the damages that you receive from winning a patent infringement lawsuit can be large, so can the costs of enforcing a patent. For example, Jet Dock Systems Inc., a company that has invented a floating dry dock, has had to engage in six lawsuits to protect its patents against infringement since its founding in 1993. The cost of enforcing the company's patents—$1.2 million since founding—is large for a company that only does approximately $15 million in sales annually.[62] This makes patent litigation a major strategic issue if you run a new and small company.

It also takes a lot of time to enforce a patent. For example, it took Ron Chasteen, the inventor of a patented snowmobile fuel injection system, 11 years to win his patent infringement lawsuit against Polaris Industries.[63]

Because new and small companies must devote a large portion of their revenues and management time to enforce their patents, large, established companies often test their willingness to do so by infringing the start-ups' patents and running the risk that they might have to pay damages. Many entrepreneurs simply run out of cash or energy before they can prevail in the several-year process of enforcing a patent and get nothing, or settle for much less than the triple damages that they would be due if they proved willful infringement.

Patent Trolls

Patent trolls are start-up companies whose business model is to buy up patents and seek royalties through licensing. Unlike large, established companies, like Lucent, IBM, and Texas Instruments, that make hundreds of millions of dollars a year on licensing, patent trolls do not produce products.

The main strategy for patent trolls is litigation. Because patents give inventors the right to sue others for making or using an invention without permission, entrepreneurs can assemble a portfolio of patents and use the threat of litigation to collect royalties from potential infringers. For example, the former chief technology officer at Microsoft, Nathan Myhrvold, has created a company that has purchased several thousand software patents and uses the potential of litigation to motivate infringers to license them.[64]

Many patent trolls buy patents cheap from the creditors of bankrupt technology companies. With those patents in hand, they seek licenses from users of the patented technology, generally asking for a small royalty from companies generating high revenues from the technology. For instance, they might seek a 1 percent royalty from a company whose product is generating $5 billion annually. Even though the trolls' odds of winning patent infringement lawsuits are small, the potential payoff is high enough to make significant profits off of their investments in the patents even if they win only occasionally.

Often companies decide that it is better to settle with patent trolls than fight them in court. Patent trolls have a powerful strategic tool—the injunction. If they can convince a court to stop a potential infringer from producing its products or services until after an infringement lawsuit is decided, then the alleged infringer will often settle and pay a royalty to the patent troll, rather than shutter its business for months, or even years. For instance, NTP won a $612.5 million settlement from RIM, the maker of the BlackBerry, because NTP got a federal judge to issue an injunction against RIM that would have caused them to shut down BlackBerry service while the case was decided.[65]

Key Points

- Effective patenting strategy often involves the creation of a picket fence of patents around a core invention and bracketing competitors' patents.
- Patents provide the right to sue others if they infringe on your patent by making, using, selling, or importing something covered by the claims of your patent.
- If a company wins a patent infringement lawsuit, it can obtain as much as triple damages and an injunction prohibiting the infringing activity.
- Triple damages are awarded for willful infringement, while lost profits or imputed royalties are common penalties when infringement is accidental.
- Because patents that are invalid cannot be infringed, a common defense against infringement is to invalidate the patent by demonstrating that the inventor had disclosed the invention prior to filing for the patent, or that the invention is obvious to a person trained in the art.
- Because small and new firms must spend a large portion of their revenues and senior management time to enforce their patent rights, large, established firms sometimes willfully infringe their patents, believing that the small, new firms will not have the money or energy to fight back.
- Patent trolls are companies whose purpose is to buy up patents and enforce their claims through litigation or threat of litigation.

SHOULD YOU PATENT?

As you might suspect, there are advantages and disadvantages to patenting. Given these pros and cons, you need to decide whether or not to patent your inventions. To help you make informed decisions, this section discusses some of the important advantages and disadvantages of patenting.

Advantages of Patenting

Companies patent for a variety of different reasons. Table 8.1 presents data from a survey conducted by faculty members at Carnegie Mellon University, which indicates the major reasons why U.S. manufacturing firms say that they patent. As the figure shows, the most common reason, given by 96 percent of respondents, is to prevent copying.

Barrier to Imitation

Under certain circumstances, patents can be an important barrier to imitation and a powerful mechanism to capture the returns to innovation.[66] For example, the mobile telephone company, Qualcomm, was founded to exploit a technology called Code Division Multiple Access (CDMA) that allows more efficient use of the radio spectrum for cellular telephones. Qualcomm obtained a patent on this technology and used the patent to protect its products as well as to earn revenues by licensing the technology to other firms. Qualcomm has grown into a $3 billion company, successfully competing against large, established firms, such as Motorola—something it would have been unable to do if CDMA had not been patented.[67]

Legal Protection

Patents also help companies to use the legal system to protect their intellectual property. As you will see in the next chapter, it is much easier to use the legal system to enforce patents than to enforce trade secrets. Moreover, having a patent helps you to defend your firm against patent litigation by allowing you to counterclaim in an infringement lawsuit.[68]

TABLE 8.1 Why Companies Say That They Patent

The table shows the proportion of R&D managers surveyed about their company's approach to patenting who responded that their company patented for each of six reasons.

Reason	Percent of Companies Giving the Reason
Prevent copying	96
Block others	82
Prevent lawsuits	59
Use in negotiation	47
Enhance reputation	48
Licensing	28

Source: Adapted from Cohen, W., R. Nelson, and J. Walsh. 2000. Protecting their intellectual assets: Appropriability conditions and why U.S. manufacturing firms patent (or not), *NBER Working Paper,* No. W7552.

Because patent litigation often results in court-mandated licensing or licensing from the settlement of lawsuits, the use of patents to protect intellectual property also has the benefit of creating a new source of revenue for many companies. In some cases, this source of revenue can be quite large. For example, much of IBM's $1.4 billion annual royalty flow is the result of patent litigation.[69]

Value Chain Leverage

Patents also give companies control over other firms in their value chain. By owning patents that are used by your customers or suppliers, you can influence their behavior and make them act more favorably toward you. For example, Nokia has obtained patents on cell phone speakers even though it does not produce phone components because the patents give the company leverage over its supplier of speakers.[70]

Markets for Knowledge

Having a patent facilitates the sale of technology to other firms. While patented technologies can be sold to others, technologies protected by secrecy cannot. Therefore, to license a technology to another company, you need to obtain a patent on it. Take the example of a podiatrist in Bedford, New Hampshire, named Howard Dananberg, who invented a "comfortable high-heel shoe" (U.S. patent number 5,782,015). Because Dr. Dananberg lacked the assets necessary to manufacture and sell shoes that were attractive to women, he sought to license his technology to companies that already produced shoes. To do this, he needed to turn over his shoe designs to shoemakers, which necessitated obtaining a patent. Otherwise, if he had tried to license his invention, he would have had no protection against an unscrupulous licensee who chose to steal his designs.[71]

Raising Funds

Patents help new companies raise money because they provide a verifiable source of competitive advantage. Investors can see the mechanism through which the new venture will deter imitation, reducing their uncertainty about the value of the venture. Moreover, patents offer salable assets if a new venture fails. Because investors can "take the patents to the bank" at the end of the day, they see their investments in start-ups with strong patents as partially collateralized, increasing the amount of money that they are willing to provide to them.[72]

Disadvantages of Patenting

While patents are valuable for many of the reasons just described, they also have several disadvantages.

Effectiveness at Deterring Imitation

Patents are not always effective at deterring imitation. Sometimes other firms can invent around your patents. (**Inventing around** is the process of coming up with something that accomplishes the same goal as the patented invention without violating the claims of the patent.) By inventing around your patents, other companies can use your invention without having to pay you royalties and without having to incur the high costs of developing the invention.

If your competitors can invent around your patents, then those patents aren't worth the cost to get. Moreover, patenting might actually be doing you more harm than good. By patenting, you have to disclose the specifications of your invention, which might be the source of information that your competitors need to copy your product.

Inventing around occurs when there are multiple ways of accomplishing the same goal. For example, an imitator can often invent around an electronic device patent by changing the design of the circuitry, allowing the imitative product to satisfy customers in the same way as the innovative one but without violating the claims on the patented invention.

Thus only for inventions for which there is a single way to accomplish a particular goal are patents very effective at deterring imitation. For example, the patent on Symantec Corporation's antivirus software, which finds computer viruses without searching every byte of data, is effective because software engineers do not believe that there is any way other than Symantec's to find computer viruses without checking all of the data in a file.[73]

Benefits of Nondisclosure

Patenting is disadvantageous when a company will gain more from nondisclosure than from a government-granted monopoly. A patent gives you a 20-year monopoly on your invention, but secrecy might allow that monopoly to last longer. For example, the chemical formula for Coca-Cola has been maintained as a secret for over 100 years. (Because the beverage is composed of complex natural substances, it is not possible to reverse engineer and duplicate it.[74]) As a result, no other companies have been able to create beverages that taste exactly like Coke. If the formula for Coca-Cola had been patented, its chemical composition would have been disclosed in the specifications of the patent. Once the monopoly on it expired decades ago, competitors would have been able to produce soft drinks with exactly the same chemical composition (and taste!) as Coke.

Pace of Change

Patents are not very helpful when the pace of technological change is very fast. When technological change is very rapid, the inventions that patents protect quickly become irrelevant. Given the time it takes to obtain patents, and the cost of patenting, you probably won't be able to earn sufficient payback to justify the investment in such patents.[75] For example, suppose your semiconductor will be obsolete in two years because of the pace of innovation in that industry. You might be better off just keeping it secret and using that proprietary knowledge to ensure that you, and not your competitors, will be able to develop the next generation of semiconductors. If you patent your invention, the semiconductor will be obsolete by the time that the patent issues. But once it issues, the invention will be public knowledge, and that might help your competitors to develop the next generation of semiconductors.

Moreover, to obtain a patent, you need to explain how the invention works in a way that enables a person trained in the art to make the invention. Sometimes, this information makes it possible for others to leapfrog your invention. If that is the case, you are better off keeping the invention a secret.

Difficulty Proving Infringement

Patents are not very useful when proving that others have infringed your patent, or defending it in a lawsuit, are very costly or difficult. If you cannot amass the evidence that it takes to prove that infringement actually occurred, or the fixed costs of defending a patent are so high that it does not pay to protect it through the court system, then

TABLE 8.2 The Advantages and Disadvantages of Patenting

Patents have many advantages and disadvantages that need to be considered when deciding whether or not to patent an invention.

Pros	Cons
Helps to raise money	Not always effective at deterring imitation
Often slows imitation	Disclosure not always worth the 20-year monopoly
Facilitates markets for knowledge	May cause competitor leapfrogging when the pace of technological change is rapid
Offers legal protection of intellectual property	Sometimes cannot be enforced
Creates value chain leverage	

you will not get enough of a return on your investment in a patent to justify its cost.[76] For example, it may be very difficult to prove that someone infringed on a surgical method that you invented, such as a type of incision used in cataract surgery, because there are no sales of physical objects, like drugs or devices, which would show that your method was used.[77] As a result, you might not want to waste your money trying to patent the surgical method.

Effectiveness of Patents in Different Industries

Patents are not equally effective in all industries. In general, they tend to be more effective in industries in which the core technology is biological or chemical, and less effective in industries in which the core technology is mechanical or electrical.[78] Why? The reason has to do with the difficulty of accomplishing the same goals through different technical means. For mechanical or electrical devices, you can make slight modification to the design and accomplish the same goal, but you cannot do this with things that are biological or chemical. For instance, a drug has a very precise molecular structure, and slight alterations will often transform the drug from something beneficial to something harmful, while relatively major changes to the structure of electronic circuitry often do not alter an electronic device's effectiveness.

Researchers have examined how effective patents are in different industries. Table 8.3 provides some data, adapted from the Yale survey on innovation, on this question. The table shows that in industries like drugs and chemicals, patents are very effective at protecting new products, but that in industries like cosmetics or pulp and paper, they are not.[79]

These differences in patent effectiveness explain why obtaining patents is crucial to generating high financial returns in industries in which patents are very effective, like pharmaceuticals.[80] It also explains why start-ups in some industries, like biotechnology, often specialize in technology development and do not build assets across the different parts of the value chain, relying, instead, on licensing to capture value from innovation.

Key Points

- Patents slow imitation, facilitate legal protection of intellectual property, make markets for knowledge possible, enhance value chain leverage, and help new firms to raise money.

TABLE 8.3
Effectiveness of Product Patents by Industry
The table shows the scores of different industries on a questionnaire given to R&D managers about the effectiveness of product patents in their industry.

INDUSTRY	PATENT EFFECTIVENESS (7-POINT SCALE— 7 EQUALS "VERY EFFECTIVE")
Drugs	6.5
Organic chemicals	6.1
Inorganic chemicals	5.2
Steel mill products	5.1
Plastic products	4.9
Medical devices	4.7
Motor vehicle parts	4.5
Semiconductors	4.5
Pumps and pumping equipment	4.4
Cosmetics	4.1
Measuring devices	3.9
Aircraft and parts	3.8
Communications equipment	3.6
Motors, generators, and controls	3.5
Computers	3.4
Pulp, paper, and paperboard	3.3

Source: Adapted from Levin, R., A. Klevorick, R. Nelson, and S. Winter. 1987 Appropriating the Returns from Industrial Research and Development. *Brookings Papers on Economic Activity*, 3: 783–832.

- However, patents allow the disclosure of information that helps competitors to invent around a technology, sometimes protect an invention for less time than secrecy, permit technology leapfrogging when the pace of technological change is rapid, and are not worth their cost when infringement is difficult to prove.
- Patent effectiveness varies substantially across industries because of differences in the nature of technology, and these industry differences affect several aspects of technology strategy.

DISCUSSION QUESTIONS

1. What can and can't be patented? What keeps some inventions from being patented? How have the criteria for patentability changed over time?
2. Patents on genes differ from patents on other things in two important ways. First, if you patent a gene, there is no way for rivals to make something better, which is not the case on other technologies (e.g., semiconductors). Second, researchers cannot identify a gene that causes a disease and develop a genetic test for it without getting DNA samples from people who have the gene, while researchers can patent other technologies without obtaining anything from other people. For these two reasons, many people think that we shouldn't be able to patent genes. Do you agree or disagree? Why?
3. What are the different parts of a patent? What purpose do they serve?
4. What strategic actions can you take to increase the effectiveness of your patents at deterring imitation? Why are these actions effective?
5. What actions should you take if you believe that your patent has been infringed? Should you engage in litigation to protect your patents? Why or why not?
6. What are the advantages and disadvantages of patents? What do these advantages and disadvantages suggest about the use of patents as part of your technology strategy?
7. Why do patents work better in some industries than in others?

Key Terms

Bracketing: The process of keeping an inventor from using his or her invention by patenting around it.

Claim: The part of a patent that states what the patent precludes others from imitating.

Infringement: The action to make, use, sell, or import something covered by the claims of a patent, without permission of the patent owner.

Invent Around: To come up with a solution that does not violate the claims of a patent but accomplishes the same goal as the patented approach.

Patent: An exclusive right given by the government to preclude others from duplicating an invention for a specified period of time in return for disclosure of an invention.

Patent Trolls: Start-up companies whose business model is to buy up patents and seek royalties through licensing.

Picket Fence: The set of patents that are obtained to offer a wall of protection around a core invention.

Pioneering Patents: The basic patents in a technical area on which a wide variety of patents build.

Specification: The part of a patent that describes how an invention works.

Putting Ideas into Practice

1. **Evaluating Patents** Go to www.uspto.gov. Do a patent search and look up the following patents: number 5,173,051—"Curriculum Planning and Publishing Method," number 5,241,671—"Multimedia Search System," number 6,080,436—"Bread Refreshing Method," number 5,443,036—"Method of Exercising a Cat," number 6,368,227—"Method of Swinging on a Swing," number 4,597,046—"Securities Brokerage Cash Management System Obviating Float Costs by Anticipatory Liquidation of Short Term Assets," and number 6,467,180—"Use of a Tape Measure to Determine the Appropriate Bra Size." Should the USPTO have granted these patents? Why or why not? (In answering these questions, think about whether the inventions are obvious, valuable, and useful, and whether they infringe on prior patents.)
2. **Basic Patent Searches** This exercise is designed to help you evaluate the potential to patent a piece of intellectual property. Please follow the steps below to figure out if you can obtain a patent.

   Step 1: Identify an invention. (It can be your own or one that someone else developed. If you can't come up with one on your own, ask your instructor to suggest one for you to investigate.) In two paragraphs, explain why you think that the invention is patentable. Make sure that you explain why the invention is novel, nonobvious, and useful.

   Step 2: List several key words that identify the invention. Now go to the U.S. government's patent Web site at www.uspto.gov and search for patents with the same key words as your invention. List all of the patents that have the same key words. Read the claims on those patents. What are other inventors claiming as their inventions? What are they not claiming? Then explain what claims you should be able to get, given what has already been claimed.

   Step 3: Evaluate the claims that you think you can get. What will they protect? What inventions will others be able to come up with that will get around your patent claims? Evaluate the nature of your patent claims. Are they broad or narrow? Are they enough to protect your product or service against imitation? Why or why not?
3. **Looking at Patent Portfolios** Go to www.chresearch.com for patent-based indicators of 1,000 leading firms. Choose two firms in the same industry. Compare their patent portfolios. How are they similar? How are they different? Looking at their patent portfolios, which company do you think has a better intellectual property position? Why?

Notes

1. Adapted from Jaffe, A., and J. Lerner. 2004. *Innovation and Its Discontents*. Princeton, NJ: Princeton University Press.
2. http://patft.uspto.gov/netacgi/nph-Parser?Sect1=PTO1&Sect2=HITOFF&d=PALL&p=1&u=/netahtml/srchnum.htm&r=1&f=G&l=50&s1=6004596.WKU.&OS=PN/6004596&RS=PN/6004596
3. Ibid.
4. McGavock, D. 2002. Intangible assets: A ticking time bomb. *Chief Executive* http://findarticles.com/p/articles/mi_m4070/is_2002_Nov/ai_94145235

5. Levin, R., A. Klevorick, R. Nelson, and S. Winter. 1987. Appropriating the returns from industrial research and development. *Brookings Papers on Economic Activity*, 3: 783–832.
6. Mansfield, E. 1985. How rapidly does industrial technology leak out? *Journal of Industrial Economics*, 34(2): 217–223.
7. Levin, Klevorick, Nelson, and Winter, Appropriating the returns from industrial research and development.
8. Ibid.
9. Ibid.
10. Jaffe and Lerner, *Innovation and Its Discontents*.
11. Lemley, M., and C. Shapiro. 2005. Probabilistic patents. *Journal of Economic Perspectives*, 19(2): 75–98.
12. Winter, S. 2000. Appropriating the gains from innovation. In G. Day and P. Schoemaker (eds.), *Wharton on Managing Emerging Technologies*. New York: John Wiley.
13. U.S. Department of Commerce. 1992. *General Information Concerning Patents*. Washington, DC: U.S. Government Printing Office.
14. Kesan, J. 2000. Intellectual property protection and agricultural biotechnology. *American Behavioral Scientist*, 44(3): 464–503.
15. Etherton, S. 2002. *Let's Talk Patents*. Tempe, AZ: Rocket Science Press.
16. Yoffie, D. 2003. Intellectual property and strategy, *Harvard Business School Note*, Number 9-704-493.
17. Schilling, M. 2005. *Strategic Management of Technological Innovation*. New York: McGraw-Hill.
18. Yoffie, Intellectual property and strategy.
19. Munoz, S. 2005. Patent No. 6,004,596: Peanut butter and jelly sandwich. *Wall Street Journal*, April 5: B1, B9.
20. Yoffie, Intellectual property and strategy.
21. Jaffe, A., and J. Lerner. 2004. *Innovation and Our Discontents*. Princeton, NJ: Princeton University Press.
22. Sandoval, G. 2005. Apple-Microsoft duke out iPod fight in patent office. *The Plain Dealer*, August 17: C2.
23. Silverman, A. 1999. The forfeiture of U.S. patent rights by placing an invention on sale. *JOM*, 51(2): 64.
24. U.S. Department of Commerce, *General Information Concerning Patents*.
25. Etherton, *Let's Talk Patents*.
26. http://patft.uspto.gov/netacgi/nph-Parser?Sect1=PTO1&Sect2=HITOFF&d=PALL&p=1&u=%2Fnetahtml%2FPTO%2Fsrchnum.htm&r=1&f=G&l=50&s1=5,380,237.PN.&OS=PN/5,380,237&RS=PN/5,380,237
27. Ibid.
28. http://patft.uspto.gov/netacgi/nph-Parser?Sect2=PTO1&Sect2=HITOFF&p=1&u=%2Fnetahtml%2FPTO%2Fsearch-bool.html&r=1&f=G&l=50&d=PALL&RefSrch=yes&Query=PN%2F3841019
29. Jaffe and Lerner, *Innovation and Its Discontents*.
30. Kesan, Intellectual property protection and agricultural biotechnology.
31. Jaffe and Lerner, *Innovation and Its Discontents*.
32. Bercowitz, L. 2000. Patent law changes: What you should know. *Research Technology Management*, 43(2): 5.
33. Fuerst, O., and U. Geiger. 2003. *From Concept to Wall Street: A Complete Guide to Entrepreneurship and Venture Capital*. New York: Financial Times Prentice Hall.
34. U.S. Department of Commerce, *General Information Concerning Patents*.
35. Flandez, R. 2005. Get a patent. *Wall Street Journal*, May 9: R9, R11.
36. U.S. Department of Commerce, *General Information Concerning Patents*.
37. Etherton, *Let's Talk Patents*.
38. Montgomery, C. 2004. Drive-in dry-dock. *The Plain Dealer*, July 20: C1, C6.
39. Etherton, *Let's Talk Patents*.
40. Regalado, A. 2004. Nanotechnology patents surge as companies vie to stake claim. *Wall Street Journal*, June 18: A1, A2.
41. Vara, V. 2006. Friendster patent on linking web friends could hurt rivals. *Wall Street Journal*, July 27: B1, B4.
42. Flandez, Get a patent.
43. Regalado, Nanotechnology patents surge as companies vie to stake claim.
44. Etherton, *Let's Talk Patents*.
45. Allen, K. 2003. *Bringing New Technology to Market*. Upper Saddle River, NJ: Prentice Hall.
46. Rivette, K., and D. Kline. 2000. Discovering new value in intellectual property. *Harvard Business Review*, January–February: 2–10.
47. Kaminski, M. 2005. Effective management of US patent litigation. *Intellectual Property and Technology Law Journal*, 18(1): 13–25.
48. Etherton, *Let's Talk Patents*.
49. Ibid.
50. Allen, *Bringing New Technology to Market*.
51. Jaffe and Lerner, *Innovation and Our Discontents*.
52. Silverman, A. 2002. I'll see you in court—Overview of a patent infringement trial. *JOM*, 54(5): 64.
53. Etherton, *Let's Talk Patents*.
54. Gomes, L. 2005. Ebay wins fresh legal victory in challenge involving patents. *Wall Street Journal*, March 30: A6.
55. Greenhouse, L. 2007. Justices alter patent landscape. *The Plain Dealer*, January 10: C2.
56. Etherton, *Let's Talk Patents*.
57. Kaminski, Effective management of US patent litigation.

58. Jaffe and Lerner, *Innovation and Its Discontents.*
59. Heinzel, M. 2005. BlackBerry maker agrees to settle patent dispute. *Wall Street Journal*, March 17: B5.
60. Bulkeley, W. 2005. Patent ruling irks inventors, aids companies. *Wall Street Journal*, March 2: B1, B2.
61. Allen, *Bringing New Technology to Market.*
62. Montgomery, Drive-in dry-dock.
63. Paris, E. 1999. David v. Goliath. *Entrepreneur*, November.
64. Varchaver, N. 2006. Who's afraid of Nathan Myhrvold? *Fortune*, July 10, http://money.cnn.com/magazines/fortune/fortune_archive/2006/07/10/8380798/index.htm.
65. Levy, S. 2006. The BlackBerry deal is patently absurd. *Newsweek*, March 13, http://www.msnbc.msn.com/id/11677343/site/newsweek.
66. Jaffe and Lerner, *Innovation and Its Discontents.*
67. Ibid.
68. Etherton, *Let's Talk Patents.*
69. Chesbrough, H. 2001. The patent and licensing exchange: Enabling a global IP marketplace, *Harvard Business School Teaching Note*, Number 5-601-124.
70. Reitzig, M. 2004. Strategic management of intellectual property. *Sloan Management Review*, 45(3): 35–40.
71. Jaffe and Lerner, *Innovation and Its Discontents.*
72. Etherton, *Let's Talk Patents.*
73. Richmond, R. 2005. Symantec patent may disturb rivals. *Wall Street Journal*, May 4: B3a.
74. Jaffe and Lerner, *Innovation and Its Discontents.*
75. Etherton, *Let's Talk Patents.*
76. Levin, Klevorick, Nelson, and Winter, Appropriating the returns from industrial research and development.
77. Miller, S. 1996. Should patenting of surgical procedures and other medical techniques be banned? *IDEA: The Journal of Law and Technology*, 255–273.
78. Levin, Klevorick, Nelson, and Winter, Appropriating the returns from industrial research and development.
79. Ibid.
80. Mullins, J. 2003. *The New Business Road Test*. London: Financial Times Prentice Hall.

Chapter 9

Trade Secrets, Trademarks, and Copyrights

Learning Objectives

After reading this chapter, you should be able to:

1. Identify the role that secrecy plays in protecting intellectual property.
2. Explain when secrecy is an effective mechanism for deterring imitation.
3. Define a trade secret and explain what characteristics are necessary for something to be a trade secret.
4. Explain why nondisclosure agreements are an important part of efforts to maintain trade secrecy.

5. Define a copyright, and explain how intellectual property can be protected by copyright.
6. Describe how a copyright is obtained.
7. Define a trademark, and explain why and how trademarks are beneficial to companies.
8. Describe how a trademark is obtained.
9. Explain the major differences in intellectual property across countries, and their effect on technology strategy.

Software Copyrights: A Vignette[1]

IBM learned a lesson about managing intellectual property the hard way. In 1980, when personal computer sales had reached $1 billion, IBM began to develop its first personal computer. The decision to develop a personal computer was an important and strategic one for the firm's management, leading to much discussion of the best way to introduce the product. IBM's senior management quickly concluded that time to market was important. So, to launch its product rapidly, IBM used off-the-shelf components, including Microsoft's operating software, and Intel's 8088 microchip.

IBM's management knew that the use of off-the-shelf components came at a risk. If the company used components available to others, then IBM's computer would be easy to copy. But IBM's management had a plan. The company would use a copyright to protect the proprietary code that its engineers were writing for the basic input-output commands that linked the computer hardware to the software.

Unfortunately for IBM, this method of protecting the company's intellectual property proved to be inadequate. A competitor then entering the personal computer business by the name of Compaq had its software programmers reverse engineer IBM's input-output code. In a few months, Compaq was able to develop comparable code without infringing IBM's copyright.

How did Compaq do this? The answer lies in the type of intellectual property protection that copyrights provide. Like most software copyrights, the copyright on IBM's input-output commands protected the code itself, but not the functions produced by the code. Consequently, as long as Compaq's programmers used different software code to create the same functions as IBM's code provided, they could copy the functionality of IBM's input-output commands without violating IBM's copyright.

Compaq's management thought up a really ingenious way to do this. First, it had a team of programmers reverse engineer the IBM software to figure out the functions that the code produced. Then the identified functions were given to a set of "virgin" programmers—programmers who had never seen IBM's code. These programmers were then told to write code for these functions. Because these programmers had never seen the IBM code, the code that they wrote was different from IBM's. The end result was input-output software that provided identical functions to IBM's but did not violate its copyright. This bit of ingenuity allowed Compaq to sell 47,000 IBM-compatible personal computers in its first year, and set the company well on its way to becoming one of the most successful companies in the personal computer business. It also taught IBM's management an important lesson about protecting intellectual property that you have now learned as well.

Introduction

While you might think that the previous chapter explained intellectual property management, protecting your company's intellectual property (IP) is more complicated than just learning how to patent your inventions. Trade secrets, trademarks, and copyrights are also important tools to protect intellectual property. For instance, Coca-Cola has made billions off of a beverage formula that is protected as a trade secret, Nike has turned its "swish" into a $7 billion trademark, and Disney derives much of its profits from animated characters and stories that it protects with copyrights.[2]

This chapter discusses trade secrets, trademarks, and copyrights. The first section explains how you can deter imitation by keeping things secret and describes how companies use trade secrets as part of their technology strategies. The second section discusses copyrights and explains how you can use them to protect intellectual property and develop competitive advantage. The third section discusses trademarks and explains their role in technology strategy. The final section discusses the issues raised by differences in intellectual property protection across countries.

Secrecy

You can deter imitation by keeping things secret and reducing the leakage of information about your products or services or how you produce them (see Figure 9.1). For example, suppose that you have discovered a chemical that makes an excellent fertilizer. If you run a fertilizer company, you might not want other people to know that you have identified this chemical. If your competitors and potential competitors do not know that the key to your fertilizer lies in the use of a particular chemical, then

FIGURE 9.1
Protecting IP by Keeping Things Secret

"These dreams of yours wherein you find great tubs of money, Mr. Croy—can you describe the spot a little more exactly?"

You will run into problems if you talk too much about your intellectual property.

Source: New Yorker Magazine. 1975. *The New York Album of Drawings, 1925–1975.* New York: Penguin Books.

they will not understand that they need to gain access to that chemical to compete with you successfully. Therefore, they will not seek to obtain access to that resource, and they will not be able to imitate your operations successfully.

When Does Secrecy Work?

Efforts to mitigate imitation by keeping information about a new product or service secret work best under certain conditions. First, they work better when there are few sources of the information about the new product or service. To imitate your product or service, a competitor needs access to the information that makes copying the innovation possible. While your competitors can obtain this information from you, they can also get it from third parties. Your efforts to keep things secret are not going to be very effective if third parties readily provide this information to your competitors. Therefore, if only you know the information necessary to imitate your product or service, then your product or service is less likely to be copied.

This is why it is easier for Coca-Cola to keep other companies from copying its soft drink formula than it is for your local dry cleaner to keep its dry cleaning formula secret. Even if your local dry cleaner never told anyone the formula for its dry cleaning solution, you could obtain it from any of thousands of other dry cleaners. However, if the few executives at Coca-Cola who know the formula to classic Coke do not tell you what it is, you are going to have no way of knowing it.

Second, secrecy is more effective when a new product or service is complex. Imitation involves understanding how to copy a new product or service, not just having access to formulas or blueprints. The more complex a product or service is, the harder it is for people to figure out how to duplicate it. Complexity affects people's understanding of the order in which tasks need to be undertaken and the difficulty of choreographing the joint efforts of different people. Take, for example, the difficulty of assembling a child's toy. Even if you have the instructions, it is much harder to make the product just as the manufacturer had intended when the product is made up of hundreds of pieces than when it is made up of only a couple of pieces.[3]

Third, secrecy is more effective when the process of creating a new product or service is poorly understood. To imitate your activities, people have to understand what you are doing. The fewer competitors that can actually understand what you are doing, the fewer that will be capable of imitating your products, and the less imitation there will be. For example, suppose that you developed a new method for keeping storm drains clean by flushing them with a chemical mixture at certain temperatures. If the process of creating this new chemical solution was poorly understood (say that very precise amounts of the chemicals have to be combined at exactly the right moments under the right temperature for unknown reasons), then few people would be able to imitate this product, and your company would capture the profits from providing it.

Fourth, secrecy works best when the information that is being kept secret involves **tacit knowledge**—knowledge about how to do something that is not documented in written form. For instance, a plant manager's knowledge of how to keep an assembly line running at high speed through a sense of where to position different workers with different skills and a salesperson's knowledge of how to close sales by timing the introduction of personal comments into a discussion are both examples of tacit knowledge.

It is easier to imitate a well-codified process than a tacitly understood one because imitation of a codified process only requires access to the document outlining the process, whereas imitation of a tacitly understood process requires the imitator to gain access to the person who holds that information in his or her head. Most of the time, it is easier to gain control of a document about a process than to gain control of a person who knows about it.[4] Take, for example, the case of expertise in boiler repair. If that knowledge is held in documentary form by a company in Michigan, then a company in Ohio could get control of that information and move it to Ohio more easily than it could if the knowledge was tacit and held in the minds of the Michigan firm's employees. To copy the tacit knowledge, competitors would need to hire the employees of the Michigan firm and get them to move to Ohio.

Moreover, when knowledge is tacit, its transfer must take place through face-to-face meetings between people. In contrast, when things are codified, knowledge can be transferred by handing a blueprint or a formula to others.[5] Because knowledge spreads much faster if the transfer is not limited to direct contact between people, codified knowledge tends to spread very quickly, and is harder to keep secret than information that is not written down.

Fifth, secrecy works better when there are limited numbers of people capable of understanding the information that is being kept secret. The fewer people who have the skills and abilities to use the information that creates your new product's value, the fewer people that can figure out how to imitate what you are doing, even if the knowledge that you are keeping secret leaks out. Researchers Lynne Zucker and Michael Darby at the UCLA business school have shown this to be true for new biotechnology companies. They learned that the new biotechnology firms founded to exploit the technical expertise of leading scientists often were successful because competition was limited to the handful of people who also had the skills to exploit the cutting-edge scientific techniques that they used.[6]

Sixth, secrecy works better for processes, inputs, and materials than for products. Why? You sell your product in the marketplace. That makes the product itself **observable-in-use**. (In fact, the more observable-in-use a product is, the less it can be kept secret. This is why it is hard to keep secret processes like techniques for providing customer service.[7]) Moreover, competitors can buy your product and reverse engineer it to figure out how it works.[8] These things make it harder for you to keep the composition of your product secret than it is for you to keep secret the production processes used to make the product. Therefore, processes make better secrets than products.

Trade Secrecy

Trade secrecy is a special case of all efforts to keep a new product or service secret. In the United States, trade secrets are governed by state law,[9] primarily the Uniform Trade Secrets Act, which is in force in 44 states.[10] This Act defines a trade secret as "information including a formula, pattern, compilation, program, device, method, technique, or process that derives independent economic value, actual or potential, from not being generally known, and not being readily ascertainable by proper means by, other persons. . ." Examples of trade secrets include: chemical processes, customer databases, food recipes, computer source code, manufacturing processes, architectural designs, vendor lists, marketing plans, sources of raw materials, design manuals, pricing policies, and blueprints.[11] For instance, one of the most valuable trade secrets today is Google's Web page ranking algorithm, which makes its search engine better than others.

Trade secrecy laws provide for legal remedies if someone benefits from your trade secret without your consent. If you believe that someone else has improperly obtained your trade secret, you can sue to collect damages for your loss and obtain an injunction to stop further use of the secret. These remedies are available to you regardless of whether the party disclosing the trade secret was bound by a duty of confidentiality, had signed a nondisclosure agreement, obtained the information illegally, obtained the information from someone who did not have authorization to disclose it, or learned the information by accident, but knew it was a trade secret.[12]

Conditions to Have a Trade Secret

Three conditions must be met for the courts to hold that something is a trade secret. First, the information must be known only by people in your company. Information that is known generally in an industry, such as standard manufacturing processes, or information that can be generated from data that are known in an industry, cannot be a trade secret.[13] Moreover, you cannot claim that the general skills that your employees learn on the job are trade secrets because that would preclude them from being able to take new jobs and use the skills that they learned working for you at their new employers.[14]

Second, the information must have economic value. For something to be a trade secret, it must generate a competitive advantage that would be lost if your competitors made use of it. This means that you must be able to document that what you term a trade secret is important to how your company derives value, and provides an advantage over your competition in the market place. You should note that this standard is stricter than for a patent, where all you have to do is prove infringement to collect damages.

Third, you must take reasonable measures to keep the information secret. This means that you have to adopt "secrecy policies" to ensure that people do not accidentally access the secret information. Your employees need to know what information is secret and that secret information is limited to only those personnel who need it. Moreover, those personnel who need access to the information must agree, in writing, to keep it confidential. Furthermore, you need to use physical mechanisms, such as limiting the access of nonemployees to your facilities, locking files, requiring computer passwords and so on, to keep the information from getting out.[15]

Take, for example, the efforts by KFC to keep the recipe for its fried chicken a trade secret. The recipe is kept in a vault at the company's headquarters, and only a few people know what it is. Those employees who know the formula are required by the terms of their employment to keep the recipe secret. Moreover, two different companies supply the herbs and spices to KFC, but each one is allowed to create only part of the ingredients, and neither company is known to the other.[16]

Secrecy as a Strategy

You might choose secrecy as your basic approach to protecting intellectual property. This choice may stem from a preference for trade secrets over patents as the basis for competitive advantage (the two sources of intellectual property protection are mutually exclusive, necessitating a choice), perhaps because trade secrecy offers a longer time horizon of protection or because it does not disclose information to your competitors. Or it may occur because you have a product for which secrecy is particularly effective: It is created through a process that is poorly understood, complex, and based on tacit knowledge for which there are few sources of information and a limited number of people who can comprehend it. You might even focus on secrecy to generate customer interest in your products and services because people are often more interested in things that they can't know about than things that they can.

Apple Computer is an example of a high-technology company that focuses very much on secrecy. (A former CEO, John Scully, was fond of using the phrase, "loose lips sink ships.") The company rarely discloses its plans for new products and compartmentalizes development efforts so that employees working on new products rarely have information about the entire product. The company vigorously maintains efforts to limit disclosure, suing employees that leak information about forthcoming products and Web sites that publish such information. It creates lists of employees that have been given access to information about new product plans, even watermarking documents with the recipient's name, and using different code numbers for different departments to better track the source of any leak. Access to buildings in Apple's headquarters is even limited to the part of the complex in which employees work.[17]

While secrecy-focused strategies, such as Apple's, have many advantages, these benefits come at a cost. As was mentioned previously, maintaining trade secrets requires the adoption of secrecy policies and reduces the level of informal exchange of information among your employees, which hinders your ability to develop new products and processes. Maintaining trade secrets also inhibits your efforts to work with other companies, which, by necessity, lack adequate information to serve as effective partners. Moreover, it hinders efforts to sell products to many business customers, who need to know about new products long in advance of their release to fit them into their own plans. Finally, maintaining trade secrets risks the independent discovery and exploitation of your inventions. Competitors who independently and legally obtain technology that you maintain as a trade secret—e.g., by reading your publications, talking to your suppliers or customers, and reverse engineering your products—are free to use it to make and sell exactly the same products as you, even though they would be barred from doing so if you patented the technology. (Table 9.1 describes the conditions when protecting intellectual property with trade secrets is better than protecting it with patents and vice versa.)

TABLE 9.1 Trade Secrets Versus Patents

This table shows when it is better to protect an invention with trade secrecy and when it is better to patent an invention.

Trade Secrets are Better When . . .	Patents are Better When . . .
The secret cannot be patented	Reverse engineering is possible
The product cycle is short	The market life of a product is close to 20 years
The patent would be hard to enforce	The patent is easy to enforce
The patent would be narrow	The patent would be broad
It is hard to identify a trade secret	It is easy to identify a trade secret

Source: Adapted from Mohr, J., S. Sengupta, and S. Slater. 2005. *Marketing of High-Technology Products and Innovations* (2nd edition). Upper Saddle River, NJ: Prentice Hall.

Nondisclosure Agreements

As Figure 9.2 shows, trade secrecy is enhanced by having people sign **nondisclosure agreements** that are crafted by lawyers who know the details of employment law. These agreements are important; you cannot make a case that you are keeping information secret unless your employees understand that they are expected to refrain from disclosing information.

Effective nondisclosure agreements must meet certain conditions. The agreements must specify exactly what information is to be kept secret, and cannot state that all information that employees learn during their employment is confidential. Moreover, the agreement must provide **consideration**. That is, employees must receive something of value, like their salaries, in return for nondisclosure. Furthermore, the agreement must specify legitimate uses for the information, including identifying those people to whom the information can be disclosed, and how the information may be used to perform a job. Lastly, the agreement must state what must be done with any documents or materials that are transferred to the employee, both during employment and after the termination of an employment relationship.[18]

Enforcing Nondisclosure Agreements

To enforce nondisclosure agreements, you need to be willing to sue your employees and others who help them because the only remedies for violation of nondisclosure agreements come through legal action. Many companies do this. For example, Biomec Inc, a Cleveland, Ohio, medical device company, sued a former employee claiming that he violated his confidentiality agreement when he moved to rival, Cleveland Medical Devices; and Wal-Mart sued Drugstore.com and the venture capital firm Kleiner Perkins when Drugstore.com hired former Wal-Mart employees who had developed that company's system for Internet retailing.[19]

FIGURE 9.2
Nondisclosure Agreements Help to Protect IP

Companies often protect their intellectual property by having people sign nondisclosure agreements that preclude them from letting anyone else know the protected information.
Source: United Features Syndicate Inc., March 16, 1996.

While the easiest case to make for violation of a nondisclosure agreement occurs when your employees take documents that belong to your company, you can make a case that they violated their nondisclosure agreements if they take only uncodified knowledge. For instance, IBM recently settled a lawsuit with Compuware Corp. in which Compuware alleged that IBM had violated Compuware's trade secrets for file management and error detection software by hiring former Compuware employees to speed the development of software for its mainframe computers. In this case, Compuware claimed that its former employees had signed confidentiality agreements and then disclosed technical knowledge and knowledge of customer preferences to IBM.[20]

Noncompete Agreements

Trade secrecy is enhanced by having your employees sign **noncompete agreements**, which bar them from working for competitors for a period of time after their employment has ended, because these agreements keep employees from moving to rivals while their company-specific knowledge still has value (see Figure 9.3 for an example of a nondisclosure and noncompete agreement). For example, Microsoft successfully forced a start-up company named CrossGain to lay off 20 former Microsoft employees until the expiration of their noncompete agreements, as a way to protect its intellectual property.[21]

Enforcing Noncompete Agreements

As with nondisclosure agreements, you need to be prepared to go to court to enforce your noncompete agreements. For example, Patio Enclosures Inc. had to take Four Seasons Solar Products to court for hiring a former Patio Enclosures employee who had signed a noncompete agreement that barred him from employment at a competing firm for two years.[22]

While noncompete agreements help you to protect your company's intellectual property, they are hard to enforce. These agreements need to be of limited length and limited geographic breadth because they will be declared invalid if they keep people from earning a living in their chosen field.[23] (For example, ExxonMobil's noncompete agreement cannot preclude a petroleum engineer from working at another oil company after leaving ExxonMobil.) Moreover, in many states, you must give employees some benefit, like a bonus or a higher salary, in return for asking them to sign a noncompete agreement;[24] in other states, like California, you cannot enforce these agreements at all.[25]

Ownership of Intellectual Property

Related to the issue of nondisclosure and noncompete agreements is the issue of who owns the rights to technologies that employees develop during the period of their employment at a company. These rights reside with employees unless you require them to assign the rights to you. Of course, most large companies do just this, which keeps many people from quitting and starting new companies to exploit technologies that they developed while working elsewhere. For instance, Jeff Hawkins, the founder of Palm computing, patented an algorithm for pattern recognition software when he was on academic leave from GRiD systems, his employer. Although he owned the patent to the algorithm, and his licensing agreement with GRiD allowed him to use it in noncompeting products, he did not have

FIGURE 9.3
Nondisclosure/Noncompete Agreement

This agreement is made as of the 23rd day of June, 2002, by and between: **ACME Inc.** located in CITY, STATE and **JOHN INVENTOR** located in CITY, STATE.

This agreement shall govern the conditions of disclosure by **JOHN INVENTOR** to **ACME Inc.** of certain "Confidential Information" including but not limited to prototypes, drawings, data, trade secrets and intellectual property relating to the "*Patent Pending*" invention named "**Mouse Trap**" invented by **JOHN INVENTOR**.

With regard to the Confidential Information, **ACME Inc.** hereby agrees:
1. Not to use the information therein except for evaluating its interest in entering a business relationship with **JOHN INVENTOR**, based on the invention.
2. To safeguard the information against disclosure to others with the same degree of care as exercised with its own information of a similar nature.
3. Not to disclose the information to others, without the express written permission of **JOHN INVENTOR**, except that:
a. which **ACME Inc.** can demonstrate by written records was previously known;
b. which are now, or become in the future, public knowledge other than through acts or omissions of **ACME Inc.**;
c. which are lawfully obtained by **ACME Inc.** from sources independent of **JOHN INVENTOR**;
4. That **ACME Inc.** shall not directly or indirectly acquire any interest in, or design, create, manufacture, sell, or otherwise deal with any item or product, containing, based upon or derived from the information, except as may be expressly agreed to in writing by **JOHN INVENTOR**.
5. That the secrecy obligations of **ACME Inc.** with respect to the information shall continue for a period three years from the date hereof.

JOHN INVENTOR will be entitled to obtain an injunction to prevent threatened or continued violation of this Agreement, but failure to enforce the Agreement will not be deemed a waiver of this Agreement.

IN WITNESS WHEREOF the Parties have hereunto executed this Agreement as of the day and year first above written.

ACME Inc.

By:________________________ Date:____________________

Title:________________________

JOHN INVENTOR and SIGNATURE

This figure provides an example of a nondisclosure and noncompete agreement.
Source: Inventnet, http://www.inventnet.com/nondisclosure.html.

the rights to improvements to the C-language enhancements he had made while a GRiD employee. As a result, he needed to work around this intellectual property to develop the Palm personal digital assistant.[26]

Key Points

- Keeping things secret is an important way to protect your intellectual property.
- Secrecy is most effective when there are few sources of information about a product, when information about it is complex, when the process of creating the

GETTING DOWN TO BUSINESS
Using Nondisclosure and Noncompete Agreements

A product development engineer who was working for a large medical device company in Minnesota had developed a new product to replace existing cardiac pacemakers. Because several aspects of the product were novel, nonobvious, and useful, it was eligible for patent protection. Moreover, the product was much better than existing cardiac pacemakers, and all the cardiologists who had seen a prototype were very enthusiastic about it.

While the previous paragraph might seem like the introduction to a tale about a technology entrepreneur who becomes fabulously wealthy, that is not the case here. You see, the engineer had signed nondisclosure and noncompete agreements as conditions of his employment at the medical device firm. Like most nondisclosure agreements, the one that this employee had signed precluded him from disclosing proprietary information that was developed during his employment at the company. And like most noncompete agreements, this one also barred him from working for a competitor for several years after leaving the medical device firm.

Even though the medical device firm was not interested in pursuing the new product and did not even want to patent it, the engineer couldn't start a company or work with another company to bring the new product to market because of the noncompete and nondisclosure agreements that he had signed. The medical device firm that owned the intellectual property behind the cardiac pacemaker could take to court him and anyone else who hired or helped him.

The threat of litigation was enough to stop the engineer from starting a company. He knew that it would be foolish to start a company that would immediately be enmeshed in a lawsuit that would take a large amount of his time and would require him to pay large legal fees that he would have no revenues to cover.

Moreover, he could not even take a job at another company working on the development of similar products. While other medical device companies were eager to hire him, believing that he would develop valuable new products, they were afraid to hire him to work on cardiac products because his former employer could then sue them, claiming that he had disclosed its knowledge to competitors.

The lesson here is that companies can often use nondisclosure and noncompete agreements to prevent their employees from using the intellectual property developed during their period of employment. This is why the creation and enforcement of strong nondisclosure and noncompete agreements are important parts of a technology strategy.

product is poorly understood, when the knowledge necessary to create it is tacit, when few people can understand that knowledge, and when value comes from a process rather than a product.

- Trade secrecy is a special case of efforts to keep information secret; to have a trade secret, a piece of intellectual property must be of value, must not be known generally, and must be kept secret.
- To protect a trade secret, you must take action to keep it secret by such activities as limiting access to facilities and having employees sign nondisclosure agreements.
- Nondisclosure agreements are legal documents in which employees agree to keep a company's information secret; noncompete agreements are legal documents that preclude employees from working for competitors for a period of time after their employment has ended.
- Noncompete and nondisclosure agreements are difficult to enforce because enforcement requires legal action, because companies cannot prevent people from earning a living in their chosen fields, because the agreements must be precise to be effective, and because employees must be given consideration in return for signing the agreements.
- Companies typically require their employees to assign them the rights to intellectual property developed during their period of employment.

COPYRIGHTS

A **copyright** is a legal protection given to the authors of original literary, musical, or artistic works.[27] It gives the right to reproduce, display, or produce derivative works from the protected item. It also gives the right to sue to collect damages if someone else infringes the copyright from the time the work was created until 70 years after the author's death (or 95 years after publication for works for hire—more about that next). Infringement occurs if another party duplicates, displays, produces, or distributes the work, or gives, rents, or lends it to others.

What Can Be Copyrighted?

Many things can be protected by copyright, including books, movies, software, music, other recordings, databases, plays, pantomimes, dances, sculptures, graphics, and architectural designs.[28] For example, the LEGO Group has used copyrights to protect the appearance of its standard, eight-studded block against imitation.[29]

The thing that you want to protect does not need to be novel or even lawful to receive copyright protection. For instance, Napster's software copyright still holds even after the file-sharing service was deemed unlawful.[30]

However, there are some limitations on what can be protected by copyright. First, the thing being protected has to be tangible. Thus, you cannot copyright an impromptu speech, but you can copyright a written one. Second, titles and names cannot be protected by copyright; these things are protected by trademarks. Third, slogans, ideas, methods, principles, discoveries, or things composed of common property, like calendars, cannot be copyrighted.[31] Fourth, the work has to be produced because the idea behind it cannot be copyrighted.

Who Gets a Copyright and How Do They Get It?

So who can get a copyright? A copyright can be obtained by the author of any completed original work, unless the work is done for hire. **Work for hire** is a technical term for work that is done under the scope of a person's employment, or under a written agreement between the author and the person contracting for the work, which requests that the work be done on the contractor's behalf. If the work is done for hire, then the copyright goes to the entity commissioning the work. For instance, you could hire your roommate to write some software for the insurance claims adjusting business that you are starting, and the copyright on that software would then belong to you.[32]

You can obtain a copyright without taking any action other than putting the intellectual property into tangible form (for instance, writing something on paper or recording it on a DVD or CD). Alternatively, as Figure 9.4 shows, you can apply for copyright protection from the U.S. Patent and Trademark Office. While applying for a copyright is not necessary, it does provide a couple of important advantages. Most notably, you need to have a registered copyright to file a lawsuit to protect your copyrighted intellectual property; so registration is useful in the event that you want to sue someone.

Enforcement Through Litigation

If you think that someone has improperly used your copyrighted materials, you can take them to court and sue them for infringement. Because plaintiffs in a copyright infringement lawsuit rarely have direct evidence of the actual incidence of copying (of course, having photos of people in the act of copying your copyrighted material and distributing it would strengthen your case!), the courts usually infer that copying has occurred if the new work is substantially similar to the copyrighted work, and the defendant had access to it.

If you win a copyright infringement lawsuit, the court will award you damages. The size of those damages depends on the intent of the infringer, how much money they made, how they made their money, and how their actions affected your business. For instance, if the infringer charged others for your copyrighted material, then the size of the damages that you can receive will be greater than if they gave away your material for free. Also, the size of the damage settlement will

FIGURE 9.4
A Copyright Application Form

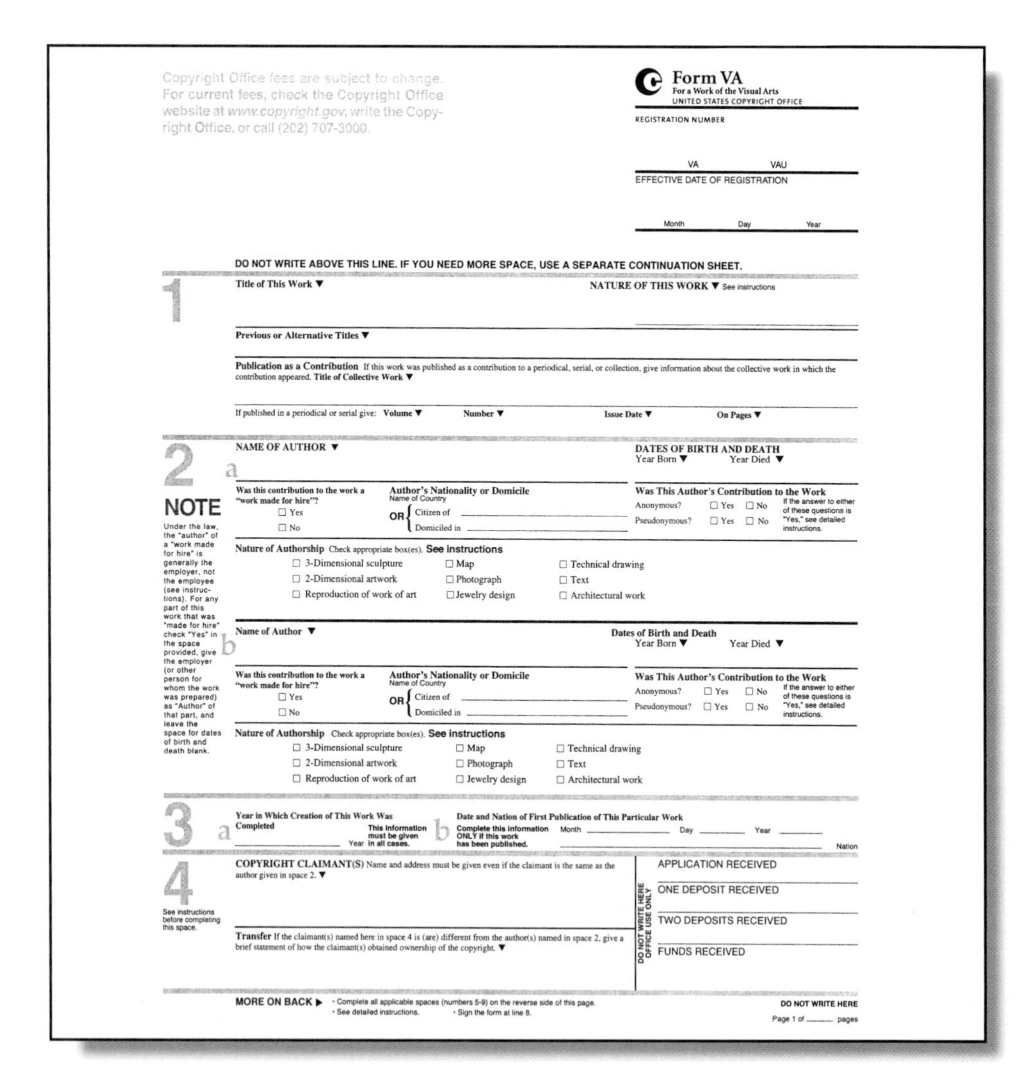

Copyright Office fees are subject to change. For current fees, check the Copyright Office website at *www.copyright.gov*, write the Copyright Office, or call (202) 707-3000.

Form VA
For a Work of the Visual Arts
UNITED STATES COPYRIGHT OFFICE

REGISTRATION NUMBER

VA VAU

EFFECTIVE DATE OF REGISTRATION

Month Day Year

DO NOT WRITE ABOVE THIS LINE. IF YOU NEED MORE SPACE, USE A SEPARATE CONTINUATION SHEET.

1

Title of This Work ▼

NATURE OF THIS WORK ▼ See instructions

Previous or Alternative Titles ▼

Publication as a Contribution If this work was published as a contribution to a periodical, serial, or collection, give information about the collective work in which the contribution appeared. **Title of Collective Work ▼**

If published in a periodical or serial give: **Volume ▼** **Number ▼** **Issue Date ▼** **On Pages ▼**

2 a

NAME OF AUTHOR ▼

DATES OF BIRTH AND DEATH
Year Born ▼ Year Died ▼

Was this contribution to the work a "work made for hire"?
☐ Yes
☐ No

Author's Nationality or Domicile
Name of Country
OR { Citizen of ______
Domiciled in ______

Was This Author's Contribution to the Work
Anonymous? ☐ Yes ☐ No
Pseudonymous? ☐ Yes ☐ No
If the answer to either of these questions is "Yes," see detailed instructions.

Nature of Authorship Check appropriate box(es). **See instructions**
☐ 3-Dimensional sculpture ☐ Map ☐ Technical drawing
☐ 2-Dimensional artwork ☐ Photograph ☐ Text
☐ Reproduction of work of art ☐ Jewelry design ☐ Architectural work

NOTE
Under the law, the "author" of a "work made for hire" is generally the employer, not the employee (see instructions). For any part of this work that was "made for hire" check "Yes" in the space provided, give the employer (or other person for whom the work was prepared) as "Author" of that part, and leave the space for dates of birth and death blank.

b

Name of Author ▼

Dates of Birth and Death
Year Born ▼ Year Died ▼

Was this contribution to the work a "work made for hire"?
☐ Yes
☐ No

Author's Nationality or Domicile
Name of Country
OR { Citizen of ______
Domiciled in ______

Was This Author's Contribution to the Work
Anonymous? ☐ Yes ☐ No
Pseudonymous? ☐ Yes ☐ No
If the answer to either of these questions is "Yes," see detailed instructions.

Nature of Authorship Check appropriate box(es). **See instructions**
☐ 3-Dimensional sculpture ☐ Map ☐ Technical drawing
☐ 2-Dimensional artwork ☐ Photograph ☐ Text
☐ Reproduction of work of art ☐ Jewelry design ☐ Architectural work

3 a

Year in Which Creation of This Work Was Completed
______ Year
This information must be given in all cases.

b

Date and Nation of First Publication of This Particular Work
Complete this information ONLY if this work has been published.
Month ______ Day ______ Year ______
______ Nation

4

COPYRIGHT CLAIMANT(S) Name and address must be given even if the claimant is the same as the author given in space 2. ▼

See instructions before completing this space.

Transfer If the claimant(s) named here in space 4 is (are) different from the author(s) named in space 2, give a brief statement of how the claimant(s) obtained ownership of the copyright. ▼

DO NOT WRITE HERE
OFFICE USE ONLY

APPLICATION RECEIVED

ONE DEPOSIT RECEIVED

TWO DEPOSITS RECEIVED

FUNDS RECEIVED

MORE ON BACK ▶ • Complete all applicable spaces (numbers 5-9) on the reverse side of this page.
• See detailed instructions. • Sign the form at line 8.

DO NOT WRITE HERE
Page 1 of ______ pages

(*continued*)

EXAMINED BY

FORM VA

CHECKED BY

☐ CORRESPONDENCE Yes

FOR COPYRIGHT OFFICE USE ONLY

DO NOT WRITE ABOVE THIS LINE. IF YOU NEED MORE SPACE, USE A SEPARATE CONTINUATION SHEET.

5

PREVIOUS REGISTRATION Has registration for this work, or for an earlier version of this work, already been made in the Copyright Office?

☐ **Yes** ☐ **No** If your answer is "Yes," why is another registration being sought? (Check appropriate box.) ▼

a. ☐ This is the first published edition of a work previously registered in unpublished form.

b. ☐ This is the first application submitted by this author as copyright claimant.

c. ☐ This is a changed version of the work, as shown by space 6 on this application.

If your answer is "Yes," give: **Previous Registration Number** ▼ **Year of Registration** ▼

6

DERIVATIVE WORK OR COMPILATION Complete both space 6a and 6b for a derivative work; complete only 6b for a compilation.

a. Preexisting Material Identify any preexisting work or works that this work is based on or incorporates. ▼

a

See instructions before completing this space.

b. Material Added to This Work Give a brief, general statement of the material that has been added to this work and in which copyright is claimed. ▼

b

7

DEPOSIT ACCOUNT If the registration fee is to be charged to a Deposit Account established in the Copyright Office, give name and number of Account.

Name ▼ **Account Number** ▼

a

CORRESPONDENCE Give name and address to which correspondence about this application should be sent. Name/Address/Apt/City/State/Zip ▼

b

Area code and daytime telephone number ( ) Fax number ( )

Email

8

CERTIFICATION* I, the undersigned, hereby certify that I am the

check only one ▶
- ☐ author
- ☐ other copyright claimant
- ☐ owner of exclusive right(s)
- ☐ authorized agent of ____________

Name of author or other copyright claimant, or owner of exclusive right(s) ▲

of the work identified in this application and that the statements made by me in this application are correct to the best of my knowledge.

Typed or printed name and date ▼ If this application gives a date of publication in space 3, do not sign and submit it before that date.

Date ____________

Handwritten signature (X) ▼

X ____________

9

Certificate will be mailed in window envelope to this address:

Name ▼

Number/Street/Apt ▼

City/State/ZIP ▼

YOU MUST:
- Complete all necessary spaces
- Sign your application in space 8

SEND ALL 3 ELEMENTS IN THE SAME PACKAGE:
1. Application form
2. Nonrefundable filing fee in check or money order payable to *Register of Copyrights*
3. Deposit material

MAIL TO:
Library of Congress
Copyright Office
101 Independence Avenue SE
Washington, DC 20559-6000

*17 *USC* §506(e): Any person who knowingly makes a false representation of a material fact in the application for copyright registration provided for by section 409, or in any written statement filed in connection with the application, shall be fined not more than $2,500.

Form VA Rev: 07/2006 Print: 07/2006—30,000 Printed on recycled paper

U.S. Government Printing Office: 2004-320-958/60,126

This figure shows a U.S. copyright application form for a copyright on a visual art.
Source: http://www.copyright.gov/forms/formva.pdf.

be larger if the infringer reduces the commercial value of your property through their actions.

If the court determines that the infringer's imitation was intentional, then you can collect triple the value of your loss as damages. So it is important to affix the copyright symbol (©) to your material. Doing so allows the court to reject any claim by an infringer that he or she did not know the material was copyrighted and innocently infringed.[33]

If you believe that someone has infringed your copyright, you can ask the court to issue an injunction, stopping that party from using your copyrighted material while the case is being decided. However, if you believe that your copyrighted material has been infringed, you need to take action quickly. The statute of limitations on copyrights only lasts three years.

Recent Developments to Strengthen Copyrights

Although copyrights were originally intended to protect written documents, in recent years, most of their growth has been as a means of protecting sound and images (as well as computer software). Now such things as video recordings of Super Bowl games and Web casts of the weather outside of college dorms are routinely copyrighted.

However, software is easier to copy than books and other printed material because duplicating and distributing multiple copies of a book takes more time and money than duplicating and distributing a piece of software. Therefore, copyright violations have been increasing in the digital age. To deter copying and illegal distribution, software companies often impose very restrictive end user license agreements (EULAs). By severely limiting how their customers can use their products, these companies strengthen their position against violators of their copyrights.

Unfortunately for copyright holders, the development of computer network technology to share digital files has made it easier to copy protected material, reducing sales of the legitimately duplicated versions. For instance, file sharing technology has made it easy to copy musical recordings, leading to declining sales of music CDs. As a result, copyright holders have become more vigilant about protecting their intellectual property. In the case of digital music files, the record labels have begun to sue anyone that does anything that lets users get around their copyrights. For example, several record labels recently sued XM Satellite Radio because XM's Inno device allows users to record, store, and create play lists of songs that they have heard on XM. The record labels claim that the use of the Inno device violates the copyrights to their songs by allowing people to obtain recordings of them without paying a royalty.[34]

Similarly, the record labels sued Napster for making it possible for people to exchange digital music files without paying royalties, charging that the company violated the copyrights of recording artists, and caused them financial loss. The record labels' argument was that Napster created a market in which other people could avoid paying royalties on copyrighted songs, thus enabling infringement.[35] (The copyright issue wasn't settled in this case because Napster was forced to shut down when the judge in the case issued an injunction banning Napster from offering the service until the courts had decided the case.[36] However, many observers believe that file-sharing networks will not be able to claim "fair use" of copyrighted material—see Figure 9.5.)

While the recording industry was able to use the court system to enforce its copyrights against the first generation of peer-to-peer networks like Napster, they face a more difficult time with second generation peer-to-peer networks that don't use a central server for file sharing. The use of second generation peer-to-peer networks spreads the copyright violation across numerous parties and makes the value of their infringement too small to justify the use of lawsuits as a way to stop it.[37]

Recently, laws have been enacted to let companies use physical tools, such as embedded authentication chips, to make it more difficult for people to copy a piece of intellectual property. For instance, the Audio Home Recording Act of 1992 requires that all digital recording devices include a Serial Copy Management System, which permits originals, but not copies, to be duplicated. And the Digital Millennium Copyright Act (DMCA) made it illegal to circumvent a technological device that is used to prevent duplication of copyrighted material.[38]

FIGURE 9.5
Fair Use

> The fair use of a copyrighted work, including such use by reproduction in copies or phonorecords or by any other means specified by that section, for purposes such as criticism, comment, news reporting, teaching (including multiple copies for classroom use), scholarship, or research, is not an infringement of copyright. In determining whether the use of a work in any particular case is a fair use the factors to be considered shall include—
>
> 1. The purpose and character of the use, including whether such is of a commercial nature or for nonprofit educational purposes;
> 2. The nature of the copyrighted work;
> 3. The whole amount and substantiality of the portion used in relation to the copyrighted work as a whole; and
> 4. The effect of the use upon the potential market or value of the copyrighted work. The fact that a work is unpublished shall not itself bar a finding of fair use if such a finding is made upon consideration of all of the above factors.

The U.S. government provides a "fair use" exemption to copyright law, which allows a user to make use of copyrighted material under the conditions listed above.

Source: Section 107 of the Copyright Act of 1976, http://www.copyright.gov/title17/92chap1.html#107.

However, the use of copy protection software to prevent sharing of intellectual property has had problematic side effects. It limits the devices that customers can use to play legitimately purchased recordings and sometimes causes damage to computers that play the recordings.[39] For instance, Sony BMG recently had to reimburse its customers more than $100 each for computers damaged by hidden antipiracy software that Sony BMG had placed on their CDs.[40]

Efforts to strengthen copyrights have had other adverse effects as well. They have diluted the concept of "fair use" of copyrighted material. As a result, it is becoming more difficult to make noncommercial use of these materials. Second, these efforts have hindered the natural process by which innovators build on the work of others by requiring them to obtain the rights to use any copyrighted material simply to build on it.

Software Copyrights

Copyrights have become an important mechanism to protect software. While the mathematical formulas and equations underlying software programs are not copyrightable, nor are the ideas or methods behind them, copyrights can be used to protect many parts of computer software, including source code, object code, microcode, and screen displays.[41] For instance, ConnectU.com, a social networking Web site, uses copyrights to protect its source code.[42]

While copyrights provide some intellectual property protection for software, they are not an ideal form of protection for this medium because they only protect the expression of ideas, not the concepts underlying those ideas. As the introductory vignette in the chapter pointed out, ideas can often be expressed in a variety of different ways, allowing someone to reverse engineer a piece of software and then write a new piece of software that works around the copyright by expressing the same idea in a different way. If a defendant in a software copyright case can show that they created a work independently and expressed an idea in a different way from the holder of the copyright, then there is no copyright violation.[43]

Moreover, demonstrating the infringement of a software copyright is not easy to do directly, making software copyrights difficult to enforce. Because it is impossible to show an exact linkage between the expression of an idea and the process of expressing it, courts have had to interpret the "look and feel" of software to determine whether copyright infringement has occurred. Of course, this reliance on "look and feel" to determine infringement makes it harder to know if infringement has actually occurred.[44]

On the other hand, software copyrights provide additional intellectual property protection to that provided by software patents. A wider variety of software programs can be copyrighted than patented because any originally authored work presented in tangible form can be protected by a copyright, while only novel, nonobvious, and useful inventions can be protected by a patent. Copyrights also are much easier to obtain than patents, and are, consequently, a much less expensive form of protection. Furthermore, copyrights offer protection until 70 years after the author's death, while patents offer protection for only 20 years after the time of invention.[45]

Key Points

- Copyrights give the authors of original works the right to distribute, duplicate, and provide derivations of that work, and to preclude others from doing the same.
- A variety of things can be copyrighted, including literary works, dramatic works, audio and video recordings, and computer software; however, intangible things, titles, names, slogans, ideas, methods, principles, and works composed of common property, cannot be copyrighted.
- Copyrights are given to the author of any original work, unless the work is done for hire.
- Copyrights can be obtained either by putting the work into tangible form or by registering the work at the USPTO; registration provides the right to sue for copyright infringement.
- Copyrights offer a negative right; they do not stop others from copying your intellectual property; they only give you the right to sue to obtain damages if your copyrighted material has been infringed.
- File-sharing software poses an important threat to copyrights on recorded music, and its rise has led to a number of infringement lawsuits.
- Recent laws have strengthened the position of copyright holders by allowing them to use physical tools to prevent duplication of their work; however, these physical tools have had problematic side effects.
- Copyrights can be used to protect the source code, object code, microcode, and screen displays in software but not the ideas, mathematical formulas, or equations behind them.
- Because it is impossible to show the exact link between the expression of an idea and the process underlying it, courts interpret the "look and feel' of software to evaluate infringement.
- Copyrights are less effective than patents at protecting software but can be obtained to protect a wider range of things.

TRADEMARKS

Trademarks are devices to identify the provider of a product or service.[46] While they offer much less intellectual property protection than patents, copyrights, or trade secrets, they do help companies to protect their brand names. For instance, the Intel Inside® trademark helps Intel build its brand by making it easier for that company to differentiate itself from competitors.

In addition, trade and service marks can be used as leverage to drive other forms of strategic advantage. For instance, Cisco recently settled a lawsuit with Apple Computer over violation of its iPhone trademark. Cisco wanted Apple to make its iPod and iPhone products compatible with non-Apple products. By blocking Apple's use of the iPhone name, Cisco forced Apple to concede on the issue of compatibility.[47]

Because consumers associate particular trade or service marks with the quality of the products or services that companies provide, some trademarks are quite valuable. For instance, the Microsoft trademark is now worth $60 billion.[48] Therefore, learning how trademarks work and how they protect intellectual property is an important part of technology strategy.

What Can Be Trademarked?

A trade or service mark can be obtained on any word, number, symbol, phrase, color, design, or even smell that distinguishes the products and services of one company from those of another. For instance, Nike has trademarked its "swoosh" symbol, while Porsche AG has trademarked the numerical sequence "911."[49]

However, not everything can be trade or service marked. For a word, number, symbol, phrase, color, design, or smell to be appropriate as a mark, it cannot describe the product or service that a company provides. For instance, a supermarket cannot trademark the word *carrot* because that word is descriptive of the products sold at a supermarket. However, an airline could trademark that word because carrots are not descriptive of what airlines do.

A common word, like *house* cannot be trademarked. However, what is a common word depends on interpretation by the courts. A Federal appeals court recently upheld *Entrepreneur* Magazine's trademark on the word *entrepreneur*, allowing that company to block the use of that word by others.

Ironically, the fact that someone else has trademarked a word, number, symbol, phrase, color, design, or smell does not mean that you can't use the same one. A trade or service mark can be used by more than one company if customers would not be confused about the identity of the provider of the product or what the product is used for, and if the use by a second party does not dilute the value of the mark. Typically, this means that a mark can be used by two companies if they sell different types of products and services (e.g., airplanes and vegetables) through different channels. For instance, Apple Computer and Apple Records are both able to have a trademark with the word *apple* in it because personal computers and Beatles songs are very different products and are sold through different marketing channels. However, as Apple Computer moves further into the music business, it may face problems using its trademarked name for that business because its name might then cause confusion among customers as to the provider of the product.

FIGURE 9.6 Many Companies Obtain a Large Number of Trademarks

A Mathematical Assistant™
ANYLITE™
AOS™
APD™
Automatic Power Down™
Avigo™
BA II PLUS™
BA Real Estate™
Calculator-Based Laboratory™
CBL™
CBL 2™
Calculator-Based Ranger™
CBR™
CellSheet™
Clear Calc™
Constant Memory™
Datamath™
Data Synchronization™
Derive™
Derive™ 5
DockMate™
EOS™
Europa™
Executive Business Analyst™
EXPLORATIONS™
Explorer™
Explorer Plus™
Financial Investment Analyst™
five2eight™
Galaxy™
GeoMaster™
Home Manager™
Hot Calc™
LearningCheck™
Math Explorer™
MathMate™
Math Star™
MBA™
muLisp™
NoteFolio™
Paper-Free™
Paperless Printer™
Personal Banker™
Phone Bank™
Pocket Dialer™
PocketMate™
Pocket Paper-Free Printer™
Pocket Speller™
Pocket Thesaurus™
Pro-Calc™
Profit Manager™
Student Business Analyst™
Study Cards™
SuperBundle™
SuperView™
T^3™
Teachers Teaching with Technology™
T^3 Europe Talk TI™
Technofiscaphobia™
Tfas™
TI-30Xa School Edition™
TI-30Xa SE™
TI-83 Plus™
TI-Cares™
TI Connect™
TI FLASH Studio™
TI-GRAPH LINK™
TI InterActive!™
TimeSpan™
TI-Navigator™
TI-Presenter™
TI-TestGuard™
ViewScreen™
Voyage™ 200

This figure shows a partial list of trade and service marks belonging to Texas Instruments.
Source: Adapted from http://education.ti.com/us/global/trdmrk.html.

Because trade and service marks are valuable tools, many large, established companies have obtained a large number of marks. Figure 9.6 shows just a partial list of the trademarks obtained by one technology company—Texas Instruments.

Obtaining a Trademark

So how do you get a trademark? In common law countries, like the United States, you get a trade or service mark by using the word, phrase, symbol, design, or smell or by registering that mark with the USPTO.[50]

The process of registering a trade or service mark is very simple. You just send an application to the USPTO along with a drawing of the mark and the payment of the fee for the relevant category of mark.

However, before you send in your application and pay the money to register a trade or service mark, you probably want to conduct a trademark search. The USPTO is not going to give you a mark that violates that of another company. Conducting a search will minimize the chances that you will select something that infringes on another mark, as well as the likelihood that you will select something that cannot be trademarked.

Although you will not get a trade or service mark right from the USPTO until you use a mark, and you don't need to register the mark to enforce it, you probably want to go down the registration route when you seek trademark protection. Registration provides a record of your claim of ownership of the mark, which is useful to signal your actions to competitors. In addition, you cannot sue to protect your trade or service mark, or collect triple damages in the case of infringement, until the mark has been registered.[51] Furthermore, registration makes it easier to obtain trade or service mark rights in other countries.[52]

Enforcing a Trademark

Once you have registered a trade or service mark, your ownership of it lasts for ten years and can be renewed as long as the mark is in use and has not been invalidated.[53] However, five years after you have obtained the mark, you will need to file an affidavit with the USPTO attesting that the mark is still in use. If you don't do this, your trade or service mark can be cancelled.

Trade and service marks can be invalidated by the USPTO in one of three ways: through cancellation proceedings, through abandonment, or through generic meaning. Cancellation occurs when the owner of the mark fails to attest to its continued use.

Abandonment occurs when someone else can show that the owner of the mark has stopped using it. For example, in the recent dispute between Cisco and Apple Computer over the trademark "iPhone," Apple Computer sought to show that Cisco hadn't sold iPhone-branded products for a period of time, thus indicating that Cisco had abandoned the trademark.[54]

The potential for abandonment is why trademark holders fight hard to protect their trademarks. For instance, Entrepreneur Magazine Incorporated fights to exert its rights to the trademark *entrepreneur* against a variety of small companies, not because it thinks it would obtain any significant royalties from enforcing the trademark, but to defend the use of the mark against other companies that claim that Entrepreneur Magazine has abandoned it.

Generic use occurs when a mark no longer represents a specific product or service and ends up representing a general category of products or services (as occurred, for example, with the once trademarked term *escalator*).[55] Once a trade or service mark becomes a generic term, it reverts to the public domain and anyone can use it. That is why Bayer works hard to ensure that Aspirin® is not used to refer to all pain medications. If that were to occur, the word could no longer be trademarked because it would no longer distinguish Bayer's product from those of other companies.

Like other forms of intellectual property protection, trade and service marks are enforced through legal action. Owners of a mark can sue to prevent both infringement and dilution of the value of the mark.[56] Infringement occurs when a competitor's use of a mark causes confusion amongst customers about the provider of a product. For example, VoIP start-up, Vonage, has sued AT&T claiming that the name of AT&T's VoIP service, CallVantage, violates its trademark because that name is too close to its own.

Dilution occurs when another party's use of a word, phrase, symbol, design, or smell lowers the value of a company's trade or service mark. For instance, American Express was able to stop a limousine service from using the name "American Express" by showing that its trademark's value was reduced by that action.[57]

You need to protect your trademarks. Failure to take legal action to enforce your rights can result in the loss of a trade or service mark through abandonment. Unfortunately, taking legal action costs money; and many organizations fail to protect valuable trademarks. For instance, the Metropolitan Transit Authority (MTA) in New York City has trademarked its circular route symbols for the A, D, F, 1, 4, and 7 trains. However, many companies frequently violate the MTA's trademarks by making unauthorized T-shirts—or, in the case of Eli Zabar's food emporium, rectangular cookies with hard icing designed to look like New York City metro cards. Although the MTA has written letters to many of the trademark violators, it lacks the legal staff to go to court to enforce its trademarks and has allowed the value of those trademarks to deteriorate.[58]

Start-ups face a greater challenge than large, established companies in developing an effective strategy toward the management of trade and service marks. Because small, new companies are often cash constrained, they face the dilemma of whether challenging—and winning—a trademark infringement lawsuit is worthwhile. The start-up might win the suit against a deep-pocketed competitor but be driven out of business by the legal effort. Take, for example, the case of Haute Diggity Dog, the maker of dog toys. They created dog toys shaped like handbags, called "Chewy Vuiton." Louis Vuitton, makers of the handbags that Haute Diggity Dog was parodying, sued them for degrading the value of Louis Vuitton's trademark. While Haute Diggity Dog won the lawsuit, it lost a lot of distributors because Louis Vuitton sent cease-and-desist letters to the retailers during the lawsuit, causing the retailers to stop carrying Haute Diggity Dog's products.[59]

Domain Names

Domain names are the names used on Web sites to identify an organization providing a good or service. As Figure 9.7 shows, domain names have become an increasingly popular form of intellectual property protection, as companies do more and more business over the Web.

Domain names are registered by the Internet Corporation for Assigned Names and Numbers (ICANN) to the first party to seek registration for a name.[60] As with trade and service marks, it is useful to conduct a search before trying to register a domain name to make sure that you can obtain the name that you'd like to use. You can do this at the ICANN Web site (www.icann.org).

The protection of domain names is similar to the protection of trade and service marks. However, two important distinctions exist. First, because geographic regions are not meaningful in cyberspace, companies in different places are not permitted to use the same domain name, though they are permitted to use the same trade or service mark.[61] Second, unlike with trade or service marks, common words can be used as domain names. For example, Procter & Gamble has obtained the domain name "cavities.com."[62]

Your domain name also cannot adversely affect another company's business. If it does, the company whose business has been hurt can sue you for control of your domain name. For example, Universal Tube and Rollerform Equipment Corporation

FIGURE 9.7
Use of Domain Names

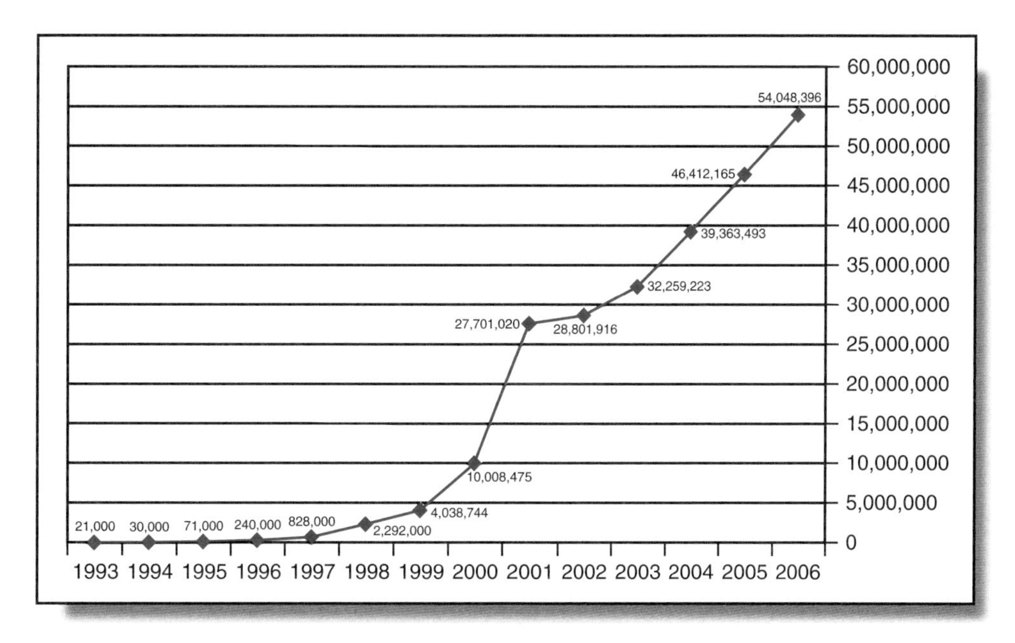

This figure shows the dramatic rise in the number of domain names in use over time.
Source: Adapted from data contained in http://www.zooknic.com/Domains/dn_length.html.

has sued YouTube for the rights to the www.youtube.com domain name because the volume of people going to Universal Tube's Web site www.utube.com when looking for www.youtube.com has caused Universal Tube's Web servers to crash repeatedly.

The enforcement of a domain name occurs in a similar way to the enforcement of a trade or service mark. If you believe that someone else has infringed on your domain name or has taken action to lower its value, you can sue the offending party.[63] For example, Sir Ratan N. Tata of Bombay, India, the leader of India's Tata group of companies (which includes Tata Steel, Tata Engineering, Tata Power, Tata Chemicals, Tata Finance, Tata Power, Tata Tea, and Tata Sons Ltd) and the Sir Rata Tata Trust, sued a New Jersey porn site in 1999, and obtained an injunction against the latter's use of the Internet domain name Bodacioustatas.com because the New Jersey company's use of the domain name harmed the reputation and, hence the value, of Sir Ratan's companies.[64]

However, enforcement of domain name infringement is often more difficult than enforcement of trade or service mark infringement because domain names operate in cyberspace. As a result, it is often difficult to determine the legal jurisdiction in which to sue an offender, and when that jurisdiction can be determined, it is often a place that does not strongly enforce intellectual property laws, making it hard for you to stop the offending action or collect damages.[65]

Key Points

- Trade and service marks can be obtained on any nondescriptive, nongeneric word, number, symbol, phrase, color, design, or even smell that distinguishes the products and services of one company from those of another.
- Trade and service marks provide a negative right and must be enforced through legal action, which is often more difficult for start-ups to undertake than for established companies to conduct.

- The same mark can be used by more than one company if that use will not cause confusion amongst consumers about the provider's identity, and does not dilute the value of another party's mark.
- You can obtain a U.S. trade or service mark by using the mark or by registering it with the USPTO; however, registration facilitates your ability to obtain similar rights in other countries and allows you to sue to enforce your mark.
- Trade and service marks are lost through cancellation proceedings, abandonment, or if they take on generic meaning.
- Domain names are names used on Web sites to identify the organization providing a good or service; they are protected through legal action.

International Issues in Intellectual Property

There is no such thing as an international copyright, trademark, or patent. You need to obtain that piece of intellectual property protection in each of the countries in which you want to use it. As Figure 9.8 shows, this means that some countries, like the United States, receive a large number of patent filings, many of which are from entities domiciled in other countries.

Moreover, this also means that the type of protection that you get in different countries on these three types of intellectual property depends on the laws of those countries and the willingness of their governments to enforce them. Just because you can obtain a patent, trademark, or copyright in one country doesn't mean that you can obtain the same protection in another. For example, Anheuser-Busch, which has owned the trademark "Bud" in the United States since 1876, was only recently able to use that trademark in Hungary. There, the name "Bud" belonged to the Czech beer maker, Budejovicky Budvar, which had objected to Anheuser-Busch's efforts to use it.[66]

Differences in Intellectual Property Regimes

While a complete discussion of the differences in intellectual property regimes across countries is beyond the scope of this book, as a technology manager or entrepreneur, you need to be familiar with three of the most important differences: first-to-invent versus first-to-file rules, policies on the timing of disclosure, and the requirement to manufacture. You also need to understand the differences between the intellectual property systems in developing and developed countries.

First to Invent

As the previous chapter indicated, the United States is one of only two countries in the world that awards a patent to the first person to invent a technology. Most countries award a patent to the first person to apply for one, regardless of who invented the technology first. Because of this rule, being the first party to file a patent application will not get you a patent in the United States. If you want a U.S. patent, you have to prove that you invented the technology before anyone else.

FIGURE 9.8 Top 20 Countries for Patent Applications

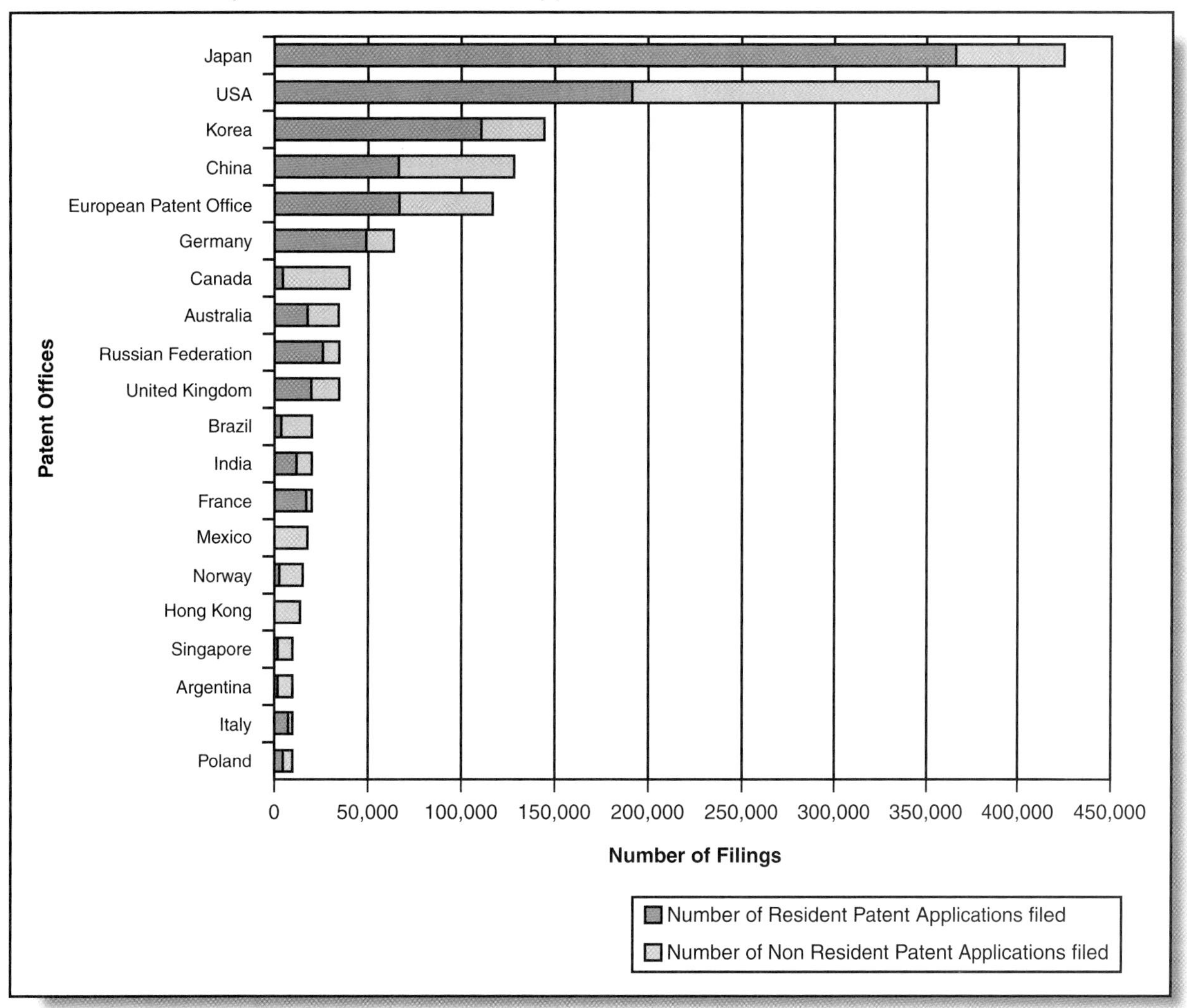

There are major differences across countries in both the total number of patent applications filed and the proportion filed by residents.

Source: WIPO Statistics Database, http://www.wipo.int/ipstats/en/statistics/patents/patent_report_2006.html#P104_9303.

The first-to-invent rule also has an important implication for obtaining worldwide patent protection: Failure to act quickly to apply for patents in other countries might keep you from obtaining patents outside of the United States.[67]

Disclosure

In the United States, inventors who publish information about an invention up to one year before applying for a patent remain eligible to obtain one. However, in many European countries, and in Japan, prior publication *at any time* will keep you from obtaining a patent.[68] So if you are in a business like pharmaceuticals, in which researchers generally publish the results of their research in scholarly journals, you need to be very careful about the timing of publication when seeking patents outside the United States.

Requirement to Manufacture

The United States imposes no requirement that an inventor actually produce a product or service that uses the patented invention. However, in many countries, you may obtain patent protection *only* if you are willing to manufacture your product in that country. If you aren't willing to do this, it doesn't make sense for you to go through the time and expense to obtain a patent in that country. It will only be invalidated.[69]

Intellectual Property in Developing Countries

Many developing countries, like China and India, do not enforce their intellectual property laws very vigorously. In fact, research has shown a positive correlation between the strength of intellectual property protection in countries and their level of per capita income.[70]

This correlation exists because companies in developing and developed countries generate value in different ways. In developed countries, companies generate much of their value from intellectual property—things like software code, drugs, automobile designs, digital music files, and so on—whereas, in developing countries, companies generate most of their value from the production of physical products—things like textiles, toys, and machinery.[71] As a result, in developing countries, few domestic companies lose anything from weak intellectual property laws, and many benefit because those laws facilitate imitation of foreign companies' IP.

Moreover, many developing country governments use weak intellectual property laws to facilitate imitation of expensive foreign products by low-cost local producers, thereby lowering the cost of those products to their citizens. Weak enforcement of pharmaceutical patents, in particular, allows local companies to reverse engineer and copy foreign drugs, reducing their cost by over 90 percent.[72] For example, the Brazilian government forced Abbott Laboratories to allow a Brazilian company to make a generic version of its patented AIDS drug in return for a 3 percent royalty. This action allowed the Brazilian government to save several hundred million dollars in medical costs by making an expensive drug available for $0.68 a pill.[73]

As a technology entrepreneur or a manager, you need to adapt your technology strategy to these weak intellectual property laws. As Figure 9.9 shows, you need to

FIGURE 9.9
Conflict over Patents

Many people in developing countries consider it unfair that biochemical companies from industrialized countries exert patent rights to genetically engineered seeds, which raises the price of the seeds used by their farmers.

Source: http://www.thimmakka.org/Newsletters/neemii.html.

be prepared for conflict with governments and businesses in developing countries over enforcing your intellectual property rights.

Moreover, you need to expect that your products will be pirated in developing countries, where counterfeiting costs American companies an estimated $20 billion to $24 billion per year.[74] This is particularly true if your company produces an intellectual property–intensive product that is also easy to duplicate, like a piece of computer software or a music video.

You might want to minimize your risk of loss by keeping valuable IP out of developing countries. For instance, you might avoid conducting R&D in countries with weak records of protecting intellectual property.[75] However, you might want to take advantage of weak intellectual property laws in developing countries by designing a strategy in which the transfer of your technology to local companies helps your company to compete better. For example, Toyota is building its Prius hybrid car in China even though the Chinese government does not enforce patents vigorously because it is counting on imitation by Chinese manufacturers to drive down the cost of batteries, which will attract buyers to its hybrids.[76]

International Agreements on Intellectual Property

Several international agreements on copyrights, trademarks, and patents exist, and facilitate the international protection of intellectual property. The major international agreement on copyrights is the **Berne Convention**, which sets a minimum level of copyright protection that has to be provided in signatory countries, and requires the same level of protection to be provided both to citizens and noncitizens.[77] (However, it does not set the actual level of copyright protection in different countries, nor does it provide for protection in multiple nations.)

The major international agreement on trademarks is the **Madrid Protocol**, which gives any individual who is a resident or citizen of one of the 70 signatories to the agreement the ability to file for international registration in as many of the signatory countries as they want through their home country's patent and trademark office.[78] Because the other country applications are given the same filing date as the home country application, this protocol has made the process of obtaining trademark protection around the world easier and less expensive.[79]

Several international agreements cover patents. The **European Patent Convention** allows an inventor to apply for patents in all participating European countries through a single application. If the patent is granted, it is then enforceable in all the participating countries in accordance with each country's laws.[80]

The **Paris Convention**, which was signed by over 160 countries, prohibits the provision of differential patent rights to citizens of a country, establishes priority in patent applications, and allows you to disclose your invention to file for another country's patents. Consequently, in any signatory country to the Paris Convention, you will obtain exactly the same patent protection as any citizen of that country, don't have to rush around the world to file your patents, and don't have to time your applications around the world to avoid disclosure problems that would preclude you from getting a patent.[81]

The **Patent Cooperation Treaty** allows inventors in the 100 signatory countries to apply for patent protection in one country and preserve the right to apply for patent protection in all other signatory countries for 30 months. Inventors can reduce their patent costs by waiting to see what happens to their patent applications at home before applying for patents in other countries because the likelihood that foreign patent applications will be granted is linked to what happens to the application in the home country.[82]

The **Trade-Related Aspects of Intellectual Property Rights Agreement (TRIPS)** makes uniform, at 20 years, the length of patents in all signatory countries, requires the countries to provide patents on chemical and pharmaceutical products,[83] and restricts mandatory licensing.[84] This agreement strengthens intellectual property laws in developing country signatories. For example, India used to recognize patents only on the process of making drugs, allowing companies to copy patented drugs by coming up with alternative ways to synthesize them, but to be part of TRIPS, India had to agree to enforce patents on pharmaceutical products.

Key Points

- There is no such thing as international copyrights, trademarks, or patents; all three of these forms of intellectual property protection must be obtained in each country where a company would like to have them.
- Major differences exist across countries in laws governing patenting, including whether they award patents to the first to invent or the first to file, their policies on the timing of disclosure, and their requirement to manufacture.
- Developing country governments often do not enforce intellectual property laws vigorously because companies in those countries generate little value from intellectual property and because weak intellectual property laws reduce the cost of many products.
- Weak intellectual property laws in developing countries lead to widespread piracy, and require companies to formulate strategies that are effective under such conditions.
- Several international agreements make it easier to obtain intellectual property protection in multiple countries; the most important of these agreements are: the Berne Convention, the Madrid Convention, the European Patent Convention, the Paris Convention, the Patent Cooperation Treaty, and the Trade-Related Aspects of Intellectual Property Rights Agreement.

DISCUSSION QUESTIONS

1. What types of legal tools can be used to protect intellectual property? How should you choose among them? When is each of them effective?
2. Why does secrecy work better for some products and processes than for others? Why does it work better in some industries than in others?
3. What are the advantages and disadvantages of using copyrights to protect software? Why do these advantages and disadvantages exist?
4. When can a company hire an employee who worked at another company to develop a new product? What factors influence when this can be done?
5. Do you think that what peer-to-peer digital file-sharing networks do is "fair use" of copyrights or an infringement of those rights? Explain your position.
6. What, if anything, should be done to protect the rights of owners of copyrighted video and audio recordings, given the development of file-sharing technologies? Explain your reasoning.
7. China has relatively weak intellectual property laws, allowing many Chinese companies to imitate products and services developed in other countries. To enter the World Trade Organization (WTO), China needs to strengthen its intellectual property laws. Is it better for China to strengthen its intellectual property laws and enter the WTO or not strengthen its laws and remain outside the WTO? What factors influence your evaluation?

Key Terms

Berne Convention: An international agreement that sets a minimum level of copyright protection in signatory countries and requires the same level of protection to be provided both to citizens and noncitizens.

Consideration: The provision of something of value in return for agreeing to take a costly action or to make a purchase.

Copyright: A form of intellectual property protection provided to the authors of original works of authorship.

Domain Name: The name used on Web sites to identify an organization providing a good or service.

European Patent Convention: An agreement that allows an inventor to apply for patents in all participating European countries through a single application.

Madrid Protocol: The major international agreement on trademarks; it gives any individual who is a resident or citizen of one of the 70 signatories the ability to file for international registration in as many of the signatory countries as they want.

Noncompete Agreement: A legal document that bars a person from working for a competitor for a period of time after his or her employment has ended.

Nondisclosure Agreement: A legal document in which a person agrees not to make private information public.

Observable-in-Use: A condition under which the intellectual property underlying a technology can be observed when it is used.

Paris Convention: An intellectual property agreement signed by 160 countries that prohibits differential patent rights to citizens and establishes priority in patent applications.

Patent Cooperation Treaty: An agreement by 100 countries to preserve the right to apply for patent protection in signatory countries for 30 months.

Tacit Knowledge: Knowledge about how to do something that is not documented in written form.

Trademark: A word, number, symbol, phrase, color, design, or even smell that distinguishes the products and services of one company from those of another.

Trade-Related Aspects of Intellectual Property Rights Agreement (TRIPS): An agreement that makes uniform, at 20 years, the length of patents in all signatory countries, requires the countries to provide patents on chemical and pharmaceutical products, and restricts mandatory licensing.

Work for Hire: Work that is done under the scope of a person's employment or under a written agreement requesting that the work be done for someone else.

Putting Ideas into Practice

1. **Deciding on a Trade Secret** Pick a company that you are interested in investigating. Identify a new product or process that the company is developing or might develop. Using the information presented in this chapter, evaluate whether the company should use trade secrecy to protect the technology. Make sure to explain the pros and cons of that choice, and why you make this recommendation.
2. **Obtaining a Trademark** The purpose of this exercise is to identify a trademark for a business. Pick a business that you know well. Then follow the steps below to identify a trademark.

   Step 1: Identify a word, number, phrase, symbol, design, or combination thereof that you would like to trademark.

   Step 2: Go to www.uspto.gov and search existing trademarks to make sure that the word, number, phrase, symbol, or design you selected is not already trademarked. If so, evaluate whether you would be allowed to use the same word, phrase, number, symbol, or design. If your evaluation is positive, explain why. If your evaluation is negative, select something else to trademark.

   Step 3: Once you have identified something that you can trademark, outline the steps that you'll take to obtain a trademark on it.

   Step 4: Explain how the trademark will help your business. What will the trademark protect? How will you enhance the value of the trademark?
3. **Obtaining a Domain Name** The purpose of this exercise is to learn how to obtain an Internet domain name for a business. Pick a business that you know well. Then follow the steps below to obtain a domain name for it.

   Step 1: Identify the word or words that you would like to use as your domain name.

   Step 2: Go to www.icann.org and search existing domain names to make sure that the one you selected is not already being used by someone else. If it is, select a different domain name.

   Step 3: Once you have identified a domain name that you can obtain, please outline the steps that you'll take to obtain the domain name for your business.

Notes

1. Adapted from Cringely, R. 1992. *Accidental Empires*. New York: Harper Collins.
2. Gleick, J. 2004. Get out of my namespace. *New York Times Magazine*, March 21: 44–49.
3. Afuah, A. 2003. *Innovation Management*. New York: Oxford University Press.
4. Nelson, R., and S. Winter. 1982. *An Evolutionary Theory of Economic Change*. Cambridge, MA: Belknap Press.
5. Teece, D. 1998. Capturing value from knowledge assets: The new economy, markets for know-how and intangible assets. *California Management Review*, 40(3): 55–79.
6. Zucker, L., M. Darby, and M. Brewer. 1998. Intellectual human capital and the birth of U.S. biotechnology enterprises. *American Economic Review*, 88(1): 290–305.
7. Winter, S. 2000. Appropriating the gains from innovation. In G. Day and P. Schoemaker (eds.), *Wharton on Managing Emerging Technologies*. New York: John Wiley.
8. Teece, Capturing value from knowledge assets.
9. Chally, J. 2004. The law of trade secrets: Toward a more efficient approach. *Vanderbilt Law Review*, 57(4): 1269–1311.
10. Yoffie, D. 2003. Intellectual property and strategy, *Harvard Business School Note*, Number 9-704-493.
11. Chally, The law of trade secrets.
12. Schilling, N. 2005. *Strategic Management of Technological Innovation*. New York: McGraw-Hill.
13. Chally, The law of trade secrets.
14. Allen, K. 2003. *Bringing New Technology to Market*. Upper Saddle River, NJ: Prentice Hall.
15. Fitzpatrick, W., S. DiLullo, and D. Burke. 2004. Trade secret piracy and protection: Corporate espionage, corporate security and the law, *Advances in Competitiveness Research*, 12(1): 57–68.
16. Schreiner, B. 2005. Colonel's recipe remains a secret. *Columbus Dispatch*, July 24: G1, 2.
17. Wingfield, N. 2006. At Apple, secrecy complicates life but maintains buzz. *Wall Street Journal*, June 28: A1, A11.
18. Allen, *Bringing New Technology to Market*.
19. http://www.crainscleveland.com/article.cms?articleId=38171
20. Karush, S. 2005. IBM settles suit over intellectual property. *The Plain Dealer*, March 23: B2.
21. Guth, R. 2005. Microsoft sues to keep aide from Google. *Wall Street Journal*, July 20: B1, B3.
22. Farkas, K. 2004. Patio enclosures awarded $8.6 million in suit. *The Plain Dealer*, August 31: D1.
23. Woolf, L. 2004. Non-competition agreements. *FDCC Quarterly*, Summer: 333–342.
24. Guth, Microsoft sues to keep aide from Google.
25. Woolf, Non-competition agreements.
26. Hart, M. 1996. Palm Computing Inc. (A), *Harvard Business School Case*, Number 9-396-245.
27. Silverman, A. 1997. Understanding copyrights: Ownership, infringement, and fair use. *JOM*, 49(8): 60.
28. U.S. Department of Commerce. 1992. *General Information Concerning Copyrights*. Washington, DC: U.S. Government Printing Office.
29. Austen, I. 2005. Block by block toy competitors build a case against Lego. *The Plain Dealer*, May 20: G1, G5.
30. Mesa, P., and R. Burgelman. 2004. Finding the balance: Intellectual property in the digital age. In G. Burgelman, C. Christiansen, and S. Wheelwright (eds.), *Strategic Management of Technology and Innovation*. New York: McGraw-Hill/Irwin.
31. U.S. Department of Commerce, *General Information Concerning Copyrights*.
32. Silverman, Understanding copyrights.
33. U.S. Department of Commerce, *General Information Concerning Copyrights*.
34. McBride, S. 2006. Music industry sues XM over replay device. *Wall Street Journal*, May 17: B1, B10.
35. Kiron, D. 2001. Napster. *Harvard Business School Case*, Number 9-801-219.
36. Moon, Y. 2005. Online music distribution in a post-Napster world. *Harvard Business School Case*, Number 9-502-093.
37. Silverman, Understanding copyrights.
38. Yoffie, Intellectual property and strategy.
39. Moon, Online music distribution in a post-Napster world.
40. Associated Press. 2007. Sony BMG to reimburse customers for CD damage. *Wall Street Journal*, January 31: B4.
41. Smith, B., and S. Mann. 2004. Innovation and intellectual property protection in the software industry: An emerging role for patents. *University of Chicago Law Review*, 71: 241–264.
42. Barnett, W., and M. Leslie. 2006. Facebook. *Stanford Graduate School of Business Case*, Number E-220.
43. Yoffie, Intellectual property and strategy.
44. Silverman, Understanding copyrights.
45. Smith and Mann, Innovation and intellectual property protection in the software industry.
46. Silverman, A. 2005. How to customize and maximize federal trademark protection. *JOM*, 57(10): 72.
47. Clark, D. 2007. At the heart of Cisco's iPhone lawsuit: A desire for open standards. *Wall Street Journal*, January 12: A9.
48. Gleick, Get out of my namespace.

49. Ibid.
50. U.S. Department of Commerce, *General Information Concerning Trademarks*.
51. Silverman, How to customize and maximize federal trademark protection.
52. Schilling, *Strategic Management of Technological Innovation*.
53. Etherton, S. 2002. *Let's Talk Patents*. Tempe, AZ: Rocket Science Press.
54. Clark, At the heart of Cisco's iPhone lawsuit.
55. Cohen, D. 1991. Trademark strategy revisited. *Journal of Marketing*, 55: 46–59.
56. Kopp, S., and T. Suter. 2000. Trademark strategies online: Implications for intellectual property protection. *Journal*, 19(1): 119–131.
57. Allen, *Bringing New Technology to Market*.
58. Chan, S. 2005. You can take the A train, but don't take its logo. You may get a warning letter. *New York Times*, June 5: 27.
59. Flandez, R. 2006. Tiny firm wins "chewy Vuiton" suit, but feels bite. *Wall Street Journal*, November 28: B1, B5.
60. Vermette, N. 2000. Domain names in the realm of trademark law. *FICC Quarterly*, 51(1): 1–15.
61. Ibid.
62. Kopp and Suter, Trademark strategies online.
63. Bagby, J., and J. Ruhnka, 2004. Protecting domain name assets. *The CPA Journal*, 74(4): 64–67.
64. Gleick, Get out of my namespace.
65. Bagby and Ruhnka, Protecting domain name assets.
66. Credeur, M. 2006. Case of Bud finished. *International Herald Tribune*, May 25: 17.
67. Kesan, J. 2000. Intellectual property protection and agricultural biotechnology. *American Behavioral Scientist*, 44(3): 464–503.
68. Yoffie, Intellectual property and strategy.
69. Schilling, *Strategic Management of Technological Innovation*.
70. Varian, H. 2005. Copying and copyright. *Journal of Economic Perspectives*, 19(2): 121–138.
71. Fishman, T. 2005. Manufacture. *New York Times Magazine*, January 9: 40–44.
72. Bellman, E. 2005. India senses patent appeal. *Wall Street Journal*, April 20: A20.
73. Benson, T. 2005. Brazil to copy AIDS drug made by Abbott. *New York Times*, June 25: B12.
74. Fishman, Manufacture.
75. Anonymous. 2005. Drug industry invests less in China, India. *Wall Street Journal*, November 10: D6.
76. Fishman, Manufacture.
77. Schilling, *Strategic Management of Technological Innovation*.
78. Ibid.
79. Haleen, I., and A. Scoville. 2003. United States ratifies the Madrid Protocol: Pros and cons for trademark owners. *Intellectual Property and Technology Law Journal*, 15(4): 1–3.
80. Kesan, Intellectual property protection and agricultural biotechnology.
81. Schilling, *Strategic Management of Technological Innovation*.
82. Ibid.
83. Yoffie, Intellectual property and strategy.
84. Kesan, Intellectual property protection and agricultural biotechnology.

Chapter 10

Capturing Value from Innovation

Learning Objectives

After reading this chapter, you should be able to:

1. Define *appropriability*, and explain why nonlegal barriers to imitation are important to high-technology firms.
2. Explain why most innovations are easy to imitate, and how this situation influences firm strategy.
3. Understand when and how controlling a key resource will allow you to appropriate the returns to innovation.
4. Explain when and how economies of scale will allow you to appropriate the returns to innovation.
5. Understand when and how moving up the learning curve will allow you to appropriate the returns to innovation.
6. Explain when and how establishing a reputation will allow you to appropriate the returns to innovation.
7. Understand when and how obtaining architectural control will allow you to appropriate the returns to innovation.

8. Explain when being a first mover is an advantage and when it is a disadvantage, and how to use moving first as an appropriability mechanism.
9. Understand the role of complementary assets in appropriating the returns to innovation, and explain when it is better to be an imitator than an innovator.

Lead Time: A Vignette[1]

A key competitive advantage in the semiconductor industry is lead time. By coming out with smaller microchips before its competitors, a semiconductor firm can sign contracts to supply computer manufacturers with a generation of microchips for their computers while competitors are not yet able to offer similar products. Because the number of transistors on a chip doubles every two years (remember the discussion of Moore's law in Chapter 2?), a six- to nine-month lead time in producing a generation of chips provides a substantial competitive advantage in this industry.

Intel has consistently managed to maintain a lead time advantage over its competitors, coming out with each new generation of semiconductors at least six months before them. Figure 10.1 shows Intel's newest semiconductor, which the company announced in January 2007. Using a new insulator, called hafnium, in place of silicon dioxide, and new metal alloys in place of polysilicon in the transistor gates, Intel was able to shrink the microchip to 45 nanometers, significantly smaller than its current 65-nanometer chips and half the size of the 90-nanometer chips that most of the industry, including rival AMD, produces.[2]

By squeezing twice as many transistors onto a chip as the typical semiconductor manufacturer, Intel can make its products smaller and get higher performance or lower power in the same space. Because Intel has already figured out how to manufacture its new chip for Windows, Mac OSX, and Linux machines, and plans to begin making the chip in 2007, it will be able to extend its lead time of six to nine months on the introduction of new semiconductors into another generation of products.[3]

FIGURE 10.1 Intel's 45 Nanometer Chip

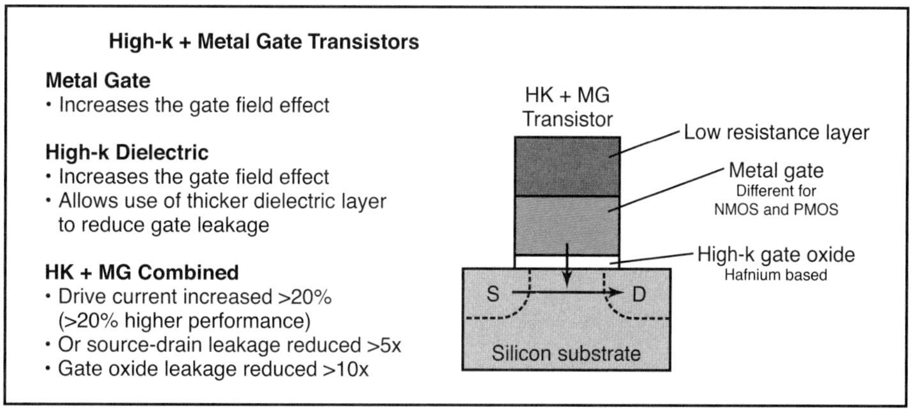

High-k + Metal Gate Transistors

Metal Gate
- Increases the gate field effect

High-k Dielectric
- Increases the gate field effect
- Allows use of thicker dielectric layer to reduce gate leakage

HK + MG Combined
- Drive current increased >20% (>20% higher performance)
- Or source-drain leakage reduced >5x
- Gate oxide leakage reduced >10x

HK + MG Transistor
Low resistance layer
Metal gate
Different for NMOS and PMOS
High-k gate oxide
Hafnium based
S
D
Silicon substrate

Intel's innovation in materials allows it to retain its lead time over rivals in producing ever smaller semiconductors.

Source: http://www.intel.com/technology/silicon/45nm_technology.htm?iid=newstab+45nm.

Introduction

Companies often use technological innovation to enhance their competitive position relative to their rivals.[4] Doing this successfully requires capturing the returns to their investment in innovation because innovating cannot provide a competitive advantage if other companies derive the benefits from it. Take, for example, the case of Palm Computing. Although the company developed an innovative personal digital assistant that customers valued a lot, the company created no barriers to imitation. As a result, competitors copied many of Palm's valuable innovations and stole many of its customers.[5]

So how can you deter imitation of your innovative new products and services? While you can sometimes use the legal mechanisms discussed in the previous two chapters, often these mechanisms are not available or not very effective, and you need to take other approaches, such as exploiting the lead time advantages discussed in the opening vignette.

This chapter focuses on the nonlegal ways that companies appropriate the returns to investment in innovation. The first section identifies several key mechanisms that you can use to capture the profits from innovation, including controlling resources, obtaining architectural control, developing a brand name reputation, moving up the learning curve, being a first mover, and taking advantage of economies of scale. The second section brings together the intellectual property issues discussed in Chapters 8 and 9 with the issues discussed earlier in this chapter to present a model of when to be an imitator, and focus on the control of complementary assets, and when to be an innovator, and focus on the introduction of new products.

Appropriability Mechanisms

As discussed in Chapter 4, developing an innovative new product or service often requires you to invest in R&D, which is costly. This investment makes sense if you can **appropriate**, or capture, the financial returns from it.[6] However, if you cannot capture those returns, then the investment in innovation makes little sense at all.

While you can block imitation and thus appropriate the returns to innovation by using the legal mechanisms described in Chapters 8 and 9, you will often need to use nonlegal barriers to imitation. First, legal barriers to imitation cannot be obtained for many products and services. Take for example, the case of a superior snowboard binding design. You will be unable to patent that design unless you can prove that it is novel and nonobvious, and even then, the patent office might deny your patent application or refuse to give you broad and enforceable claims. You also will be unable to protect the design as a trade secret because it could be easily identified through reverse engineering. Therefore, to capture the returns to your investment in developing this new design, you would need some type of nonlegal barrier to imitation.

Second, most nonlegal barriers to imitation are more effective at deterring innovation than legal barriers.[7] As Table 10.1 shows, managers in high-technology industries believe that patents deter imitation a little more than a third of the time; whereas complementary assets in manufacturing and marketing block imitation over 43 percent of the time, and lead time does so approximately half of the time.[8]

TABLE 10.1 Effectiveness of Barriers to Imitation

This table shows the percentage of companies whose R&D managers believe that different barriers to imitation are effective at protecting new products and processes; it indicates that patents are less effective than lead time, secrecy, and complementary assets in manufacturing and sales and service, in deterring imitation.

Mechanism	Products	Processes
Lead time	53	38
Secrecy	51	51
Complementary Assets in Manufacturing	46	43
Complementary Assets in Sales and Service	43	31
Patents	35	23

Source: Cohen, W., R. Nelson, and J. Walsh. 2000. Protecting their intellectual assets: Appropriability conditions and why U.S. manufacturing firms patent (or not). *NBER Working Paper,* No. 7552.

Third, legal and nonlegal barriers are not mutually exclusive. You can combine the two simultaneously, as would occur if you developed a trade secret and moved up the learning curve, or sequentially, as would occur if you used a patent to protect your new product until you had established a brand name.

Fourth, the degree of protection provided by legal mechanisms in certain technical areas, such as genetics and Internet business methods, are uncertain. Changing legislation and unresolved legal issues can make relying on legal mechanisms to deter imitation in these areas risky. You might invest heavily in obtaining patent protection only to find that it doesn't have the effect that you thought that it would have because legislation or court decisions subsequently limited the strength of that protection.

Fifth, technological change itself has weakened the value of the legal barriers to imitation in many high-technology industries. For instance, the creation of technology to facilitate the exchange of video and audio content on the Internet, such as that used by video-sharing Web sites or peer-to-peer music networks, makes the enforcement of copyrights on video and audio content much more difficult. By dramatically increasing the number of people who violate copyrights, and dramatically reducing the average size of each violation, these technological changes have essentially reduced the value of lawsuits as a deterrent to imitation. Violators know that the odds they will get caught and sued is very small.

Because you cannot rely only on legal barriers to imitation, you need to learn how to employ nonlegal mechanisms to protect your innovations, including controlling key resources, creating a brand name reputation, establishing architectural control, exploiting of economies of scale, moving up the learning curve, and exploiting first mover advantages.

Controlling Key Resources

One way to deter imitation and appropriate the returns to your investment in innovation is by obtaining control over the key resources needed to create and sell your product or service (see Figure 10.2). These key resources will vary across industries and can include such things as manufacturing facilities (e.g., a billion-dollar semiconductor fab), distribution channels (e.g., shelf space in a supermarket),[9] key inputs (e.g., part of the communication spectrum),[10] and the patents, copyrights,

FIGURE 10.2
Controlling Key Resources

By controlling the sources of supply of the star anise plant, a key raw ingredient in the production of the medication Tamiflu, the pharmaceutical firm Roche makes it difficult for competitors to imitate its product.
Source: Alamy.com.

trademarks, and trade secrets that were discussed in the previous two chapters. For example, if you run an oil company, you might want to purchase, or sign long-term contracts with, all of the low-cost sources of oil. This would create a barrier to imitation by insuring that you had a lower cost of oil than your competitors.[11] Similarly, if you run a biotechnology company, you might want to buy the patents necessary to produce the drug that you are developing from the university where it was invented to make sure that your competitors cannot get access to this resource.

The use of resource control as a barrier to imitation does not require the key resources to be natural resources, or even physical resources.[12] They could easily be human resources. While you can't buy up all of the key human resources needed to create and distribute your product, you can still obtain control over those resources through contracting. For example, if you run a biotechnology company, you could sign long-term employment contracts with all of the top genetic engineers in the world so that you had control over their talent, which is a rare resource in drug development.[13]

Controlling resources is most effective at deterring imitation when the key resource is rare and also a **rival good**, which keeps it from being used by two companies simultaneously.[14] If the resource meets these two conditions, you can deter imitation by gaining control over just a few sources of supply. For example, you can deter imitation in cellular communications by licensing part of the radio spectrum from the government, which will keep other people from using that part of the spectrum to compete with you.[15]

Establishing a Reputation

A second way that you can appropriate the returns to your investment in innovation is by developing a reputation for satisfying customers.[16] When value of a new product's attributes is unknown, customers often look to the reputation of the seller as an indication of that value.[17] As a result, companies with stronger brand names can attract customers to their new products more easily than companies with weaker reputations. For example, when IBM developed its first personal computer, it was able to attract business customers much more easily than its competitors because of its reputation as a leading computer manufacturer.[18]

In addition, by creating a strong brand name, you can generate the perception in the minds of customers that your product or service is better than those offered by your competitors. This mitigates the tendency for your customers to shift to competitors' products or services, even if they can obtain those products or services at a lower price. For instance, Apple Computer has sought to convince customers that its products are better than those of its competitors by stressing its brand and its reputation for design excellence.

Furthermore, you can use a brand name to cross-sell products into new markets. If your brand becomes known for something in one area, such as value or excellent engineering, you can leverage that customer perception in other markets. For example, Microsoft has used its brand name to sell video games.[19] Even though Microsoft had no proven ability in making video games when it first introduced those games, its brand name led consumers to believe that its video games would be very good.

Of course, the effectiveness of using reputation to capture the returns to investment in innovation varies greatly across industries. Reputation matters more in industries that serve consumers than industries that serve businesses because businesses are less likely to be swayed by perceptions than by the economics of a transaction. Moreover, among industries that serve consumers, brand names are most effective in industries, such as fashion, in which customer behavior is more heavily influenced by perceptions, and in industries in which advertising is more important in affecting buying decisions.

Developing a brand name is expensive. To build one, you have to invest in advertising, which provides information to customers about the qualities of your new product or service and persuades them that these qualities make the product or service better than those offered by competitors. Because the price of developing and running a radio, television, or print advertisement tends to be fixed, regardless of how many units you produce, advertising is subject to considerable economies of scale. Therefore, advertising is very expensive on a per unit basis if you produce and sell very few units of a product, but falls as your volume increases.

Moreover, it takes a long time to build a reputation though advertising. The nature of the human mind is such that it can only process a certain amount of information at a time, whether that information comes from advertising or some other source. As a result, people do not absorb much from an ad each time they see it. For advertising to be truly effective, repeated messages need to be sent over a long period of time. This means that you have to invest in advertising for a while before you can see any benefits from it.

The use of advertising to build a brand name means that reputation is not a very effective way for start-ups to prevent imitation. Usually, it takes too much time and costs too much money for a new company to develop a strong enough brand name to be competitive. Take, for example, the case of Pets.Com Inc., which spent $25 million on advertising in 1999 and 2000, yet failed to develop enough of a brand name reputation to keep its customers from going to competitors.[20]

Moreover, large, established companies can often use investments in brand names to offset the other competitive advantages of start-up firms. For example, Intel succeeded in taking market share in the notebook computer microchip business away from Transmeta, which had a three-year head start in chip design, because Transmeta could not match Intel's $300 million investment in advertising its Centrino chip. In the end, customers were swayed by brand name reputation in their purchasing decisions.[21]

Obtaining Architectural Control

A third way that you can deter imitation and appropriate the returns to your investment in innovation is by developing **architectural control**, or control over the operation and compatibility of a product or service. Architectural control allows you to determine what products and services work with your own, making it possible to bias compatibility toward your own products. Moreover, it allows you to manage the type and pace of improvements to your technology to ensure that improvements benefit you and not your competitors. For example, Microsoft has been able to influence the development of computer software to favor its products at the expense of its competitors' products because it has architectural control over the dominant interface between computer hardware and software, the Windows operating system.[22]

Architectural control also permits you to maximize your profits from the sale of older versions of products before the introduction of newer ones. Microsoft's architectural control in personal computer operating systems allows it to time the introduction of new generations of operating system software to maximize customer upgrades and minimize the cannibalization of its sales.

When companies have architectural control, they are generally very successful at bundling. As you probably remember from Chapter 6 bundling allows companies to combine old products of known value to customers with new products of unknown value to increase the odds that customers will adopt the new products.

When a company has architectural control, the effectiveness of a bundling strategy is enhanced because the old product that is part of the bundle is not only known, it is also critical to customers. For example, Microsoft used its architectural control over the Windows operating system to push its Web browser forward at the expense of Netscape's Navigator. By bundling its Web browser with its operating system, Microsoft essentially forced computer manufacturers to adopt its Web browser as the one that they would offer on the computers that they sold.[23]

Exploiting Economies of Scale

Another way to deter imitation and appropriate the returns to your investment in innovation is by exploiting **economies of scale**, the reduction in unit costs that occurs as production volume increases. When economies of scale exist, larger firms have a lower cost of production than smaller firms, which allows them to deter imitation by keeping their prices low. For example, the capital intensity of the semiconductor business means that Intel, which is larger than other firms in the industry, can produce new generations of semiconductors at much lower unit costs than its competitors.

Larger firms can also deter imitation by investing in so much capacity that entry into the industry by other firms would be unprofitable. Because volume production reduces costs dramatically, new entrants have to enter on a large scale to match the cost structure of existing competitors. This means that their entry will create a large amount of excess capacity, dragging down prices, and reducing profits.[24] Faced with the prospect that entry will be unprofitable, imitators often choose not to enter.

For example, Monsanto is the world's largest producer of glycophosate, the active ingredient in herbicides, like Roundup. Because the production of glycophosphate requires high levels of capital investment, production is subject to scale economies. By operating at a very large scale, Monsanto has driven production costs so low that it is cheaper for companies to source glycophosate from Monsanto than to produce it on their own, which keeps other companies from entering the glycophosphate business.[25]

In general, economies of scale provide more of a barrier to imitation for large, established firms than for small, new ones. The advantages of scale economies go to those companies that have more scale, which tend not to be the new ones. While new companies can sometimes be created at a larger size than established competitors, most of the time, the cost, risk, and difficulty of creating a new company on a large scale means that new companies usually are smaller than established companies in the businesses that they enter.

Moving Up the Learning Curve

You can deter imitation and appropriate the returns to your investment in innovation by moving up the learning curve. (A **learning curve** is a graphical depiction of how well someone does at something as a function of the number of times that they have done it.[26]) Moving up the learning curve helps you to deter imitation for two reasons. First, as Figure 10.3 indicates, by doing more of something than your competitors, you become better at it than them. (You produce more output at a lower input cost.[27]) As a result, your competitors cannot produce a new product with the same efficiency as you, and so choose not to copy what you are doing.

For example, as semiconductor firms produce more semiconductors, they learn how to solve problems that lead to poor yields. As a result, their production yields—the proportion of products that meet performance standards—improve with the volume produced. Because of the capital intensity and complexity of the process of making semiconductors, the advantage that innovators have in increasing production yields makes copying their efforts uneconomical, making the learning curve an important barrier to imitation in this industry.[28]

Second, as you gain experience making and selling a product, you learn how to create features that your competitors cannot match. The inability of your competitors to match your product deters imitation because customers do not find their alternatives as appealing as yours. For example, many consumer electronics firms have learned how to make their products smaller and more robust to wear-and-tear through the process of producing and selling the devices. Other companies have been unsuccessful at copying these products because their lack of experience keeps them from providing similar features.

FIGURE 10.3
The Learning Curve

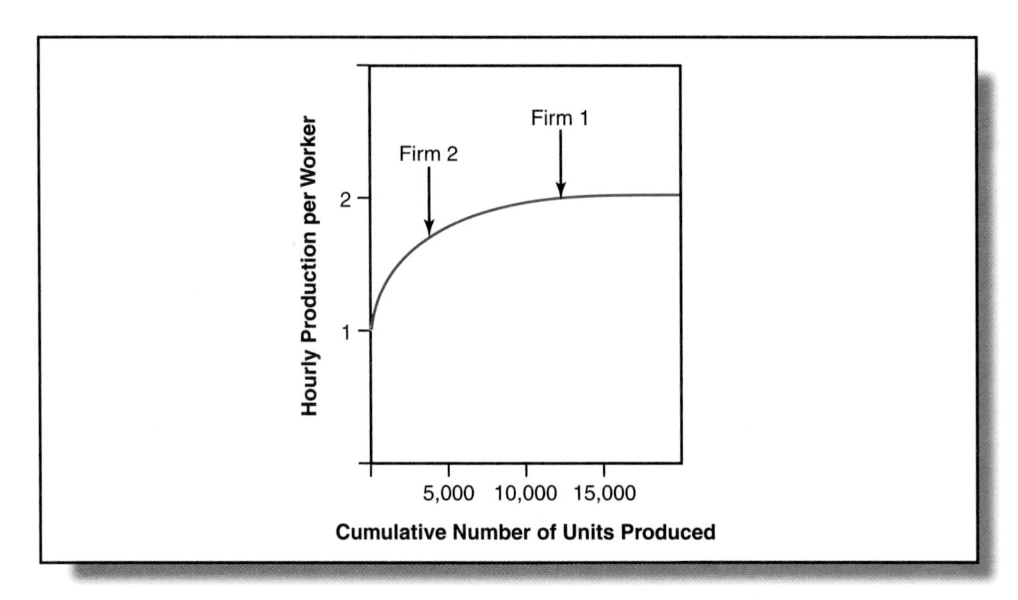

In this example, Firm 1 has a learning curve advantage over Firm 2; because of its experience at producing the product, it can produce more units per worker per hour than its competitor.

Your ability to use the learning curve to deter imitation depends on your ability to learn from experience. If your organization has a lot of employee turnover; poor mechanisms for capturing learning; an inability to access information that already has been learned; or has poor mechanisms for transferring knowledge internally, then it will be poor at learning from experience. As a result, your organization will be unable to use the learning curve to appropriate the returns to innovation.[29]

In addition, you can only use the learning curve to appropriate the returns to innovation if learning is proprietary. If what you have learned seeps out to other companies in your industry, then your competitors will know whatever you know, equalizing their performance with yours, even if they have less experience. For instance, many early Internet clothing start-ups were unable to generate learning curve advantages because later start-ups learned what those entrants had figured out about how to get people to buy clothing online simply by observing the efforts of the early entrants.

Learning that is easily **codified** (it can be written down in drawings or words) is less proprietary than learning that is tacit because tacit knowledge cannot be transferred easily and can only be gained by doing.[30] Because tacit knowledge is only available to firms with operating experience in an industry, while codified knowledge is available to anyone, companies exploiting tacit knowledge can stay ahead of their competitors on the learning curve more easily than companies exploiting codified knowledge.

The learning curve is more effective at deterring imitation in some industries than in others because more of the knowledge that provides value in those industries is tacit. For example, much of the knowledge about how to make aircraft is tacit, and held in the brains of aerospace engineers, while most of the knowledge about computer networking is codified in books, articles, and patents.[31] As a result, the learning curve is a stronger deterrent to imitation in aerospace than in computer networking.

As you have probably already figured out, the learning curve is a much better mechanism to appropriate the returns to investment in innovation for established companies than for new ones. By definition, any learning curve advantages that exist go to the companies that have produced more of a product. Therefore, established companies, which have been operating in an industry longer than new companies, tend to have the learning curve advantages. The only exception to this occurs when all of the learning resides in the head of individuals, as might be the case in, say, IT consulting. In that industry, if all of the experienced employees quit an established consulting company to start their own, the new company might end up higher on the learning curve than the old one because it would have the more experienced consultants.

Exploiting a First Mover Advantage (Lead Time)

A sixth way that you can deter imitation and appropriate the returns to your investment in innovation is by exploiting a **first mover advantage**, or the advantage that accrues to a company from being the first to enter a market. Several studies have shown that first movers often are able to obtain a higher market share[32] and earn higher profits than later entrants.[33]

First Mover Advantages

Being a first mover can help you to appropriate the returns to investment in innovation in several ways. First, by moving early, you are more likely to obtain control over key resources. For instance, you can acquire assets that are in limited supply,[34] such as the most attractive physical locations, the best distribution channels, or the least expensive sources of raw materials and other inputs.[35] In addition, you can work to reduce your costs of production by moving up the learning curve unfettered by competition before other firms can reap the benefits of learning.[36] Furthermore (as will be explained in greater detail in Chapter 13), you can tip the industry toward your product before competitors can catch up, as Apple appears to have done in the music download business with its early introduction of the iPod.[37]

Second, by moving first, you can target the best customers in a market. By selling your products to the innovators and early adopters, you will leave only potential customers who are negatively disposed to adopting new products, making it hard for imitators to sell their products successfully.[38] Moreover, as a first mover, when you first enter a market, you will have a monopoly on promoting your products. Therefore, you can advertise without customers hearing competing messages,[39] making your advertising more effective.

Third, as a first mover, you can exploit **switching costs**, or the cost to customers of changing suppliers. While moving first doesn't guarantee that you can create switching costs, it is a necessary condition. If you are a late mover, then someone else will have the opportunity to create switching costs that create obstacles for customers to change from its products to yours.

Switching costs can take a variety of different forms. Sometimes, they take the form of the additional expense necessary to purchase a replacement product. These costs can be particularly high when products are bundled and other products must also be changed if suppliers are replaced.[40] For example, for a long time, customers were deterred from switching from Apple to Microsoft as a supplier of graphics software because changing to Windows-based software also required changing computer hardware.

Other times, switching costs exist because it is costly for customers to learn a new system. For example, many hospitals use particular intravenous solution delivery systems because nurses have learned to use those systems, and it is too costly for the hospitals to retrain them to use different systems.[41]

Switching costs also exist when breaking contractual arrangements is expensive, as occurs when those contracts are written to impose large penalties for changing suppliers. For example, most cellular telephone companies require their customers to pay stiff penalties and return free telephone hardware if they change carriers, as a way to deter them from doing so.[42]

Switching costs can also be psychological.[43] People are biased in favor of the status quo because they dislike searching for alternatives.[44] As a result, they treat first movers as the default option and the standard against which other alternatives are compared.[45] As long as customers have had a positive experience with a first mover's product, they prefer to continue with it and demand a significantly better experience from a follower's product to switch.[46] Therefore, the sellers of a follower product have to spend more money to get the same level of customer adoption as the first product receives,[47] making the cost of advertising and promotion per unit sold lower for first movers than for followers. For instance, Amazon.com is the default choice for most customers who buy books online. Customers switch suppliers only when Amazon.com fails to satisfy their needs.[48] As a result, Amazon.com has to spend less money to keep its customers than its competitors have to spend to woo them away.

Late Mover Advantages

Despite the advantages listed previously, moving first is not always a good strategy. As Table 10.2 shows, many late movers have been more successful than early movers. For instance, Google was a late entrant in Internet search but has done better than Lycos, the first mover. Similarly, many other first movers have been less successful than their late moving competitors, including Netscape in the Web browser business,[49] VisiCalc in the spreadsheet business,[50] and Apple in the PDA market.[51] In fact, one study showed that first movers have a high—47 percent—failure rate and achieve only about 10 percent market share.[52]

Why are late movers sometimes more successful than first movers? The answer is that being a late mover provides several advantages to companies. First, as a late mover, you benefit from the ability to free ride on the investments that the first mover makes in creating supply infrastructure and distribution channels.[53] For instance, when DEKA Research developed its IBOT wheelchair, it had to create its own ball bearings because no company could supply it with the ones it needed.[54] By the time later movers had entered the industry, however, other companies had figured out how to provide these ball bearings, reducing the cost of sourcing supply.

Second, as a late mover, you can design products that correct the mistakes that the first mover has made in meeting customer needs.[55] First movers often face customers who do not know what features they want in new products or services, or what they will pay for them, while later movers face customers that are better educated about the product category.[56] The lower uncertainty about customer preferences faced by later movers allows them to design products that better fit customer needs.[57]

Third, as a late mover, you can leapfrog ahead of the first mover's technology.[58] Often, first movers cannot adapt to changing supply or demand conditions, and are unable to develop products based on a new generation of technology because they are locked into earlier product designs or process technologies.[59] As a result, later movers can often develop better products and production processes than first movers. For example, later movers in personal computers benefited from the

TABLE 10.2
Successful Companies That Were Not First Movers
This table shows several examples of companies that were the most successful firms in their product markets but were not first movers.

Product Market	Successful Company	First Mover
Mainframe computer	IBM	Atansoff's ABC Computer
DNA synthesis	Genentech	Biologicals
Pocket calculator	Texas Instruments	Bowmar
Fax machine	Sharp	Xerox
VCR	JVC	Ampex
Personal computer	IBM	Osborne
Video games	Magnavox	Nintendo
Spreadsheet	Microsoft	VisiCalc
35 mm camera	Canon	Leica
PC operating system	Microsoft	Digital Research
Food processor	Black & Decker	Cuisinart
Disposable diaper	Procter & Gamble	Chicopee Mills
Web browser	Microsoft	Netscape
CT scanner	GE	EMI
Personal digital assistant	Palm (U.S. Robotics)	Apple
Internet search	Google	Lycos
Online bookselling	Amazon.com	Computer Literacy Bookstore
Online brokerage	Charles Schwab	Howe Barnes Investments
8 mm video camera	Sony	Kodak
Microwave	Samsung	Raytheon
Word processing software	Microsoft	MicroPro
Computer workstation	Sun Microsystems	Xerox

Source: Created from information contained in Markides, C., and P. Geroski. 2005. *Fast Second: How Smart Companies Bypass Radical Innovation to Enter and Dominate New Markets*. San Francisco: Jossey-Bass; Schilling, M. 2005. *Strategic Management of Technological Innovation*. New York: McGraw-Hill.

changes in customer preferences to which Osborne Computer, the industry pioneer, could not respond because it had already committed to a particular technological approach that was incompatible with these changes.[60]

Fourth, as a late mover, you can benefit from the investments in R&D that the first mover has made. Because knowledge spills over from the first mover to later movers, the cost of innovation is higher than the cost of imitation. As a result, first movers spend more money than later movers to develop comparable new products.[61] Therefore, you may be better off entering a market late, particularly if the technology underlying a new product is costly to develop and inexpensive to imitate.[62]

Fifth, as a later mover, your entry into the market is often better timed to take advantage of the development of complementary technology than the first mover's entry, which sometimes occurs before the complementary technology has had a chance to develop. For example, to have hydrogen fuel cell vehicles, someone first needs to figure out how to store compressed hydrogen in pressurized tanks at fueling stations. A first mover that enters the market before this refueling problem is resolved would face very slow adoption of its vehicles, and might be worse off than a later mover that enters after the problem had been solved.

First Mover or Late Mover?

So how do you know if you should be a first mover or a late mover? The answer depends on your industry. The following dimensions of industry affect this choice:

- First, you are better off being a first mover in industries in which products and services are expensive, cannot be valued easily prior to purchase, are durable, and are infrequently purchased.[63] In these industries, once customers have adopted a product, they are not likely to repurchase it very soon. As a result, first movers are likely to remove the best customers from the market before later movers have a chance to sell to them.
- Second, you are better off being a first mover in advertising-intensive industries, and industries in which customers learn very little or very slowly about new products.[64] In these industries, it is very difficult to persuade customers to switch from the first mover's product to the later mover's version. As a result, first movers face a much lower cost to attract customers than later movers.
- Third, you are better off being a first mover in industries in which products require distributors to hold large stocks, additional parts, or complementary products to satisfy the needs of end users.[65] In these industries, later movers often find it difficult to obtain access to distribution channels because the distributors lack the space to handle the later mover's product in addition to the first mover's.
- Fourth, as will be discussed in greater detail in Chapter 13, you are better off being a first mover in industries where network externalities exist.[66] In these industries, the first mover has the opportunity to build an installed base and tip the market to its product before later movers have a chance to compete.

TABLE 10.3
Mechanisms to Appropriate Returns
This table shows some examples of the different mechanisms that are used to appropriate the returns to investments in innovation.

MECHANISM	EXAMPLE	WHY IT WORKS
Control over resources	Providing cellular telephone service	Without access to the wireless spectrum purchased at a government auction, companies cannot provide cell phone service
Economies of scale	Selling herbicide	Because of economies of scale in production, the company that produces the largest volume has the lowest cost
Reputation	Providing accounting software	Reputation keeps customers from defecting to the competition
Learning curve	Selling MP3 players	By making the devices, a company reduces its defect rate and has higher quality products than its competition
First mover advantage	Establishing an Internet auction house	Because of increasing returns, the company that enters first builds a larger installed base and tips the market to its product
Architectural control	Offering a computer operating system	Architectural control allows a company to determine which products are compatible and which are not

- Fifth, you are better off being a first mover in industries in which patents are more effective.[67] In these industries, first movers have the chance to develop pioneering patents that later movers cannot get around.
- Sixth, you are better off being a first mover in industries in which scale economies are very large.[68] In these industries, moving first will allow you to grow large and drive your costs down before your competitors can introduce an alternative product.

Key Points

- Companies need to deter imitation to appropriate the returns to investment in innovation.
- Nonlegal mechanisms—controlling key resources, exploiting economies of scale, moving up the learning curve, being a first mover, building a brand name reputation, and establishing architectural control—help companies to do this.
- Nonlegal barriers to imitation are valuable because legal and nonlegal barriers are not mutually exclusive, and legal barriers cannot always be obtained, are of uncertain value in some fields, are less effective than nonlegal barriers in most industries, and because technological change has weakened their value in some industries.
- Controlling key resources is most effective when resources are rare and are rival goods.
- By exploiting economies of scale, companies can reduce their costs below their competitors', and make entry by other firms unprofitable.
- By moving up the learning curve, companies become more efficient at making products and develop product features that competitors cannot match.
- Learning curve advantages exist only if learning is proprietary and firms are good at learning.
- By building a reputation, companies can attract customers more easily and keep them from shifting suppliers; however, brand names are more effective at appropriating the returns to investment in innovation in industries that serve consumers, particularly those that are strongly affected by perception.
- Architectural control allows firms to limit compatibility of their products to companies that are not a competitive threat, to bias compatibility to their own products, and to control the type and pace of product improvement.
- Being a first mover offers a variety of advantages and disadvantages to firms; whether it is better to be a first mover or a later mover depends on which industry you are in.

Teece's Model

Many companies do not succeed with efforts to introduce innovative products because their competitors imitate those products and offer their versions at a lower cost, taking customers (and profits) away from the innovators.[69] This means that you need to figure out when you are better off being an innovator and developing new products, and when you are better off being an imitator and letting your competitors develop them. David Teece, a professor at the Haas School of Business at the University of California at Berkeley, has developed a model that you can use to make this decision. The model is based on the idea that imitators are more successful than innovators when innovations are easy to imitate, a dominant design has

FIGURE 10.4
Teece's Model

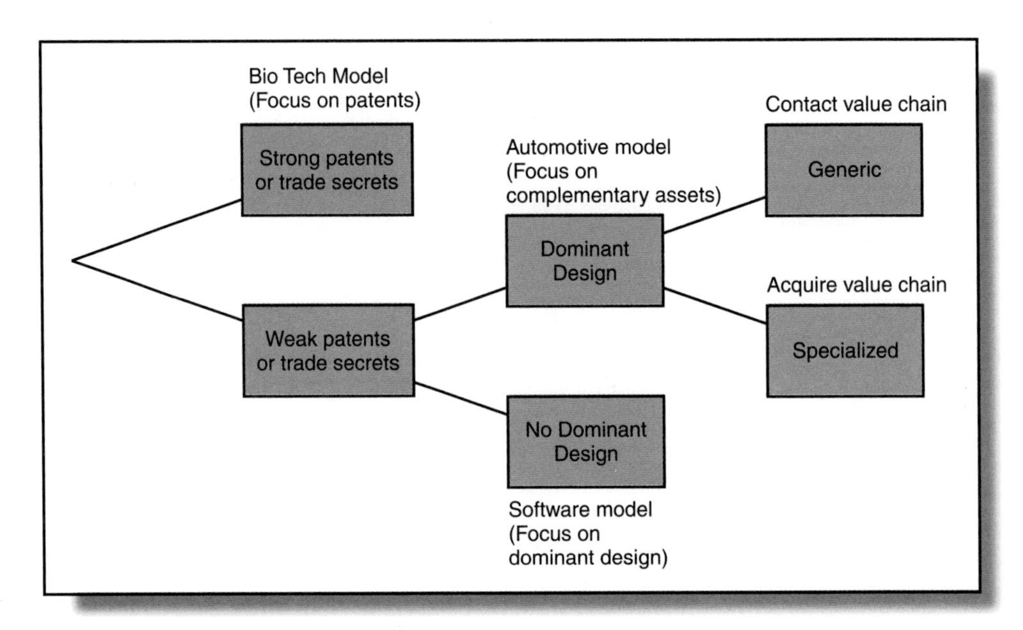

When using Teece's model, decision makers first need to determine if the industry is one in which patents or trade secrets are effective, then determine if convergence on a dominant design has occurred, and then if the value chain assets are specialized or generic.

emerged in an industry, and imitators control the key complementary assets in the industry.

Previous chapters explained when a dominant design exists, and when innovations are easy to imitate. So we just need to define **complementary assets**, and then we can explain Teece's model. Complementary assets are upstream or downstream assets that are used to develop, produce, or distribute an innovative new product or service.[70] For example, Merck's pharmaceutical sales force, which numbers in the thousands and can reach most doctors in the world; Sony's specialized facilities for producing high-definition televisions; and Bristol Myers Squibb's process of conducting drug clinical trials and obtaining FDA approval are all complementary assets because they are all used along with the innovations themselves to generate and capture value from innovating.

Difficult to Imitate

Innovators usually do not lose out to imitators in industries in which imitation of new products and services is difficult. Therefore, the first step in applying Teece's model is to determine how difficult it is to imitate new products and services in your industry. To do this, you want to look at the effectiveness of patents and trade secrets in your industry (perhaps by examining data like that in Table 8.3, which shows patent effectiveness in different industries). In industries in which patents or trade secrets are very effective at deterring imitation, being an innovator is a good idea. The barriers to imitation afforded by patents or trade secrets will allow you to keep your competitors from copying your new products, adjust the design of your new products if market feedback suggests that such changes are necessary,[71] and build or contract for the complementary assets needed to exploit the innovations.[72] For example, being an innovator is a good idea in biotechnology because companies can exploit the effectiveness of drug patents to compete successfully with imitators.

Easy to Imitate, No Dominant Design

However, your strategy needs to be different in industries in which innovations are easy to imitate. This is the case in industries like consumer electronics, in which patents are very easy to get around. Here your decision of whether or not to be an innovator depends a lot on whether or not your industry has converged on a dominant design.[73] (Remember from Chapter 2 that a dominant design is a common form that all products or services in an industry take).

If your industry has not yet converged on a dominant design, then it is hard to say whether or not you should be an innovator. The success of innovators in industries in which imitation is easy, but no dominant design exists, depends largely on what product designs are favored by different niche markets, and what design ultimately becomes dominant. If you come up with a design that appeals to a valuable niche market, or better yet, a design that ultimately becomes dominant, then your company can capture the returns to innovation through lead time or first mover advantages.[74] For example, in the computer industry, Sun Microsystems succeeded as an innovator by developing a product that appealed to the niche market of computers for network servers in the period before the industry converged on a dominant design.

Easy to Imitate, Dominant Design

The story is very different if your industry has already converged on a dominant design, but new products are easily imitated. Under these conditions, you need to control complementary assets in marketing and manufacturing to be successful. As Chapter 2 explained, once a dominant design is in place, companies make products that are very similar to one another. Consequently, success goes to the firms that control manufacturing and marketing assets because they can produce products at a lower cost.[75]

To succeed in industries in which the imitation of new products and services is easy, and convergence on a dominant design has occurred, innovators need to gain control over complementary assets quickly. Typically, this requires contracting for manufacturing and marketing assets because building these assets from scratch usually takes too long, and because contracting minimizes the cost and risk of capital investment.[76] For instance, when they first began, cellular telephone companies contracted with traditional telecommunications companies to gain access to the telephone companies' switching networks.[77]

The success of this strategy depends on whether the complementary assets are generic or specialized. Complementary assets are **generic** if they do not need to be modified to fit the innovation, and **specialized** if they need to be modified. If the complementary assets are generic, then an innovator can obtain access to the needed complementary assets through contracting.[78] However, if the complementary assets are specialized, then contracting will not work. Specialized assets cannot be redeployed cheaply to other uses. Therefore, any investment that a company undertakes to modify a specialized complementary asset to work with another company's innovation is vulnerable to the second company's opportunistic actions to strike a better deal by exploiting the first company's dependence. This threat of opportunistic action deters firms from contracting. As a result, when complementary assets are specialized, companies usually have to own them to use them.[79]

This need for ownership is what dooms the innovator's effort. Innovators often cannot obtain enough capital fast enough to build complementary marketing or manufacturing assets to exploit their innovations (if they do not already own them). As a result, in industries in which specialized complementary assets are important to

GETTING DOWN TO BUSINESS

Should You be an Innovator or an Imitator?[80]

In 1973, the central research laboratory of EMI Ltd., the dominant player in the movie and recording industry in the United Kingdom, developed a new medical imaging technology called computerized axial tomography (CAT), which allowed doctors to take three-dimensional images of the human body. To capture the value of this invention, EMI set up a separate subsidiary, called EMI Medical Inc., and began to manufacture and sell CAT scanners.[81]

The technology proved to be very popular with doctors and hospitals. Medical professionals saw CAT scans as a significant improvement over alternative imaging technologies, such as X-rays. Because demand for the product was strong, EMI's sales grew rapidly, and, by 1976, sales of CAT scanners were contributing 20 percent of that company's profits.[82]

The profitability of the product attracted competitors, like General Electric (GE), which had a strong medical device operation. GE developed its own CAT scanner, which it introduced to the market in 1976. Because patent protection on the EMI scanner was weak, GE's scanner used the same core technology as EMI's scanner and had most of its important features.[83]

At the time that GE entered the CAT scanner market, EMI had a commanding position, accounting for a 75 percent market share. However, one year later, its market share had dropped to 56 percent, and in subsequent years, it continued to fall. By the end of the 1970s, EMI was forced to exit the CAT scanner business as GE took control of the market.[84]

So what happened? Why did an innovator like EMI fail to benefit from the introduction of a valuable and innovative product like the CAT scanner? The answer lies in the importance of controlling complementary assets in an industry in which a dominant design is present and intellectual property protection is weak. Weak patent protection allowed GE to copy the key features of the EMI scanner. This, combined with the dominant design present in the industry, meant that success in the medical imaging business depended on the efficiency of sales and marketing. Because GE had a large medical products sales force and service network in place, and EMI did not, GE quickly drove EMI out of the CAT scanner business.[85]

exploit an innovation, new products and services are easy to imitate, and convergence on a dominant design has already occurred, it is better to be an imitator than an innovator.[86] For example, being an imitator is a better strategy than being an innovator in the auto industry in which patents tend to be relatively ineffective, the dominant design of the internal combustion engine exists, and complementary assets in manufacturing are highly specialized.

In short, being an innovator or an imitator depends on industry conditions. Table 10.4 summarizes the different conditions under which firms will be more successful if they are innovators and the conditions under which they will be more successful if they are imitators.

TABLE 10.4 Imitator Strategy

This table shows when you are better off being an imitator and when you are better off being an innovator.

Be an Imitator	Be an Innovator
Patents and trade secrets are not effective	Patents and trade secrets are effective
A dominant design exists	A dominant design does not exist
Complementary assets are important and specialized	Complementary assets are unimportant and generic

Key Points

- Innovators do not always capture the returns to innovation; these returns sometimes go to imitators who control complementary assets in manufacturing and marketing.
- In industries in which products and services are difficult to imitate, innovators tend to be successful.
- In industries in which new products and services are easy to imitate and dominant designs have not yet been established, innovators' success depends on their ability to make their technologies the dominant designs.
- In industries in which new products and services are easy to imitate and dominant designs have been established, innovators' success depends on control of complementary assets.
- If complementary assets are generic, then innovators can contract for them.
- If complementary assets are specialized, then contracting is undermined by problems of opportunistic renegotiation.
- When innovators do not control specialized complementary assets, intellectual property protection is weak, and a dominant design exists, being an imitator is a better strategy than being an innovator.

Discussion Questions

1. How can a firm appropriate the returns to investment in innovation? Is one approach to appropriating returns better than another? Why? What factors influence the effectiveness of different approaches to appropriating the returns to investment in innovation?
2. What are the limitations to legal mechanisms for appropriating the returns to investment in innovation? What are the implications of these limitations for technology strategy?
3. What are the advantages and disadvantages of being a first mover? When should you be a first mover and when should you be a late mover?
4. Are any mechanisms to appropriate the returns to investments in innovation more effective for start-ups than for established firms? What about the opposite? Why?
5. How does Teece's model help an entrepreneur or a manager formulate a technology strategy? Are there differences in exploiting the model if you run a large established company or a small start-up? Why or why not?
6. When is it better to be an innovator and when is it better to be an imitator? Why?
7. Why is it easier to contract for generic complementary assets than specific ones?

Key Terms

Appropriate: To capture the financial returns from investment in innovation.

Architectural Control: Control over the operation and compatibility of a product or service.

Codified: Written down in drawings or words.

Complementary Assets: The assets that are used in conjunction with an innovative new product or service to capture value.

Economies of Scale: A situation in which unit costs decline as production volume increases.

First Mover Advantage: The advantage that accrues to a company from being the first to enter a market.

Generic Complementary Assets: The assets that are used in conjunction with an innovative new product or service to capture value that do not need to be modified to fit the innovation.

Learning Curve: A graphical depiction of how well someone does at something as a function of the number of times that they have done it.

Rival Good: A good that cannot be used by two companies simultaneously.

Specialized Complementary Assets: The assets that are used in conjunction with an innovative new product or service to capture value that need to be modified to fit the innovation.

Switching Costs: The cost to customers of changing suppliers.

PUTTING IDEAS INTO PRACTICE

1. **Understanding Value Creation Strategies in High-Technology Industries** The purpose of this exercise is to develop an understanding of what strategies are effective and ineffective at capturing value from innovation in a high-technology industry. Pick a company. Using the information in this chapter on mechanisms to appropriate the returns to investment in innovation, as well as those in the chapters on patents, trade secrets, copyrights, and trademarks, identify all of the mechanisms that the firm uses to deter imitation. Explain why they use these mechanisms and not others. Next, indicate whether these barriers to imitation are likely to change in the future. Explain why or why not. Then, specify whether the barriers to imitation would be different if the firm were small and new or if it were large and established, and explain why or why not.

2. **Learning Curve** This exercise[87] is designed to help you understand the learning curve. (The learning curve measures the relationship between the cumulative amount produced and the cost of producing that product.) First, calculate the cumulative number of gigabytes of DRAM produced by Intel by the end of each year. Next, calculate the cost per gigabyte in each year. Then, calculate the natural log of the cumulative number of gigabytes of DRAM produced in each year and the natural log of the cost per gigabyte in each year. Next create a scatter plot of the natural log of the cost per gigabyte against the natural log of the cumulative number of gigabytes produced. Does the scatter plot show evidence of a learning curve? Why or why not? What does this scatter plot tell you about Intel's ability to appropriate the returns to its investment in innovation in DRAM? Does Intel have a learning curve advantage?

TABLE 10.5
Data for Exercise 10.2: Learning Curve
DRAM production at Intel from 1974 to 1997.

YEAR	GIGABYTES PRODUCED	COST OF SALES (MILLIONS)
1974	2	68
1975	15	67
1976	78	117
1977	180	144
1978	441	196
1979	964	313
1980	2,128	399
1981	3,026	458
1982	7,735	542
1983	19,899	624
1984	45,915	883
1985	58,775	943
1986	1,28,564	961
1987	1,71,488	1,044
1988	3,23,469	1,506
1989	4,30,596	1,721
1990	6,46,964	1,930
1991	10,30,197	2,316
1992	18,97,935	2,557
1993	31,28,925	3,252
1994	48,03,048	5,576
1995	88,26,830	7,811
1996	1,54,59,917	9,164
1997	3,00,10,120	9,945

Source: Adapted from Schmidt, G., and S. Wood. 1999. The growth of Intel and the learning curve. *Stanford Graduate School of Business Case,* Number S-OIT-27.

TABLE 10.6
Data for Exercise 10.3
Some first movers remain market leaders, but others do not.

Product Market	First Mover	Remained Market Leader?
Antivirus software	Symantec	Yes
CAD software	Autodesk	Yes
HDTV	Zenith	Yes
Ink-jet printer	Hewlett-Packard	Yes
Camcorder	Kodak	No
DVD player	Toshiba	No
Laser printer	IBM	No
Home VCR	Sony	No

Source: Adapted from Srinivasan, R., G. Lilien, and A. Rangaswamy. 2004. First in, first out? The effects of network externalities on pioneer survival. *Journal of Marketing* 68(1): 41–58.

3. **First Mover Advantage** The purpose of this exercise is to help you understand when being a first mover is an advantage, and when it is not. From Table 10.6, select one of the first movers that remained a market leader and one of the first movers that did not. Investigate the introduction of the new product by each of the competitors. For the first mover that remained a market leader, explain why being a first mover was an advantage. For the first mover that did not remain a market leader, explain why being a first mover was not an advantage.

Notes

1. Adapted from Markoff, J. 2007. Intel says chips will run faster, using less power. *New York Times*, January 27: A1, B9.
2. Ibid.
3. Ibid.
4. Narayanan, V. 2001. *Managing Technology and Innovation for Competitive Advantage*. Upper Saddle River, NJ: Prentice Hall.
5. Cyran, R., and E. Hadas. 2007. Learning from Palm's pain. *Wall Street Journal*, March 6: C2.
6. Kesan, J. 2000. Intellectual property protection and agricultural biotechnology. *American Behavioral Scientist*, 44(3): 464–503.
7. Cohen, W., R. Nelson, and J. Walsh. 2000. Protecting their intellectual assets: Appropriability conditions and why U.S. manufacturing firms patent (or not). *NBER Working Paper*, No. 7552.
8. Ibid.
9. Shane, S. 2003. *A General Theory of Entrepreneurship: The Individual-Opportunity Nexus*. Cheltenham, United Kingdom: Edward Elgar.
10. Lieberman, M., and C. Montgomery. 1988. First mover advantages. *Strategic Management Journal*, 9: 41–58.
11. Ibid.
12. Ibid.
13. Afuah, A. 2003. *Innovation Management*. New York: Oxford University Press.
14. Kim, W., and R. Mauborgne. 1999. Strategy, value innovation, and the knowledge economy. *Sloan Management Review*, Spring: 41–54.
15. Schilling, M. 2005. *Strategic Management of Technological Innovation*. New York: McGraw-Hill.
16. Rao, P. 2005. Sustaining competitive advantage in a high-technology environment: A strategic marketing perspective. *Advances in Competitiveness Research*, 13(1): 33–47.
17. Calantone, R., and K. Schatzel. 2000. Strategic foretelling: Communication-based antecedents of a firm's propensity to preannounce. *Journal of Marketing*, 64(1): 17–30.
18. Afuah, *Innovation Management*.
19. Yoffie, D., D. Mehta, and R. Seseri. 2006. Microsoft in 2005. *Harvard Business School Case*, Number 9-705-505.
20. Tam, P., and R. Buckman. 2007. Tech start-ups have money to burn, but choose thrift. *Wall Street Journal*, January 18: B1.
21. Markides, C., and P. Geroski. 2005. *Fast Second: How Smart Companies Bypass Radical Innovation to Enter and Dominate New Markets*. San Francisco: Jossey-Bass.

22. Schilling, *Strategic Management of Technological Innovation*.
23. Foster, J., and P. Brennan. 2005. Qualcomm sued by smaller rival for unfair trade. *The Plain Dealer*, July 6: C3.
24. Afuah, *Innovation Management*.
25. Urban, T. 1995. Monsanto Company: The coming age of biotechnology. *Harvard Business School Case*, Number 9-596-034.
26. Yelle, L. 1979: Historical review and comprehensive survey. *Decision Sciences*, 10: 302–328.
27. Suarez, F., and G. Lanzolla. 2005. The half-truth of first-mover advantage. *Harvard Business Review*, April: 1–8.
28. Yoffie, D., R. Casadesus-Masanell, and S. Mattu. 2004. Wintel (A): Cooperation or Conflict? *Harvard Business School Case*, Number 9-704-419.
29. Argote, L., and D. Epple. 1990. Learning curves in manufacturing. *Science*, 247(4945): 920–924.
30. Rao, Sustaining competitive advantage in a high-technology environment.
31. Levin, R., A. Klevorick, R. Nelson, and S. Winter. 1987. Appropriating the returns from industrial research and development. *Brookings Papers on Economic Activity*, 3: 783–832.
32. Robinson, W. 1988. Sources of market pioneer advantages: The case of industrial goods industries. *Journal of Marketing Research* 25(1): 87–94.
33. Boulding, W., and M. Christen. 2003. Sustainable pioneering advantage? Profit implications of market entry order. *Marketing Science*, 22(3): 371–392.
34. Dyer, B., A. Gupta, and D. Wilemon. 1999. What first-to-market companies do differently. *Research Technology Management*, 42(2): 15–21.
35. Kerin, R., P. Varadarajan, and R. Peterson. 1992. First mover advantage: A synthesis, conceptual framework and research propositions. *Journal of Marketing*, 56(4): 33–52.
36. Ibid.
37. Fleetwood, C. 2006. Microsoft's Zune falls off sales pace for media players. *Wall Street Journal*, November 28: B2.
38. Kerin, Varadarajan, and Peterson, First mover advantage.
39. Ibid.
40. Lieberman and Montgomery, First mover advantages.
41. Ibid.
42. Ibid.
43. Robinson, Sources of market pioneer advantages.
44. Lieberman and Montgomery, First mover advantages.
45. Mellahi, M., and M. Johnson. 2000. Does it pay to be a first mover in e-commerce? *Management Decision*, 38(7): 445–452.
46. Kerin, Varadarajan, and Peterson, First mover advantage.
47. Ibid.
48. Mellahi, and Johnson, Does it pay to be a first mover in e-commerce?
49. Heinzel, M. 2005. With its BlackBerry a big hit, RIM is squeezed by all comers. *Wall Street Journal*, April 25: A1, A5.
50. Eisenmann, T., and F. Suarez. 2003. Symbian: Setting the mobility standard. *Harvard Business School Case*, Number 9-804-076.
51. Allen, K. 2003. *Bringing New Technology to Market*. Upper Saddle River, NJ: Prentice Hall.
52. Golder, P., and G. Tellis. 1993. Pioneer advantage: Marketing logic or marketing legend? *Journal of Marketing Research*, 20: 65–75.
53. Lieberman and Montgomery, First mover advantages.
54. Schwartz, E. 2002. The inventor's playground. *Technology Review*, 105(8): 69.
55. Dyer, Gupta, and Wilemon, What first-to-market companies do differently.
56. Robinson, W., and S. Min. 2002. Is the first to market the first to fail? Empirical evidence for industrial goods businesses. *Journal of Marketing Research*, 34(1): 120–128.
57. Ibid.
58. Adapted from Cusumano, M., Y. Mylonadis, and R. Rosenbloom. 1992. Strategic maneuvering and mass-market dynamics: The triumph of VHS over Beta. *Business History Review*, 66: 51–94.
59. Lieberman and Montgomery, First mover advantages.
60. Suarez and Lanzolla, The half-truth of first-mover advantage.
61. Dyer, Gupta, and Wilemon, What first-to-market companies do differently.
62. Katz, M., and C. Shapiro. 1986. Technology adoption in the presence of network externalities. *Journal of Political Economy*, 94: 822–841.
63. Kerin, Varadarajan, and Peterson, First mover advantage.
64. Boulding and Christen, Sustainable pioneering advantage?
65. Kerin, Varadarajan, and Peterson, First mover advantage.
66. Robinson and Min, Is the first to market the first to fail?
67. Boulding and Christen, Sustainable pioneering advantage?
68. Kerin, Varadarajan, and Peterson, First mover advantage.
69. Kim and Mauborgne, Strategy, value innovation, and the knowledge economy.
70. Teece, D. 1987. Profiting from technological innovation: Implications for integration, collaboration,

licensing and public policy, in D. Teece (ed.), *The Competitive Challenge.* Cambridge, MA: Ballinger.
71. Ibid.
72. Gans, J., D. Hsu, and S. Stern. 2004. When does start-up innovation spur the gale of creative destruction? *RAND Journal of Economics*, 33: 571–586.
73. Teece, 1987, Profiting from technological innovation.
74. Ibid.
75. Teece, D. 1986. Profiting from technological innovation: Implications for integration, collaboration, licensing and public policy. *Research Policy*, 15: 285–305.
76. Teece, 1987, Profiting from technological innovation.
77. Roethaermel, F. 2002. Technological discontinuities and interfirm cooperation: What determines a start-up's attractiveness as alliance partner? *IEEE Transactions on Engineering Management*, 49(4): 388–397.
78. Roethaermel, F., and C. Hill. 2005. Technological discontinuities and complementary assets: A longitudinal study of industry and firm performance. *Organization Science*, 16(1): 52–70.
79. Teece, 1987, Profiting from technological innovation.
80. Adapted from Bartlett, C. 1983. EMI and the CT Scanner (A) and (B), *Harvard Business School Case*, Numbers 5-383-194 and 5-383-195.
81. Ibid.
82. Ibid.
83. Ibid.
84. Ibid.
85. Ibid.
86. Roethaermel and Hill, Technological discontinuities and complementary assets.
87. Adapted from Schmidt, G., and S. Wood. 1999. The growth of Intel and the learning curve. *Stanford Graduate School of Business Case*, Number S-OIT-27.

Chapter 11

Competitive Advantage in High-Tech Industries

Learning Objectives

After reading this chapter, you should be able to:

1. Analyze the attractiveness of an industry.
2. Explain how the five forces model can be used to figure out industry attractiveness.
3. Define the value chain and explain why understanding an industry's value chain is important to technology strategy.
4. Analyze a value chain and identify the strategy decisions that can be made on the basis of a value chain analysis.
5. Explain why some industries operate through dynamics of creative destruction, while others operate through dynamics of creative accumulation.
6. Identify the industry conditions that make small and new firms more innovative, compare them to the industry conditions that make large and established firms more innovative, and explain why these industry conditions have these effects.

7. Describe a competitive advantage and explain how firms develop competitive advantages.
8. Define *resources*, *capabilities*, and *core competencies*, and explain the role of all four of them in firm strategy.
9. Explain how core competencies lead to core rigidities, and how the latter hinder firm performance.
10. Define *strategic dissonance* and explain how companies can redirect their strategies when dissonance occurs.

Core Competence: A Vignette[1]

Should your company focus on a single core technical capability across all of the markets in which it operates, or will such a focus lead to core rigidities that will keep it from responding to changes in the market and technology? This question is central to technology strategy and the answer to it has profound effects on how your company will create competitive advantage. To decide which approach you think is better, take a look at the experience of Nippon Electric Corporation (NEC).[2]

In the late 1970s, NEC was a world leader in three industries: telecommunications, semiconductors, and computers. Because the company's senior management believed that all three industries were likely to converge in the future, they developed a strategy to concentrate on the exploitation of a core capability in semiconductor manufacture. The company formed a team of top managers whose job was to ensure that products developed in all three industries focused on this core capability. Moreover, it established a large number of strategic alliances with companies around the world to source technology that would help it build on this capability.[3]

During the 1980s, the company's strategy appeared to be a success. Its operations expanded in all three industries, making NEC the world's leader in semiconductors and Japan's leader in personal computers and telecommunications. However, in the 1990s, this strategy ran into trouble. As the focus of the semiconductor industry shifted from the production of memory chips to the production of microprocessors that ran personal computers, NEC lost its dominant position to Intel. In telecommunications, the company had trouble keeping up with rapid deregulation worldwide and lost ground to more nimble competitors, like Motorola. In personal computers, NEC was stunned by Compaq's entry into the Japanese market with low-priced, Wintel-based, personal computers that halved NEC's share of the Japanese personal computer market.[4]

The company's focus on its semiconductor core capability seemed to straightjacket the company and keep it from coming up with new products. Critics charged that the company's core capability had become a core rigidity as markets and technology had changed. Although the company undertook a series of joint ventures and alliances to try to right its operations, those efforts failed to avert a financial crisis in the late 1990s.[5]

As the new millennium emerged, the company dramatically changed its strategy. It refocused on businesses tied to the Internet, restructured its alliances and joint ventures to fit this new focus, and exited many lower growth markets. To do this, the company had to end its focus on a core capability in semiconductor manufacture. Going forward it would treat each business as a separate financial unit, rather than as part of a larger effort to exploit a common capability.[6]

The experience of NEC raises an important question: Was the focus of the company on a single core capability the right way to go, or was its later plan to maximize financial returns through a less coordinated strategy a better approach? This chapter helps you to answer this question by exploring the role of core capabilities and core rigidities in firm strategy.

INTRODUCTION

Many aspects of general business strategy apply to technology-intensive companies. In particular, for your company to be successful, it needs to occupy a favorable position in the value chain in an attractive industry. It also needs to develop the capabilities to transform resources into valuable products and services, and create competitive advantages relative to business rivals.[7] While these activities are sometimes serendipitous, often they are part of a planned effort to develop the right strategy, which you need to learn to undertake was wordy (see Figure 11.1).

This chapter focuses on the ways that companies create and sustain competitive advantage in high-technology industries. The first section provides an overview of the use of industry analysis to identify favorable positions in the value chain in attractive industries. The second section discusses firm resources and capabilities, focusing on the core competencies that help firms to innovate repeatedly in their lines of business.

FIGURE 11.1
The Right Strategy?
It is important for a company to have the right strategy.

Source: United Press Syndicate Inc., May 24, 1998.

One important part of technology strategy is to analyze the attractiveness of your industry. Some industries are more attractive than others, making the companies in them consistently more profitable than those in other industries. In addition to identifying the attractiveness of different industries, which can be gathered from data such as that in Table 11.1, it is important to understand the factors that make one industry more attractive than another. One important and widely used model for doing that is the five forces model.

Five Forces Model

Professor Michael Porter of the Harvard Business School created a model that many entrepreneurs and managers use to assess the level of attractiveness of an industry. According to Porter's five forces model (see Figure 11.2), the level of industry attractiveness depends on five dimensions:[8]

1. *Buyer Power,* which measures the degree of power that customers have over companies in the industry. Industries are faced with low buyer power if buyers are fragmented, purchase in low volume, have little information about alternatives, cannot integrate backward to supply themselves, and face high costs to switching suppliers.
2. *Supplier Power,* which measures the degree of power that suppliers have over companies in the industry. Industries are faced with low supplier power if suppliers are fragmented, supply in low volume, cannot integrate forward into production themselves, offer supplies that add little value to later stages in the production process, and can be replaced easily.

TABLE 11.1
Profit as a Percentage of Sales for Manufacturing Industries, 2001–2005
Over these five years, the electrical equipment, appliance, and component industry was more than five times as profitable as the plastics and rubber products industry.

Industry	Profit as a Percent of Sales
Beverage and tobacco products	14.10
Electrical equipment, appliances, and components	10.94
Chemicals	10.46
Miscellaneous manufacturing	8.36
Petroleum and coal products	7.50
Apparel and leather products	5.84
Food	4.34
Fabricated metal products	4.16
Furniture and related products	3.42
Wood products	3.28
Printing and related support activities	2.98
Machinery	2.96
Primary metals	1.98
Paper	1.98
Transportation equipment	1.96
Plastics and rubber products	1.94
Textile mills and textile product mills	1.64
Nonmetallic mineral products	0.14
Computer and electronic products	−0.58

Source: Adapted from U.S. Census data, http://www.census.gov/compendia/statab/business_enterprise/profits.

FIGURE 11.2 Porter's Five Forces Model

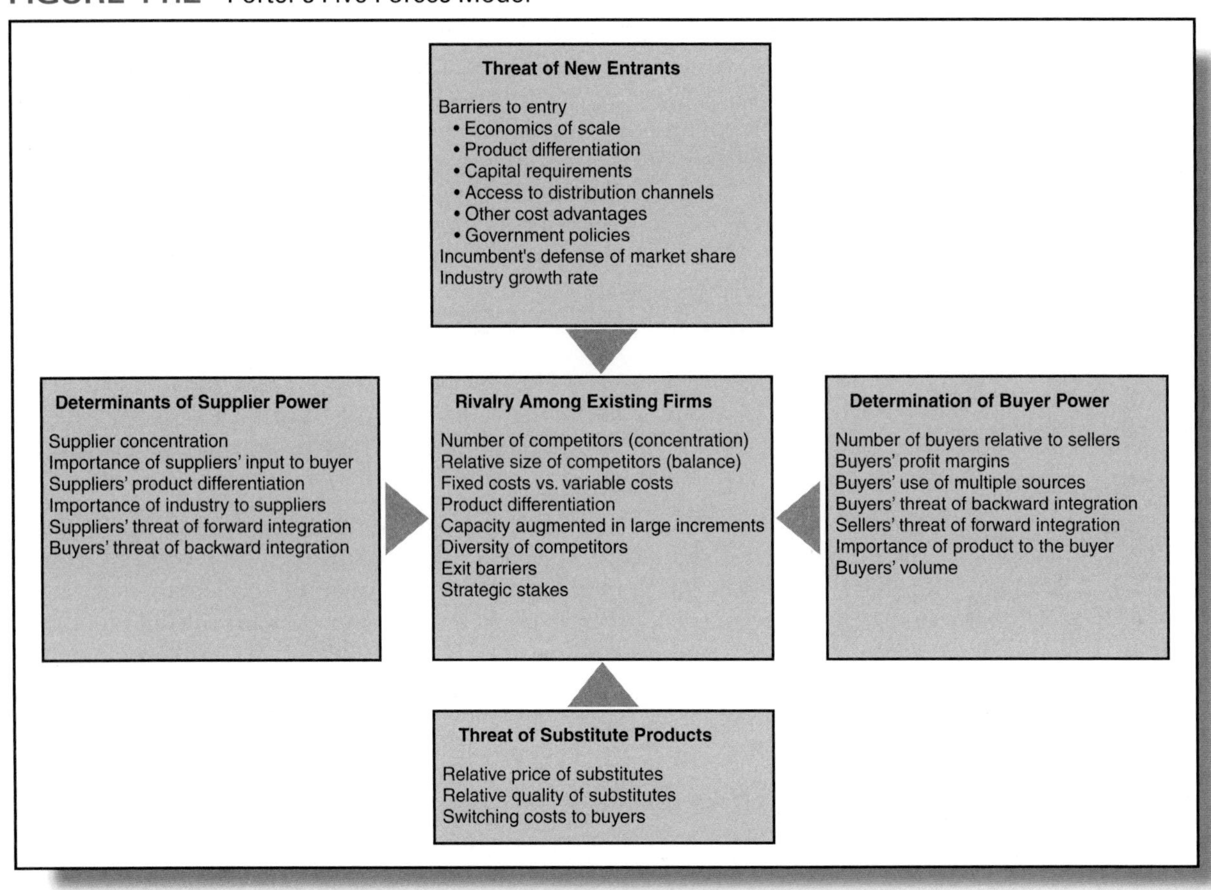

The five force's model of competition focuses on rivalry between existing firms, supplier power, customer power, the threat of substitutes, and the threat of new entrants.

Source: http://images.google.com/imgres?imgurl= http://www.strategy4u.com/assessment_tools/porters_five_forces/porters_five_forces_lg.gif&imgrefurl=http://www.strategy4u.com/assessment_tools/porters_five_forces/five_forces_popup.shtml&h=690&w=958&sz=23&hl=en&start=1&tbnid=ggYl_OvrVA0bHM:&tbnh=107&tbnw=148&prev=/images%3Fq%3D%2522five%2Bforces%2Bdiagram%2522%26gbv%3D2%26svnum%3D10%26hl%3Den.

3. ***Threat of New Entrants,*** which measures ease of entry into the industry. The threat of new entrants is low if economies of scale and switching costs are high, learning curves are steep, capital costs are large, brand name reputation is strong, and key resources are scarce.
4. ***Threat of Substitutes,*** which measures the likelihood that new products or services will substitute for those supplied by the industry. The threat of substitutes is low if alternative products and services will not serve the same function as well as existing products and services, there are few alternative products and services available, the price-value trade-off for the alternatives isn't very good, and the cost of switching to other alternatives is high.
5. ***Degree of Rivalry,*** which measures the degree of competition between firms in the industry. The degree of rivalry is high if the growth of sales is low, products are commodities and cannot be differentiated easily, the number of firms in the industry is large, and the firms are of comparable size.

An important step in formulating a technology strategy is to conduct an industry analysis using the five forces model. This analysis will help you to understand the attractiveness of your industry. For an entrepreneur, this information might affect your decision to start a business to compete in the industry. For a manager of an existing company, it might affect whether you diversify your company into other, more attractive, industries.

In addition, a five forces analysis will help you understand the sources of industry attractiveness. Knowing whether industry attractiveness comes from, say, entry barriers or the level of industry rivalry will help you to predict what conditions are necessary to keep the industry attractive in the future.

The Value Chain

Another important dimension of industry analysis is to examine the value chain. The **value chain** is a description of activities that are used to produce and deliver a product to customers. It includes such things as the sourcing of raw materials and inbound logistics, the manufacture of products, their distribution to entities that sell them to customers, marketing, and after sales service.

Figure 11.3 shows the value chain in the wireless phone business and the different companies that operate at the different stages on the chain. The key strategic issues facing firms at different stages of the value chain vary substantially.

Examining the value chain will help you with your technology strategy in several ways. First, it will help you to figure out where most of the value added is created in an industry. Industries differ in the proportion of value added that comes from manufacturing and marketing activities, as opposed to product development and innovation. In the automobile industry, for instance, much of the value comes from manufacturing and marketing. In the software industry, by contrast, manufacturing and marketing account for a much smaller proportion of the value added. Knowing where the value is added in an industry is important for deciding how to compete with other firms.

Second, examining the value chain will help to determine whether it makes sense to focus on a different stage of the value chain if the locus of value creation in an industry changes. For example, Millenium Pharmaceuticals, a biotechnology firm that was founded to use novel technologies to research genes, has shifted from a focus on contract research to the testing and manufacture of molecules because the locus of value creation in biopharmaceuticals has shifted since the human genome was mapped, and a lot of genetic information has become publicly available.[9]

FIGURE 11.3
The Value Chain in Mobile Phones

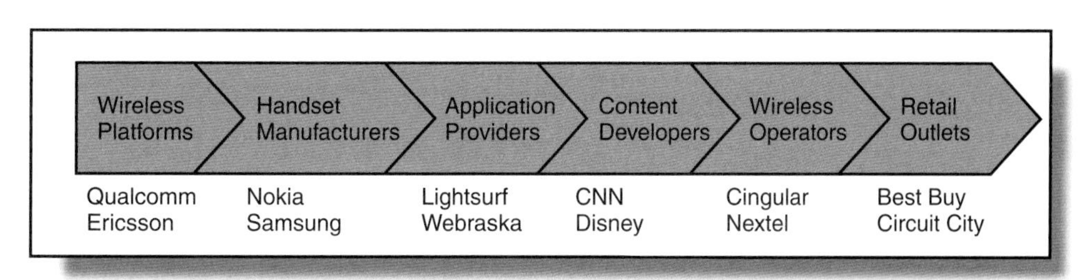

A variety of different companies are part of the wireless phone value chain, which includes the makers of platforms, handsets, applications, and content, as well as operators and retailers.
Source: Adapted from http://www.immr.org/Slides/P2.2_Wireless%20Value%20Chain.png.

Third, understanding the value chain offers insight into whether new or established firms will be more effective at innovation. New firms tend to perform poorly in industries in which manufacturing and marketing account for most of the value added. When firms create new products and services, they often need manufacturing and marketing assets to exploit those innovations. As a result, established firms figure out how to market and manufacture effectively and make those activities routine. Therefore, new firms, which have not yet figured out how to do these things are at a disadvantage when competing with established firms because their manufacturing and marketing activities are often quite inefficient.

Moreover, manufacturing assets are often specialized, which makes them difficult to outsource. As discussed in Chapter 10, when specially designed machines are needed to produce products, it is difficult to find someone willing to provide those assets on a contractual basis. As a result, established firms tend to own the necessary manufacturing assets, making it difficult for new firms to get access to these assets to compete with established firms in industries in which these assets are specialized.

Fourth, understanding the value chain helps to identify key performance drivers at different stages. Because technology is often a key performance driver, value chain analysis helps to identify ways to use technology to develop competitive advantages. For instance, a company might be able to use communications technology to develop an advantage in after sales service, or use materials technology to improve manufacturing.

Fifth, value chain analysis helps with decisions about ownership of different parts of the chain. As we will discuss in greater detail in Chapter 14, companies can vertically integrate and own all stages of the value chain or they can use contractual arrangements, such as licensing and outsourcing, to transfer those activities to other firms. Effective decisions about ownership of parts of the value chain can help companies enhance financial performance, minimize risk, and focus on key competencies.

Regimes of Creative Destruction and Creative Accumulation

A third dimension of industry analysis is to determine the innovation regime under which your industry operates. Some industries operate through dynamics of creative destruction.[10] In these industries, entrepreneurs enter with new firms; challenge established firms on the basis of new ideas; disrupt the old ways of production, organization, and distribution; and replace the old firms. Examples of industries in which this process of creative destruction appears to operate are chemical processes, computer disk drives, machine tools, and lighting.[11]

Other industries operate through dynamics of creative accumulation. In these industries, entrepreneurs enter and challenge established firms on the basis of their new ideas. However, established firms defend their old ways of production, organization, and distribution; and the new firms tend to fail. Examples of industries in which this process of creative accumulation tends to be found are organic chemicals, telecommunications, and electronics.[12]

Examining the innovation regime is particularly important for entrepreneurs who need to know whether entry into an industry is likely to be effective. Table 11.2 provides evidence of large industry differences in the performance of new firms. Specifically, it shows the proportion of start-ups that made the *Inc.* 500 list of the fastest growing young private companies from 1982 to 2000. Those industries in which a higher proportion of new firms made the *Inc.* 500 list are more favorable to new firms than those in which a smaller proportion of new firms made the list.

TABLE 11.2 Proportion of New Firms That Become *Inc.* 500 Firms by Industry
Start-ups in some industries are hundreds of times more likely to become *Inc. 500* firms than start-ups in other industries.

SIC	INDUSTRY	*INC.* 500 FIRMS	STARTS	PERCENT OF STARTS
357	Computer and office equipment	99	2,359	4.2
335	Nonferrous rolling and drawing	14	581	2.4
474	Railroad car rental	3	136	2.2
382	Measuring and controlling devices	49	2,482	2.0
262	Paper mills	3	152	2.0
381	Search and navigation equipment	6	310	1.9
366	Communications equipment	29	1,543	1.9
283	Drugs	20	1,092	1.8
384	Medical instruments and supplies	55	3,025	1.8
316	Luggage	3	172	1.8
314	Footwear, except rubber	4	271	1.5
623	Security and commodity exchanges	2	141	1.4
356	General industrial machinery	26	2,173	1.2
386	Photographic equipment and supplies	7	646	1.1
276	Manifold business forms	3	281	1.1
363	Household appliances	4	390	1.0
362	Electrical industrial apparatus	11	1,080	1.0
811	Legal services	10	129,207	8/1,000 of 1%
581	Eating and drinking places	34	494,731	7/1,000 of 1%
175	Carpentry and floor work contractors	4	66,383	6/1,000 of 1%
651	Real estate operators	5	90,042	6/1,000 of 1%
701	Hotels and motels	2	39,177	5/1,000 of 1%
172	Painting and paper hanging contractors	2	43,987	5/1,000 of 1%
546	Retail bakeries	1	22,165	5/1,000 of 1%
541	Grocery stores	5	112,473	5/1,000 of 1%
593	Used merchandise stores	1	24,442	4/1,000 of 1%
753	Automotive repair shops	5	124,725	4/1,000 of 1%
723	Beauty shops	3	79,081	4/1,000 of 1%
836	Residential care	1	27,710	4/1,000 of 1%
784	Video tape rental	1	27,793	4/1,000 of 1%

Source: Adapted from data contained in Eckhardt, J. 2003. *When the Weak Acquire Wealth: An Examination of the Distribution of High Growth Startups in the U.S. Economy*, Ph.D. Dissertation, University of Maryland.

So what makes some industries follow regimes of creative destruction and others follow regimes of creative accumulation? Research has shown that industry structure influences these patterns. (**Industry structure** is the set of factors that affect the nature and composition of firms in an industry.)

Capital Intensity

New firms tend to be better at innovation in industries that are not capital intensive.[13] (**Capital intensity** is a measure of the importance of capital as opposed to labor in the production process. Some industries, like aerospace, involve a great deal of capital and relatively little labor. Other industries, like textiles, involve relatively little capital and a great deal of labor.) Why? At the time that they are founded, new firms lack cash flow from existing operations. Yet new firms need to expend capital to create production and distribution assets. Because new firms must expend capital before they have internal cash flow, they must obtain capital from financial markets. Capital obtained from financial markets is more expensive than internally generated capital because investors demand a premium for bearing the risk that comes from the gap of information between investors and entrepreneurs. The magnitude of this premium is related to the size of the capital requirement necessary to create the business. The larger the capital requirement, the greater the disadvantage faced by new firms in the industry.

Advertising Intensity

New firms are disadvantaged relative to established firms in more advertising intensive industries. (**Advertising intensity** is a measure of the amount that companies spend on advertising as a percentage of their sales.) As the previous chapter discussed, advertising is a mechanism through which companies develop the reputations that help them to sell their products and services. To build reputations, companies need to undertake a lot of advertising over a long period of time, because, as Figure 11.4 shows, advertising initially has little effect on sales. This disadvantages new firms, which usually have undertaken advertising for less time and can amortize the fixed cost of advertising across fewer units.[14] Of course, this advertising disadvantage is more problematic the more important advertising is for an industry, making new firms less competitive with established firms in more advertising-intensive industries than in less advertising-intensive ones.

Concentration

New firms are disadvantaged at innovation relative to established firms in more concentrated industries.[15] (**Concentration** is a measure of the market share that is held by the largest companies in an industry. Typically, it is measured by the four-firm

FIGURE 11.4
The Effect of Advertising on Sales

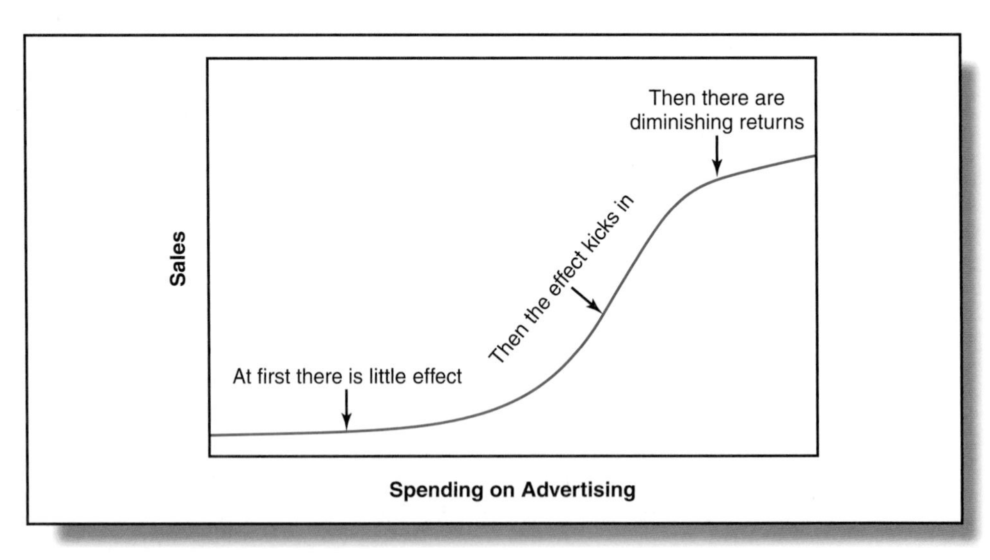

Because advertising initially has little effect on sales, new companies often have trouble in advertising-intensive industries.

concentration ratio, which is the percentage share of the market held by the four largest firms.) New firms perform relatively poorly at innovation in concentrated industries because industry concentration provides large, established firms with the resources to keep new firms from establishing a beachhead in the industry. Moreover, entry can be deterred more easily in concentrated industries than in fragmented ones for two reasons. First, in fragmented industries, there are small, vulnerable firms that can be challenged more successfully than the large, powerful firms that are the only competitors in concentrated industries. Second, in concentrated industries, established firms can collude to keep other firms from entering. For instance, they can collectively cut prices when a new entrant comes into the industry until that entrant is driven out of business, and then raise prices again. Because collusion only works if all of the colluders participate, it is much easier to pull off when there are few players in an industry than when there are many.

Average Size of Firms

As Figure 11.5 shows, some industries tend to have smaller firms than others. New firms are better innovators in industries in which the average size of firms is small.[16] New firms tend to begin small as a way to minimize the risk of entrepreneurial miscalculation. In industries in which most firms are small, this does not create much of a disadvantage. In contrast, in industries where the average firm size is large, starting small creates a number of disadvantages, such as the inability to purchase in volume and higher average manufacturing and distribution costs due to the absence of economies of scale. As an example, think of the difference between Web site developers and steel

FIGURE 11.5
Average Firm Size by Industry over Time

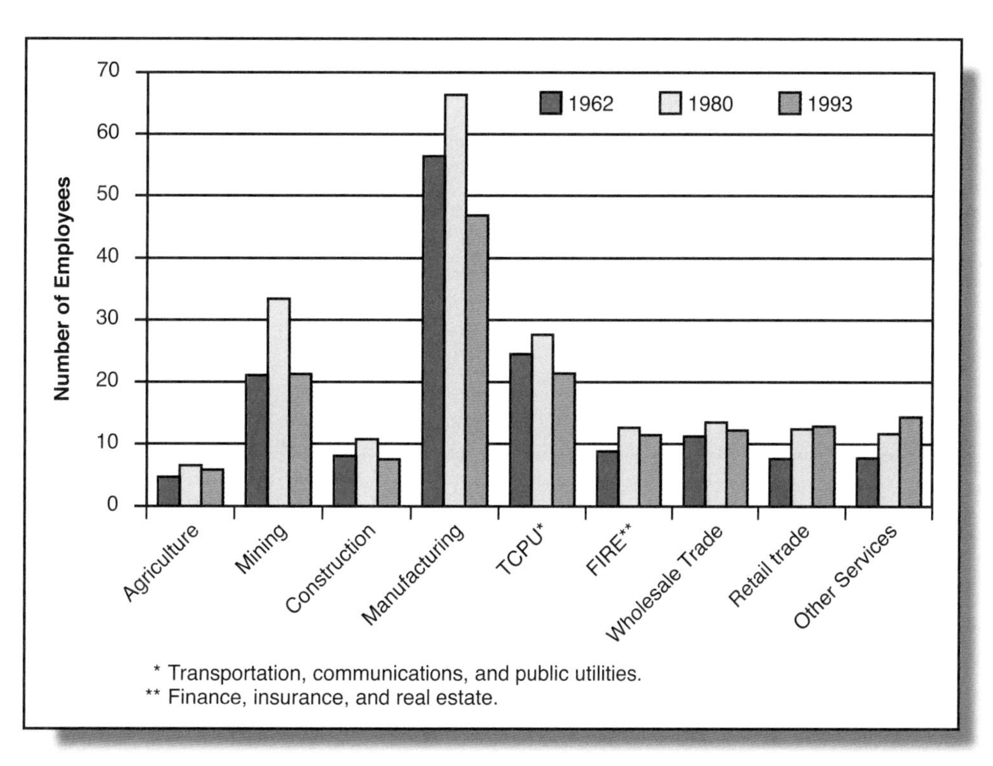

The average size of firms has changed relatively little over time but varies a lot across industries.
Source: http://www.ncpa.org/bg/bg146/bg146a.html.

mills. Because the average Web site developer is small, a new small Web site developer is able to operate at almost the same scale, if not the same scale, as the established players. However, the average steel mill is quite large. So, if a new steel mill is started small, it is initially at a great disadvantage relative to the established firms with which it needs to compete.

Key Points

- Technology strategy involves analysis of industry attractiveness because some industries are consistently more profitable than others.
- The five forces model analyzes the attractiveness of industries by looking at the power of suppliers and customers, the threat of substitution and new entry, and the degree of rivalry among existing firms.
- The value chain is a description of the activities that are used to produce and deliver a product to customers.
- Value chain analysis helps to determine where most of the value lies in an industry, offers insight into the relative advantage of new or established firms at innovation, suggests how companies can create competitive advantage at different stages of the value chain, and helps make decisions about the ownership of different stages in the value chain and movement up or down the chain.
- Some industries operate through dynamics of creative destruction, while others operate through dynamics of creative accumulation.
- Established firms are better than new firms at innovation in advertising-intensive industries because advertising is subject to economies of scale and takes time to have an effect.
- Established firms are better than new firms at innovation in capital-intensive industries because new firms need to finance innovation through external capital markets.
- New firms are worse than established firms at innovation in concentrated industries because concentrated industries provide established firms with market power.
- New firms are better than established firms at innovation in industries where the average size of firms is small because the small size of start-ups is more of a disadvantage in industries where the average firm is larger.

A Resource-Based View

Companies are rarely successful unless they can create a **sustainable competitive advantage (SCA)**, or a way of generating economic value that is difficult for other firms to replicate, purchase, or replace,[17] and that persists over time.[18] SCA often emerges from two key components: **resources**, or the assets that a firm employs in its efforts to generate economic value, and **capabilities**, or the skills and abilities used to provide the products that generate that value.

Resources and Capabilities

The resources that you use to create and deliver products to generate economic value fall into three major categories: tangible, intangible, and human.[19] Tangible resources include such things as plant and equipment, raw materials, and financial reporting systems; whereas intangible resources are composed of such things as trade secrets and relationships with customers; and human resources are made up of such things as employees' knowledge, skills, and abilities.[20]

Resources are not a complete explanation for competitive advantage. They can, but do not have to, give you an SCA.[21] Your business will develop a competitive advantage only when its effort to transform resources into products is valuable, rare, nonsubstitutable, difficult to imitate, and durable; otherwise, your firm's effort will not be superior to that of other firms.[22] Your approach is valuable if it satisfies customer needs. It is rare if few other firms take the same approach. It is nonsubstitutable if there are few alternative ways to accomplish the same goal. It is difficult to imitate if others cannot easily copy your approach. And it is durable if the value of your approach persists over time.[23]

The approach that your business takes to transforming resources into products and services is affected by its capabilities, or knowledge and skills about how to undertake a particular activity. For instance, Merck has developed a capability in molecular design, which allows it to develop a variety of drugs from small molecules made of organic matter.[24] This helps Merck to make money from the manufacture and sale of drugs. (Figure 11.6 shows the role that capabilities play in firm strategy.)

Capabilities can be found in all parts of a company. Some capabilities reside in the skills, knowledge, and ability that employees have accumulated over time in the process of doing their jobs.[25] For example, Chaparral Steel developed a capability in quality control by tapping the knowledge that its employees gained as the company moved up the learning curve in manufacturing steel.

FIGURE 11.6
The Importance of Capabilities

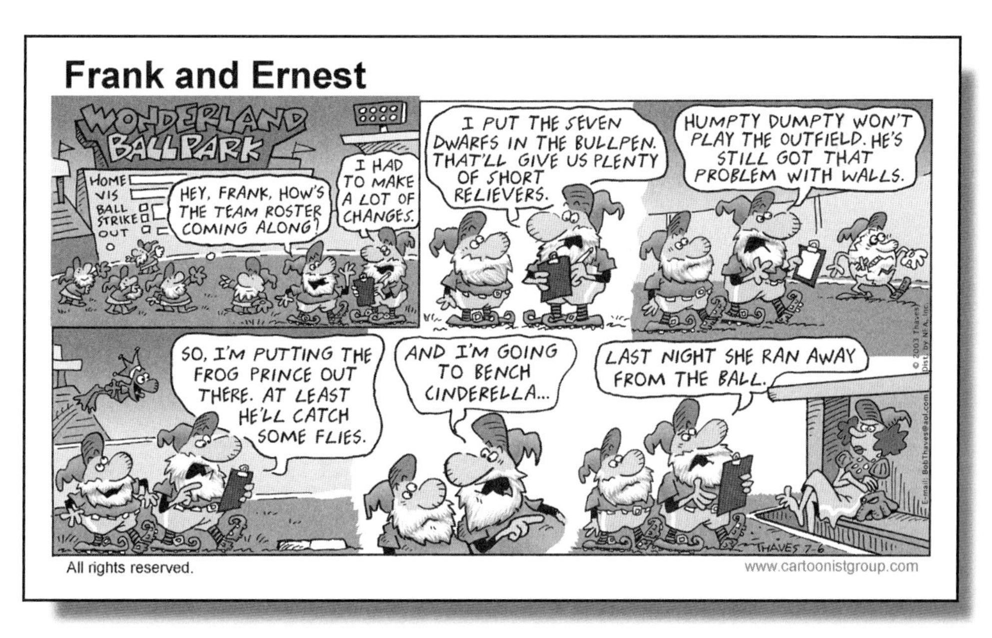

Firm strategy depends on capabilities.
Source: Cartoonist Group. Reprinted with permission. All rights reserved.

Other capabilities reside in an organization's processes, such as those for product development, production, purchasing, supply chain management, and marketing.[26] For example, Monsanto has developed a capability in obtaining government approval through its work with the Environmental Protection Agency to get new chemicals approved.[27]

In technology-intensive industries, one very important capability is the set of organizational routines and processes that help companies to create, combine, and reconfigure resources to create new value through innovation.[28] These might include the product development skills in an automobile company,[29] drug discovery techniques at a pharmaceutical firm, aircraft design processes at an aerospace firm, or software code writing skills at a software firm.[30]

Capabilities make your company's approach to the transformation of resources into products and services valuable, rare, nonsubstitutable, difficult to imitate, and durable in several ways. First, past experience might allow a company to do things that its competitors cannot do.[31] For instance, the company might have moved up the manufacturing learning curve or developed a reputation for quality, and so it can design more durable products than its competitors.[32] Alternatively, a company might have grown large already and can take advantage of economies of scale that competitors cannot exploit.[33]

Second, your company's approach to making products might depend on tacit knowledge, be **causally ambiguous** (the relationship between cause and effect is not well understood) or **socially complex** (the approach requires the interactions of multiple people in complex ways). Consequently, other companies might see the value in the approach but not know how to imitate it. For example, your company might have developed a culture of innovation. While other companies might like to have a similar culture, they might not be able to figure out what combination of organizational structure, human resource policy, and employee profile creates that culture.[34]

Third, your company's approach might depend on having undertaken specific transactions.[35] As a result, companies that have not undertaken these transactions may be unable to follow it. For example, a company might have excellent quality control because it has engaged in a variety of efforts to train its employees in that activity. Other companies cannot achieve the same level of quality control without engaging in the same training processes.

Fourth, your company's approach to making its products might be protected by a property right, such as a patent.[36] For example, it might hold a patent on the process for making fertilizer. If this were the case, then other companies would be precluded by law from making fertilizer in the same way.

Fifth, your company's approach might depend on access to resources that are specific to the firm. As a result, other firms cannot easily obtain them.[37] For example, the production of very small semiconductors might require access to equipment that the company has developed in-house and is not available on the open market.

The capability to innovate is one of the best capabilities for creating sustainable competitive advantage. This capability is difficult to replicate because the routines that lead to its development are usually tacit and require learning by doing. Moreover, the capability is often created through the accumulation of knowledge that occurs as firms gain experience.[38] Furthermore, the complementary relationship between knowledge of technologies and product markets that is necessary for technological innovation means that the replication of this capability requires the joint accumulation of both types of knowledge.[39]

Clearly, companies in technology-intensive industries benefit from the development of capabilities, particularly the capability to innovate. So how do you create these capabilities? You need three things:

- First, you need access to resources that can be transformed into innovative products and services, including intellectual property, complementary assets in manufacturing, marketing and distribution, cash, and brand name reputation.
- Second, you need to make a systematic investment in the management, technical, and other skills that help you to understand different technical areas, to identify opportunities, to develop new products and processes, to improve product quality and performance, and to lower production costs.[40]
- Third, you need to establish the right organizational boundaries (ownership of the right parts of the value chain) to take advantage of the opportunities that present themselves.[41]

It is important to note that the benefits of capabilities in technology-intensive industries illustrate a major difference between start-up and established companies. By definition, new firms cannot have the capabilities for generating competitive advantage in an industry. This means that, initially, new firms are disadvantaged vis-à-vis established firms. It also means that established companies can make themselves vulnerable to competition by new firms by failing to develop the kinds of innovative capabilities that lead to the creation of competitive advantages that new firms cannot easily overcome.

Core Competencies

Capabilities can be used in a wide range of firm activities or in only a few. **Core competencies** are capabilities that that are used to generate value across many firm activities.[42]

SCA depends largely on the quality and quantity of core competencies. Take, for example, the experience of the pharmaceutical firm Pfizer. Over many years, this company has developed two core competencies: the ability to manufacture drugs and the ability to sell them. By applying tacit knowledge of the drug manufacturing process, honed over years of running pharmaceutical plants, Pfizer has become expert at making drugs; by exploiting the socially complex process of selling drugs, developed through many years of intense interaction with doctors and hospitals, Pfizer has become expert at marketing them.

Pfizer can compete successfully against biotechnology firms and other pharmaceutical firms by applying these core competencies to the manufacture and sale of drugs that the company has developed or acquired from other firms. Thus, Pfizer's core competencies provide it with a sustainable competitive advantage that generates a continuing stream of positive cash flow.

Core competencies are often created through the coordination of different activities or technologies.[43] For instance, a company might have a core capability in persuading customers, which it has developed by bringing together advertising, marketing, and customer service; in infrastructure management, which it has created by joining supply chain management, inventory logistics, and information technology; in technology development, which it has built by combining basic research, product design, and prototyping; or in manufacturing, which it has developed by linking scale-up, quality control, and production processes.[44]

Core competencies often can be applied in more than one product market.[45] As a result, you can expand into new markets by using your core competencies in them. For instance, 3M has used its core competence in adhesive technology to produce surgical tape, electrical tape, and pet wraps. Similarly, Sony has exploited its core competence in the use of high-precision equipment and clean rooms, developed in the semiconductor business, to expand into the production of liquid crystal displays, which require similar production processes.[46]

The ability to leverage core competencies across markets is important because companies that expand into new markets in which they can exploit their core competencies perform better than companies that expand into markets in which they cannot. For example, in the early television market, radio producers performed better than other entrants because they could exploit their core competence in electronic production.[47]

In fact, many established companies perform quite poorly when they expand into markets in which their core competencies do not apply. For instance, Samsung's move into movies, video games, and music—in which it had little expertise—almost drove the Korean electronics company out of business.[48]

Core Rigidities

While core competencies provide many benefits, those benefits come at the cost of **core rigidities**, or the inability to do new things in areas outside of the firm's core competencies.[49] Organizational capabilities reinforce a company's current routines and activities by supporting ways of learning, corporate culture, and incentive systems. They become core rigidities because these things often limit the way in which people work together or solve problems, and what activities they believe are acceptable and unacceptable.[50] Existing capabilities lead managers to think in terms of "what the organization does best" so that they can take advantage of its existing ways of doing things and existing assets.[51] But this can hinder efforts to do new things. For example, Hewlett-Packard's core competencies allowed it to develop generation after generation of successful scientific instruments but hindered its ability to develop personal computers.[52]

Core competencies are particularly problematic when companies face radical technological change. A core competency can quickly turn into a core rigidity when the technological base in an industry changes, and the organization's expertise ends up in the wrong technology.[53] For example, Kodak's capability in chemical-based photographic imaging became a core rigidity when digital cameras replaced chemical film.[54] To develop digital cameras, Kodak needed to dramatically change the expertise of its workforce from chemical-based imaging to electronics and computer science.

Because companies develop core rigidities, new firms often have advantages over established firms. For example, when Dell Computer first began, the company was able to develop a direct distribution model that reduced inventory costs far below those of competitors. However, its competitors, who all had existing retail outlets, could not replicate this business model, which required changing their core competencies. As a result, Dell was able to compete quite successfully with established computer companies.

Strategic Dissonance

Because technologies and the environments in which companies operate change over time, managers and entrepreneurs need to align strategic intent—what they want their companies to accomplish—with their strategic actions—what the company actually does. In high-technology industries, in which change occurs at a very high pace, CEOs and entrepreneurs cannot formulate 10-year plans and then try to develop the capabilities to achieve those plans. Because the rate of change is so great, these leaders simply cannot predict the future with enough accuracy to develop long term plans.

For example, Intel was originally a computer memory company. However, over time, technological and competitive changes made the memory business unattractive. To remain successful despite these changes, Intel had to evolve into a microprocessor company. Had Andy Grove, the CEO of Intel at the time, adhered to the original strategic intent of the company, it would probably have been driven out of business by Japanese competitors back in the 1980s.[55]

Some authors call **strategic dissonance** the situation that occurs when the intent and action of entrepreneurs and managers get misaligned.[56] The potential for strategic dissonance raises an important question: How can managers and entrepreneurs ensure that their companies develop the capabilities necessary to exploit opportunities in their industries, given the changes that occur in the technology and environment of the industries? The answer is to establish an effective organizational process for developing strategy. Such a process involves:

1. A mechanism for having employees who are in contact with customers and suppliers identify information about changes that are occurring outside of the company that might lead to a misalignment of intent and action, and a way to bring it to the attention of senior managers who can change strategy. (See Figure 11.7 for a humorous look at the role of senior managers in this context.)
2. A framework for evaluating whether or not the information gathered indicates that the company's goals, actions, and competencies are in alignment with each other and the environment in which the firm operates.
3. A willingness to invest organizational resources that were accumulated during periods of alignment in making changes to the company's strategy and competencies, and a process for making that happen.[57]

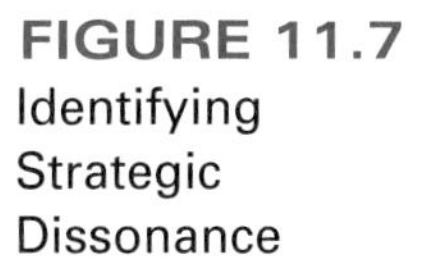
FIGURE 11.7 Identifying Strategic Dissonance

"I'm disappointed; if anyone should have seen the red flags, it's you."

One of the jobs of senior managers and entrepreneurs in high-technology industries is to act like the bull in a rodeo and notice the "red flag" of a shift in the environment that signals the need to change strategy.

Source: http://www.andertoons.com.

GETTING DOWN TO BUSINESS
Microsoft and Netcentric Computing[58]

The rise of the Internet challenged Microsoft's dominance in computing by altering the role of the PC and the importance of the operating system that runs it. The Internet made possible applications on an open Web-delivered platform instead of relying on the proprietary Microsoft operating system.[59]

However, in early 1994, Microsoft was not paying much attention to the Internet. Even though approximately 20 million people were surfing the Internet using a Web browser provided by a start-up company called Netscape, and the Web was becoming a new platform for computing, using Sun Microsystems's Java as the programming language, Microsoft's executives were focused on Windows 95.[60]

By 1996, all of this had changed. Bill Gates had announced that Microsoft would undertake a dramatic shift and embrace the Internet, creating its own Web browser and Web server and making all of its products Web-compatible.[61] Now more than a decade later, Microsoft remains the dominant software company in the world, demonstrating the success of its strategic change. So how did Microsoft respond successfully to the strategic dissonance created by the rise of the Internet?

First, the company had a process for getting employees who had contact with the outside world to identify changes that would lead to a misalignment of strategic intent and action, and bring them to the attention of senior managers. In the case of the Internet, Steven Sinosky, Bill Gates's technical assistant, noticed, while on a recruiting trip to Cornell, that students were making widespread use of the Internet, and brought that information to the attention of Bill Gates and other senior executives.[62]

Second, the company had a framework for evaluating whether or not new information indicated that the company's goals, actions, and competencies were out of alignment with each other. In this specific case, the company held an executive retreat at which senior managers discussed a 300-page brief about the Internet and Microsoft's strategic direction that Sinofsky had prepared.[63]

Third, the company had a willingness to invest organizational resources to change its strategy and competencies, and a process for making that happen. In this case, in December 1995, Microsoft announced a plan to create a Web browser, a Web server, a Web-based version of MSN, and a strategy to in-and-out license technologies that would facilitate Web-based computing. Then, it immediately tried to buy Excite, a search engine, and Vermeer, the maker of Web page–creating software. Shortly thereafter, it signed a deal with AOL to make Internet Explorer AOL's Web browser in return for putting AOL into Windows 95.[64]

IBM is one company that has successfully responded many times to the problem of strategic dissonance. When it became clear that its intent and action were misaligned in the mechanical calculator business, the company moved into the mainframe computer business. Then when IBM's strategic intent and action again became misaligned in the mainframe computer business, the company made the transition to personal computers. And when IBM's intent and action became misaligned in the personal computer business, the company became a provider of computer services.[65]

Key Points

- The creation of sustainable competitive advantage depends on resources and capabilities.
- Resources can be tangible, intangible, or human.
- Capabilities affect the process through which resources are turned into products and services; they create competitive advantage if they are rare, valuable, inimitable, durable, and nonsubstitutable.

- Capabilities are core if they are used to generate value across a wide range of firm activities.
- Core competencies are often created through the coordination of different activities or technologies.
- Core competencies allow firms to expand successfully into new product markets.
- Core rigidities, which result from the exploitation of core competencies, make it difficult for companies to do things in new areas.
- Strategic dissonance occurs when what managers want to accomplish and what companies are doing are misaligned; it indicates the need to change strategy.
- Companies are most successful in responding to strategic dissonance by evaluating information on the misalignment, gathering information from frontline employees, and devoting organizational resources to the new direction.

Discussion Questions

1. Is the five forces model a complement to or a substitute for value chain analysis in analyzing industries? Why? How does each of these tools help entrepreneurs and managers?
2. What are regimes of creative accumulation and creative destruction and how do they affect technology strategy?
3. What are resources, capabilities, and core competencies? What role does each have in firm strategy?
4. Describe something about a technology company that provides a competitive advantage and something that does not provide a competitive advantage. Why does the former provide an advantage but not the latter?
5. What are the benefits and drawbacks of having core competencies at a company? On balance is a company better off having a core competence than not having one? Are there conditions under which a company would be better off and conditions when it would be worse off? Why?
6. What is strategic dissonance? What problems does it create for technology strategy? How can entrepreneurs and managers overcome the problems created by strategic dissonance?

Key Terms

Advertising Intensity: A measure of the amount of money that companies spend on advertising as a percentage of their sales.

Capabilities: The knowledge and skills necessary to undertake a particular activity.

Capital Intensity: A measure of the importance of capital, as opposed to labor, in the production process.

Causal Ambiguity: A situation in which the relationship between cause and effect is not well understood.

Concentration: A measure of the market share that is held by the largest companies in an industry.

Core Competencies: The knowledge and skills necessary to undertake the central activity of a firm.

Core Rigidities: The routines and activities of a firm that support its core competencies and make it difficult to do new things.

Industry Structure: The set of factors that affect the nature and composition of firms in an industry.

Resources: The assets that a firm employs in its efforts to generate economic value.

Socially Complex: A situation in which the approach to creating a product or service requires the interaction of multiple people in complex ways.

Strategic Dissonance: A situation in which strategic intent and action are misaligned.

Sustainable Competitive Advantage (SCA): A way of generating economic value that is difficult for other firms to replicate, purchase, or replace, and that persists over time.

Value Chain: The set of activities, from the input of raw materials to the distribution of products to customers, through which value is created in an industry.

Putting Ideas into Practice

1. **Value Chain** The purpose of this exercise is to understand how to analyze the value chain in a technology business. The following entities are parts of the value chain for delivering music to customers over wireless phones: record labels (they record and compile the songs in a digital format available for transmission to customers), wireless application service providers (they operate the wireless platform on which the phones operate), providers of mobile delivery (they offer connectivity and transmit music over the radio waves to devices), content creators (they write and sing the songs listened to by the customers), and handset makers (they provide the phones that are used to play music).

   Please place these elements in the order that they would appear in a value chain that begins with the initiation of the process and ends with the customer. Second, evaluate the amount of value that you think is provided at each stage of the value chain. What stages generate a lot of value? What stages generate relatively little? Explain your reasoning. Third, identify the nature of the relationships between stages of the value chain (are they vertically integrated or contracted out)? Explain why the dominant form of the relationship between each stage tends to be vertical integration or contracting. Fourth, identify at least two supporting aspects of the value chain that do not provide a competitive advantage to firms in the industry. Explain why these aspects of the value chain do not provide firms with a competitive advantage.

2. **Strategy for Sony** Go to Sony's Web site www.sony.net. Read through the different Web pages. Are there any common themes among its products in electronics, music, games, movies, and financial services that might identify the company's core competencies? For example, does the company have capabilities in product design that cut across a variety of products? What about manufacturing? What about miniaturization of products? List all of the core competencies that Sony has, and explain why you think that they are core competencies. Then identify a new technology that you think that Sony could use to create a product that would meet customer needs and draw on these core competencies. Specify which of the company's core competencies will help it to exploit the new technology successfully and explain why those competencies will help.

3. **Interview a Manager or Entrepreneur** Identify a manager or an entrepreneur in an industry in which you are interested in working about the challenges of developing a strategy to generate competitive advantage. (If you don't know any entrepreneurs or managers, ask your instructor for the name and contact information of someone to speak to.) Set up a time when you can call the person on the telephone and can speak for 30 minutes. When you are conducting the interview, ask the manager/entrepreneur how companies in the industry generate and capture value from innovating. Also ask the manager/entrepreneur to identify the firm's core capabilities and how it uses those capabilities to create a competitive advantage. For both issues, ask the manager/entrepreneur to justify his or her arguments.

   Now evaluate the interviewee. Does the manager/entrepreneur understand how companies in the industry generate and capture value from innovating? Does the manager/entrepreneur understand the company's capabilities and see how they create competitive advantage? If so, why do you think this is the case? If not, why not? What suggestions can you make to improve the quality of the manager/entrepreneur's understanding of how value from innovation is generated and captured in the industry, and how the company's capabilities can be used to create competitive advantage?

Notes

1. Adapted from Enright, M. 2002. Core competence at NEC and GTE. *Centre for Asian Business Cases*, Number HKU213; Prahalad, C., and G. Hamel. 1990. The core competence of the corporation. *Harvard Business Review*, 68(3): 79–91.
2. Ibid.
3. Ibid.
4. Ibid.
5. Ibid.
6. Ibid.
7. Narayanan, V. 2001. *Managing Technology and Innovation for Competitive Advantage*. Upper Saddle River, NJ: Prentice Hall.
8. This section is adapted from Porter, M. 1980. *Competitive Strategy*. New York: Free Press.
9. Champion, D. 2001. Mastering the value chain: An interview with Mark Levin of MilleniumPharmaceuticals. *Harvard Business Review*, June: 109–117.

10. Schumpeter, J. 1942. *Capitalism, Socialism and Democracy*. New York: Harper and Row.
11. Malerba, F., and L. Orsenigo. 1997. Technological regimes and sectoral patterns of innovative activities. *Industrial and Corporate Change*, 6: 83–117.
12. Ibid.
13. National Academy of Engineering. 1995. *Risk and Innovation*. Washington, DC: National Academies Press.
14. Shane, S. 2003. *The Individual-Opportunity Nexus Approach to Entrepreneurship*. Aldershot: UK: Edward Elgar Publishing.
15. Eisenhardt, K., and K. Schoonhoven. 1990. Organizational growth: Linking founding team, strategy, environment, and growth among U.S. semiconductor ventures, 1978–1988. *Administrative Science Quarterly*, 35: 504–529.
16. Audretsch, D., and T. Mahmood. 1991. The hazard rate of new establishments. *Economic Letters*, 36: 409–412.
17. Amit, R., and P. Schoemaker. 1993. Strategic assets and organizational rent. *Strategic Management Journal*, 14: 33–46.
18. Barney, J. 1991. Firm resources and sustained competitive advantage. *Journal of Management*, 17(1): 99–120.
19. Afuah, A. 2003. *Innovation Management*. New York: Oxford University Press.
20. Barney, Firm resources and sustained competitive advantage.
21. Peteraf, M. 1993. The cornerstones of competitive advantage: A resource-based view. *Strategic Management Journal*, 14: 179–191.
22. Reed, R., and R. DeFillipi. 1990. Causal ambiguity, barriers to imitation, and sustainable competitive advantage. *Academy of Management Review*, 15(1): 88–102.
23. Barney, Firm resources and sustained competitive advantage.
24. Galambos, L., and J. Sturchio. 1998. Pharmaceutical firms and the transition to biotechnology: A study in strategic innovation. *Business History Review*, 72: 250–278.
25. Leonard-Barton, D. 1992. Core capabilities and core rigidities: A paradox in managing new product development. *Strategic Management Journal*, 13: 111–125.
26. Ibid.
27. Urban, T. 1995. Monsanto Company: The coming age of biotechnology. *Harvard Business School Case*, Number 9-596-034.
28. Eisenhardt, K., and J. Martin. 2000. Dynamic capabilities: What are they? *Strategic Management Journal*, 21(10/11): 1105–1121.
29. Ibid.
30. Teece, D., G. Pisano, and A. Shuen. 1997. Dynamic capabilities and strategic management. *Strategic Management Journal*, 18(7): 509–533.
31. Barney, Firm resources and sustained competitive advantage.
32. Peteraf, The cornerstones of competitive advantage.
33. Ibid.
34. Barney, Firm resources and sustained competitive advantage.
35. Reed and DeFillipi, Causal ambiguity, barriers to imitation, and sustainable competitive advantage.
36. Peteraf, The cornerstones of competitive advantage.
37. Barney, Firm resources and sustained competitive advantage.
38. Zollo, M., and S. Winter. 2002. Deliberate learning and the evolution of dynamic capabilities. *Organization Science*, 13(3): 339–351.
39. Nerkar, A., and P. Roberts. 2004. Technological and product-market experience and the success of new product introductions in the pharmaceutical industry. *Strategic Management Journal*, 28: 779–799.
40. Sathe, V. 2003. *Corporate Entrepreneurship*. Cambridge, UK: Cambridge University Press.
41. Teece, Pisano, and Shuen, Dynamic capabilities and strategic management.
42. Burgelman, R., C. Christiansen, and S. Wheelwright. 2004. *Strategic Management of Technology and Innovation*. New York: McGraw-Hill/Irwin.
43. Prahalad and Hamel, The core competence of the corporation.
44. Gallon, M., H. Stillman, and D. Coates. 1995. Putting core competency thinking into practice. *Research Technology Management*, 38(3): 20–28.
45. Prahalad and Hamel, The core competence of the corporation.
46. Dvorak, P., and E. Ramstad. 2006. TV marriage. *Wall Street Journal*, January 3: A1, A6.
47. Klepper, S., and K. Simons. 2000. Dominance by birthright: Entry of prior radio producers and competitive ramifications in the U.S. television receiver industry. *Strategic Management Journal*, 21: 997–1016.
48. Dvorak and Ramstad, TV marriage.
49. Christiansen, C. 1999. *Innovation and the General Manager*. New York: McGraw-Hill.
50. Leonard-Barton, Core capabilities and core rigidities.
51. Cohen, W., and R. Levin. 1989. Empirical studies of innovation and market structure. In R. Schmalensee and R. Willig (eds.), *Handbook of Industrial Organization*, Vol II. Amsterdam: Elsevier.
52. Leonard-Barton, Core capabilities and core rigidities.
53. Ibid.
54. Gallon, Stillman, and Coates, Putting core competency thinking into practice.

55. Burgelman, R., and A. Grove. 1996. Strategic dissonance. *California Management Review*, 38(2): 8–28.
56. Ibid.
57. Ibid.
58. Adapted from Carroll, P. Inside Microsoft: The untold story of how the Internet forced Bill Gates to reverse course. In R. Burgelman, C. Christensen, and S. Wheelwright, *Strategic Management of Technology and Innovation*, 587–592.
59. Lohr, Preaching from the Ballmer pulpit.
60. Carroll, P. Inside Microsoft: The untold story of how the Internet forced Bill Gates to reverse course.
61. Ibid.
62. Ibid.
63. Ibid.
64. Ibid.
65. Lohr, S. 2007. Preaching from the Ballmer pulpit. *New York Times*, January 28: Section 3, 1, 8–9.

Chapter 12

Technical Standards

Learning Objectives

After reading this chapter, you should be able to:

1. Define *technical standards*, explain why they are important, and identify the industry conditions that make them useful.
2. Explain the different mechanisms through which technical standards develop, and explain why the best technology does not always emerge as the industry standard.
3. Understand the relationship between technical standards and customer adoption in industries in which technical standards exist.
4. Describe a "standards battle" and explain what you can do to increase the odds that your technology will win one.
5. Outline a strategy to adopt if your technology loses a standards battle.
6. Explain the dilemma of differentiating your product and adhering to technical standards, and explain how you should manage this dilemma.

7. Explain how to defend your technical standard once it has been established.
8. Understand how technical standards shift competition from firms to systems, and explain how that affects firm strategy.
9. Describe how technical standards affect new and established firms and explain what strategy start-ups should take toward technical standards.
10. Define *open* and *closed standards*, explain the pros and cons of the two types of standards, and figure out when to make your technical standard open and when to make it closed.

Technical Standards: A Vignette[1]

Working with the movie producers, Warner Brothers, MGM, and Columbia Pictures, the consumer electronics companies, Sony, Toshiba, and Panasonic, created the DVD Forum, an organization designed to establish technical standards for digital video recorders (DVDs). The DVD Forum's goal was to avoid a battle over technical standards for the digital formats that would replace videocassettes.

The problem with a standards battle is that it would make customers unsure of the future existence of the version that they selected and so afraid to buy a DVD player. Moreover, the providers of complementary technology, in this case the movie producers, would be reluctant to offer their movies for DVD because they wouldn't know which format to put them in, and would not want to incur the cost of putting them in more than one. So they too would hang back and wait to see what would happen.

In September 1996, the DVD Forum released its specifications for a common standard for DVDs. The specifications were designed to ensure that both DVD players and movies recorded in the DVD format would be compatible. By including movie producers in the arrangement, the electronics companies that developed the standard were able to ensure that 40 movies were available in DVD when the first players were sold to the public in 1997.

However, in September 1997, Circuit City, the leading seller of home electronics in the United States, announced that it would launch a rival product, Digital Video Express (DIVX), which was based on a different technology. Circuit City also announced that it had commitments from major movie studios to release DIVX disks. While Circuit City did not give a product launch date for DIVX, and said that the product launch would be at least a year away, the company explained that DIVX players would be designed to play DVDs. Moreover, DIVX recordings would be protected by encryption so that they could only be shown on DIVX players. (See Figure 12.1 for pictures of the DVD and DIVX players.)

Circuit City's preannouncement was part of a strategy to gain control over the technical standard for digital video recorders. The preannouncement made sense because there would be no market for DIVX technology if customers adopted the DVD technology first, and it became the de facto digital recording standard.

The market reacted strongly to the Circuit City announcement. Many customers decided to delay their purchase of digital video recorders until after one of the technologies emerged as the industry standard. As a result, sales of DVDs were slow in the months following Circuit City's announcement.

In January 1998, Circuit City demonstrated a DIVX prototype and announced that it would release the product in the fall of 1998. However, by the time that DIVX was released, many of the movie studios had already opted for the DVD standard, and Circuit City was unable to get many of them to support its format. As a result, few retailers expressed an interest in carrying the DIVX product. In June 1999, having sold only 165,000 units (versus 1.32 million DVD players on the market), Circuit City stopped selling DIVX players.

So what happened to Circuit City's effort to make DIVX the technical standard in digital video recording devices? Why wasn't the company able to get its DIVX product adopted? Reading this chapter will help you to answer these questions. It discusses the importance of getting the producers of complementary products on board, forming standard-setting bodies, making preannouncements, and launching products quickly in battles to set technical standards.

FIGURE 12.1 Firms Often Battle to Control Technical Standards

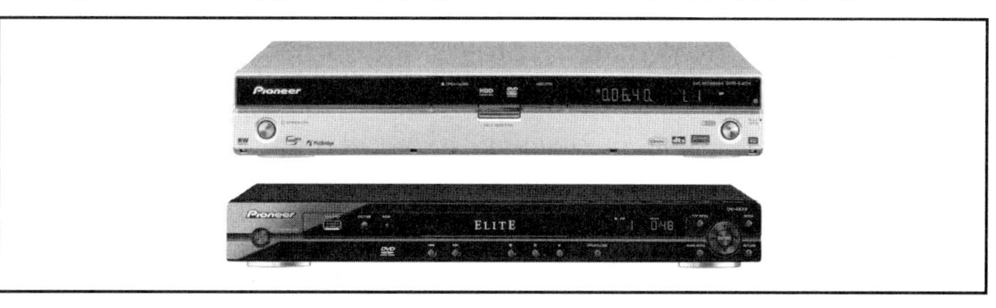

The DIVX (above) and DVD (below) digital video recorders were the subject of the standards battle described in the text.

Source: Pioneer North America, Inc.

INTRODUCTION

Because many industries converge on technical standards, understanding their role in business is important to the accurate formulation of technology strategy. For instance, you need to understand how technical standards affect customer adoption of new products. You may need to fight a standard-setting battle. And you certainly will need to adopt strategies that are appropriate to the outcome of the standard-setting process to make your business successful.

This chapter discusses the creation and exploitation of technical standards. The first section defines a technical standard, and explains how standards develop, why they aren't always the best technology, and how their formation relates to customer adoption. The second section examines standard-setting battles. It explains how to win a standards battle, and what to do if you lose one. It also explains why technical standards shift the locus of competition from firms to systems of firms and identifies the way that the shift affects strategy. The final section compares open and closed technical standards and explains how technology strategy differs depending on the openness of the industry standard.

TECHNICAL STANDARDS

In Chapter 2, we discussed the concept of a dominant design, or form of a product on which all producers converge, when we considered the Abernathy-Utterback model. A related, but broader, concept is that of **technical standards**, which are specifications to ensure that different components of the same system are compatible.[2] For example, the electrical plug is designed to specifications that ensure that it will work in all electrical outlets.

Technical standards are important in a wide variety of industries, from computers to transportation to telecommunications, because they permit independent companies to produce different components for the same product. They exist when components made by one company need to be used with components made by another because companies cannot ensure that components made by others will be compatible, unless they are made to conform to a standard. For instance, in computers, technical standards permit hardware, printers, software, memory chips, and peripheral devices that are produced by different companies to all work together.[3] As Figure 12.2 highlights, technical standards in measurement are particularly important.

FIGURE 12.2
Key Technical Standards

One of the most important technical standards is that of measurement; compatibility depends a lot on whether one is using the English or metric system.

Source: http://www.iso.org/iso/en/networking/pr/cartoons/cartoons.html#home.

Technical standards are particularly important to start-up firms. While large, established companies can make all of the components in a product and avoid problems with the lack of established technical standards, start-ups usually lack the capital to do this. Because start-ups usually cannot produce all of the components in a product, they need technical standards to ensure that their products are compatible with complementary products.

The Development of Technical Standards and Dominant Designs

Research has shown that technical standards and dominant designs develop in a variety of ways. Sometimes, chance occurrence leads them to emerge. For example, some researchers argue that the all-steel body in automobiles emerged as a dominant design just because aluminum was relatively costly in comparison to steel in the 1920s.[4]

The emergence of a technical standard or dominant design also depends, at least in part, on the nature of technology. Some technical solutions are better than others, and the best technical solution sometimes emerges.[5] For example, nylon and polyester emerged as the dominant designs in synthetic fiber because they have the chemical composition to produce long fibers, which is useful in making fabric.[6] Similarly, suspension-preheating became the dominant design in cement manufacture because it was the most fuel-efficient alternative.[7]

Sometimes governments impose technical standards and dominant designs.[8] For example, the European Union adopted the GSM wireless telephony standard to ensure that people in all the countries of the Union had compatible mobile phones.

Other times, the government does not impose a technical standard or a dominant design, but influences its adoption nonetheless. Typically, this happens when the government serves as an early adopter of a product and creates a large **installed base**—the number of users of a product at a point in time—for a particular variant.[9] For example, the U.S. government could make a particular type of fuel cell vehicle the technical standard by purchasing those vehicles for the military, postal service, and other government agencies.

Governments often get involved in setting technical standards or dominant designs when companies have little incentive to adopt a new technology. Take, for example, the case of high-definition television. Broadcasters have little incentive to switch to the new technology because it is unlikely that the audience for television, and hence advertising revenues, will increase as a result.[10] And television users do not want to adopt the new technology until the price of high-definition televisions comes close to that of traditional televisions, and programs are produced in high definition. Because the market has no mechanism to spur a shift to high-definition television, government mandate is the only way to put the new technology in place.[11]

Industry trade associations or standard-setting organizations sometimes create technical standards and dominant designs, usually through votes of committee members that are selected by their employers to represent them on working groups. For example, an international organization called the International Telecommunication Union created the H.320 standard for videoconferencing.[12]

Because the choice of a technical standard has huge ramifications for companies, battles often break out in standard-setting bodies. For example, Sony, Matsushita,

HP, and Dell battled with Toshiba and NEC over the standard for high-definition television. Each group wanted a standard based on a technology that was more favorable to its members.[13]

Companies sometimes cooperate with other firms to create a technical standard or dominant design.[14] The logic of this approach is that if companies cooperate with others to jointly develop a common standard, they will attract customers more quickly. If the cooperating companies are large enough, other companies will often go along with their approach, and the common standard will be adopted. For instance, Philips and Sony worked together to develop a standard for compact disk technology that was favorable to both of them; the size of the two companies led other companies to go along, rather than pursue a rival standard, and convinced the music labels to provide content that adhered to the Philips-Sony standard.[15]

However, many firms, even large firms, fail at strategic efforts to make their technology the industry standard, as occurred when IBM backed a standard for local area networking, but lost out to Ethernet.[16] This raises the question: Is investing in efforts to establish and control a technical standard worthwhile? Given the significant chance that you will fail at this effort, investing a lot of money in trying to do so might not be a good strategy. You might be better off simply accepting the evolution of a technical standard and adopting a strategy that best fits it.

Not Always the Best Technology

As you can probably tell from the description of the different ways that technical standards develop, the technology that becomes the industry standard is not always the "best" technology from the perspective of functionality.[17] Often, it is technically inferior to the other alternatives available.[18]

The "best" technology doesn't always become the industry standard for many reasons, including government mandate, the strategic actions of firms, agreement by standard-setting bodies, or social support for a particular alternative.[19] For example, in automobiles, the internal combustion engine emerged as the dominant design, rather than the electric battery engine, which many observers thought was technically superior, in part because the electric battery engine needed recharging, which made it less popular for touring, a major social use of the automobile at the beginning of the twentieth century.[20]

Nevertheless, the most common way that industries converge on technically inferior alternatives is through the natural workings of the market.[21] If a technology achieves a large installed base, later adopters will tend to choose this technology even if it is inferior to alternatives that emerge later.

The QWERTY typewriter keyboard (named for the order of letters on the top line) is a good example. The QWERTY keyboard was designed initially to *slow* typing, which was important with the jamming-prone typewriters of the 1880s. Once this technical problem was resolved, however, typing performance could be enhanced by the adoption of different keyboard designs. In fact, one of these keyboards, patented by Dvorak and Dealey in 1932, was so good that the costs of retraining typists to use it could be amortized in 10 days. Yet, despite the performance advantages of the alternative keyboard designs, they have never been adopted in large numbers because the installed base of QWERTY keyboards has always been too large to make widespread switching possible.[22]

Technical Standards and Customer Adoption

The emergence of a technical standard is important to attract customers in large volume. Customers are risk averse and do not want to adopt products that might be abandoned or discontinued because the products are incompatible with a technology that becomes the industry standard.[23] As a result, customers are often slow to adopt new products until a technical standard is in place.[24] For instance, adoption of high-definition DVDs has been slow because their producers have been unable to agree upon a common standard, with two groups battling over competing Blu-ray and HD-DVD alternatives.

Technical standards are particularly important to the customer adoption decision when products are systemic. Because systemic products are composed of components that need to be used with other components, having a common standard is important in ensuring that components produced by different companies can be used together. For example, the adoption of digital cameras did not take off until technical standards for the different camera components had been established because these cameras are made up of an image gathering device, a processing device, a storage device, and a reproduction device, all of which are produced by different companies.

Technical standards also enhance customer adoption by making many products more functional.[25] For example, imagine how difficult it would be to use a telephone if the people you wanted to call had a telephone that operated on a different technical standard from yours. You wouldn't be able to speak to them. And if there were multiple telephone standards, you would probably get pretty frustrated because you would only be able to call a small portion of the people that you wanted to reach.

Standardization also enhances customer adoption because it facilitates the creation of complementary products.[26] The existence of a technical standard permits companies to make products that work with those developed by other companies, increasing the number of products available to customers. Take Web sites, for example. Because all Web sites are created in accordance with a common technical standard, consumers have Web sites that are produced by millions, if not billions, of entities when they browse on their computers. The greater availability of Web sites makes Web surfing a more useful, and enjoyable, activity.

Key Points

- In some industries, products must conform to a technical standard or specifications that ensure that different components are compatible.
- Standards are of particular importance to start-up firms, which generally cannot produce all of the components needed to make a product.
- Technical standards develop because of chance occurrence, because one technology is superior to others, because the installed base of one technology is far ahead of the installed base of others, because companies take strategic action, because industry trade associations or standard-setting bodies establish them, and because governments mandate them.
- Industries sometimes converge on standards that are technically inferior to other alternatives.
- Technical standards influence customer adoption because customers desire compatibility, particularly for systemic products.

Where technical standards exist, your company will have a tremendous advantage and the potential to earn high profits if your proprietary technology is the industry standard. There are several reasons why. Products that conform to the technical standard can be sold at a premium given their greater value to customers. These higher prices create higher profit margins and give you the opportunity to generate outsized financial returns. Moreover, if your technology becomes the industry standard, your suppliers will have to adhere to it, giving you leverage over them, and allowing you to capture a large portion of industry profits. Furthermore, your competitors will have to adopt your technology, which puts them at a competitive disadvantage.

Because gaining control over a standard is valuable to companies, battles for technical standards have long been a part of technology strategy. In the nineteenth century, for instance, railroads fought over whether the technical standard for railway gauges should be 4 feet 8.5 inches wide or 5 feet wide; and in the 1940s, CBS and NBC fought over the technical standard for color televisions.[27]

How to Win a Standards Battle

So what can you do to win a standards battle? First, you can gain the support of the producers of complementary products. If all other things are equal, one technical standard will be preferred to another if more complementary products are available for that standard. Why? Because the availability of complementary products makes a given product more valuable to customers.

In the high-definition DVD war, for example, the Blu-ray faction, led by Sony, and the HD-DVD faction, led by Toshiba, are each working to gain the support of the movie studios, which provide content for their version of the high definition DVD player (see Figure 12.3). Each faction offers different advantages to those complementors, which it stresses to convince them to support its standard. Blu-ray offers greater recording capacity and protection against illegal copying, while HD-DVD offers a manufacturing process that requires few changes from conventional DVDs.[28]

FIGURE 12.3 Complementary Products in the High-Definition DVD War

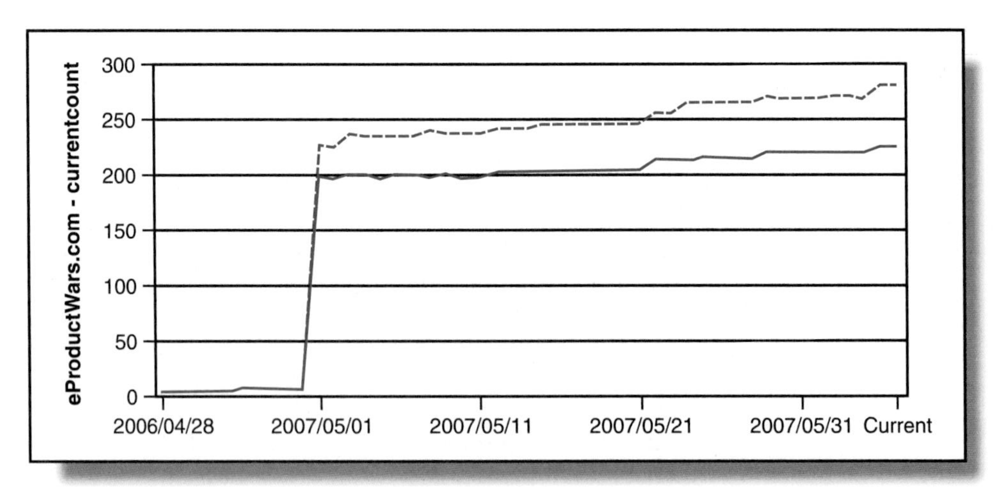

As of the beginning of June 2007, Blu-ray had opened up a small lead over HD-DVD in terms of the availability of complementary products—the number of high-definition DVDs that have been shipped in each format.

Source: http://www.eproductwars.com/dvd.

The two groups are also seeking the support of other complementors, like video game manufacturers. In this case, the Blu-ray group has an advantage because Sony is vertically integrated into the production of both high-definition DVDs and video game consoles. Because Sony controls the production of PlayStation 3, it can ensure that the device uses a Blu-ray disk drive. Thus, Blu-ray's integration of hardware with software negates any advantage that HD-DVD software has. However, the HD-DVD group has been able to use the rivalry between Sony and Microsoft to get Microsoft to incorporate the HD-DVD in its competing product to the PlayStation 3, the Xbox 360, to offset Blu-ray's integration advantage.[29]

A company will likely win a standards battle if there is a complementary product—a "killer application"—that that makes one technology much more attractive than another. For many physical products, killer applications lie in software or content. For instance, Lotus 1–2–3 was the "killer application" for personal computers. That is, in the early days of the personal computer business, many companies adopted personal computers to gain access to the Lotus 1–2–3 spreadsheet software.

Second, you can make your product **backward compatible** so that it works with a previous generation of products. (If products are not backward compatible, they won't work with previous generations of technology, as is the case with MP3 players and CDs.) For example, the backers of both competing standards for high-definition DVDs, Blu-ray, and HD-DVD, have made their products compatible with standard DVD players.[30] Backward compatibility helps you win a standards battle by facilitating customer adoption of your new product. As was mentioned earlier in the chapter, customers often do not adopt a new technology product because they are afraid that the product will be useless if the new technology does not become the industry standard. By making your product backward compatible, you reduce the cost to customers of choosing the wrong technology—they can always use it with the previous generation of technology—and enhance their willingness to buy your new product.

However, making a new product backward compatible is expensive, and therefore, not always worthwhile. It also reduces the incentive for the producers of complementary products to make their products unique. If complementary products are not unique to your product, then customers will derive less benefit from switching to your new product, and will be less likely to do so.[31]

Furthermore, making your product backward compatible might undermine its unique advantages over other products on the market. For instance, Intel made its 486 processor using complex instruction set computer (CISC) architecture as opposed to reduced instruction set computer (RISC) architecture so that it would be backward compatible with earlier generations of personal computers. While this approach made its customers more comfortable adopting the new processor, it also meant that Intel could not take advantage of the greater scalability, faster product development time, and better performance of the RISC processor.[32] As a result, the 486 processor offered less of an advantage over rival processors than it could have.

Third, you can manage customer and competitor expectations.[33] Customers are influenced not just by reality but by their *perceptions* about products, their installed base, and the availability of complementary products. Because a small advantage in the early life of a market can lead customers to tip to a particular product, you can often win a standards battle by changing customer perceptions through the management of information.

You can manage customer expectations by making pronouncements about the size of your installed base. Because potential customers favor products with a larger installed base, companies frequently try to convince customers that they have a larger market share than their competitors. For example, in the early 1990s, Sega and Nintendo each claimed they controlled over 60 percent of the video game market, knowing that future customer adoptions would go disproportionately to the product that customers perceived as dominant.[34]

You can also manage customer and competitor expectations by making product **preannouncements**, or announcements of product release dates before the actual release occurs. Preannouncements help to manage expectations in several ways:

- First, they help convince competitors that they are so far behind the announcing firm that they should not launch a competing product.[35] For example, Sony widely publicized its investment in the development of the PlayStation video game product to convince other companies not to launch competing products.[36]
- Second, they lead customers to believe that the announcing company's product will be better than its competitors', which causes them to delay their purchase of competing products and wait for the announcing company's product to be released.[37]
- Third, they signal that the announcer's product will be on the market before competitors' products, and so will become the industry standard because of its greater installed base.[38] For example, the preannouncement of the DIVX standard slowed the rate of adoption of DVD technology, enhancing the probability that DIVX would become the industry standard in digital video recorders.[39]
- Fourth, they signal the announcer's intentions to suppliers or distributors with whom it would like to form agreements[40] and encourages other companies to produce complementary products.[41] For example, Microsoft used product preannouncements to signal that it would successfully enter the Web browser and video game businesses as a way to encourage suppliers and distributors to work with it.[42]

However, preannouncements are not without costs. They provide information to competitors, who might respond strategically by accelerating their development of competing products. Moreover, they motivate customers to delay purchases of your company's product until the newer, and presumably better, versions have been released.[43] Furthermore, preannouncements are anticompetitive and can provoke antitrust action if your firm is large and has a dominant market position.[44] Finally, preannouncing, which can involve making statements that do not come true, undermines the trust that your customers have in the statements that you make.[45]

Despite these costs, preannouncements are a very common part of technology strategy in industries in which customers are biased toward the adoption of products with the largest installed base. For instance, one research study showed that software firms engage in preannouncements almost half the time that they release products.[46]

A final way that companies manage customer expectations is through criticism of their competitors' technology.[47] By criticizing your competitors' products, you can convince customers not to adopt their products and to adopt yours instead. Perhaps the most famous example of an effort to criticize competitors' products was Thomas Edison's electrocution of a dog, using his rival, George Westinghouse's, electrical standard. Edison's goal was to persuade customers that his electricity standard was superior to Westinghouse's, and his gruesome act did just that.

While all of the tactics described previously will help you to win a standards battle, companies cannot always predict or control the emergence of a standard, as the earlier part of the chapter explained. Therefore, your strategy should hedge against the loss of a standards battle. For example, you will be better off if you can make your technology work equally well with multiple standards so that your firm will not be greatly disadvantaged if one standard emerges rather than another.

What If You Lose?

Suppose you fight and lose a standards battle. What options are open to you?[48] If the industry only has room for one standard, then you might have to exit the market. Quitting will be the best choice if your company would be at too much of a competitive disadvantage to serve that market effectively.[49] This is what Circuit City was forced to do when it lost the digital video recorder standards battle.

Another option is to give up the fight, conform to the standard, and try to compete on some other dimension, like cost, service, or features. This is what Amdahl and Hitachi did when they lost the mainframe computer standards battle to IBM. The advantage of this strategy is that it gives you the chance to come up with a new technology in the future that leapfrogs the standard holder's technology. For example, Apple first conformed to the MP3 player standard and then, having come up with a better design, introduced a new standard with the iPod.

However, pulling off this strategy isn't easy. You will need to change your technology to conform to the industry standard and will have to abandon all of the capital, reputation, and learning that you have invested in it.[50] As a result, your company may suffer significant losses. For example, Rockwell and Lucent were badly damaged in their loss to 3Com in the battle over the 56K modem standard.

If your industry doesn't have an all-or-nothing standard, then you face a different strategic choice. Do you conform to the technical standards that have emerged, or do you try to sell products that don't conform to industry standards (see Figure 12.4)?

You might not conform to the standard and focus on a niche where you can use your technology to meet the needs of a market segment better than the alternatives that conform to the technical standard.[51] For example, Apple Computer was able to survive for many years without conforming to the PC standard by providing computers tailored to the needs of the desktop publishing segment of the personal computer market (graphics-oriented applications). Ultimately, however, it had to conform to the Intel standard.

Moreover, adhering to the technical standard will hinder your ability to differentiate your products from those of your competitors. The aspects of your technology that conform to the industry standard cannot be a source of competitive advantage because they must be the same as those offered by your competitors. If you can't be the low-cost producer, then customers will have no reason to select your product. For example, many personal computer manufacturers who adopted the Wintel standard lost out to Dell and Compaq because all of the companies that adhered to the PC standard offered virtually identical products, and Dell and Compaq were the low-cost producers.[52]

On the other hand, by conforming to industry standards, you will have many more customers to target. And you will find selling your products to be easier because customers prefer products that conform to the standard.[53]

Moreover, it is often difficult to focus on a niche over a long period of time because your competitors will have the advantage of scale economies. This is why

FIGURE 12.4
Fighting a Standard

Advanced Micro Devices, the maker of semiconductor products such as the one shown here, has successfully fought the Intel standard and has taken market share away from that company.
Source: AP Wide World Photos.

Apple finally gave in and adopted the industry standard, letting its machines run Windows software. Over time it became difficult for Apple to support a different standard than Microsoft because the latter's much larger installed base (1 billion versus 30 million) meant that it took Microsoft only eight weeks to break even on the $1 billion cost of developing a new operating system, as opposed to four years for Apple. Apple also could not benefit from the creation of complementary software for its operating system because software developers needed to write separate software for the Macintosh, and had less incentive to write for the small installed base. Therefore, Apple earned lower profits than Microsoft on its operating system, and had a less desirable product in terms of software availability.[54]

Furthermore, customers put pressure on companies to adopt the dominant industry standard if they cannot continue to offer the advantages that make differentiation valuable. For instance, Sun was pushed to abandon its more powerful and more expensive microchips, and adopt the Intel standard, when the company could not provide its network server customers enough benefits to offset the lack of compatibility and higher cost of its chips.[55]

Given the pros and cons of conforming to the industry standard, should you do it, or should you focus on a niche? The answer to this question depends on the stage of the product life cycle. If your market is young, and most of the customers are innovators or early adopters, then you should differentiate your product from existing technical standards. At this point in the life cycle, so little of the market has adopted the technology that convergence on a single standard is unlikely, especially if you can offer a viable alternative. Given the advantages of not adhering to the standard described previously, the balance falls in favor of differentiating your product.

However, if the market is already mature, and most of the users are from the early or late majority, then you should make your product compatible with existing technical standards. At this point in the life cycle, so much of the market has adopted the technology that convergence on a single standard is quite likely. Given the advantages of adhering to the standard described previously, the balance falls in favor of conforming.[56]

Defending a Technology Standard

If your proprietary technology becomes an industry standard, then you'll want to defend this privileged position against the efforts of other firms to dislodge you. So how should you do that? The answer lies in making it difficult for customers to switch to the products and services offered by your competitors. For example, Microsoft has made it difficult for customers to switch to competitive products by maintaining a very wide range of products that work with the Windows operating system, even though some of those products are not very profitable for the company. Until recently, the range of software that was compatible with Windows locked many users into this system because there were one or two programs that they used—niche software for games or hobbies or custom software for their industry—that did not work with the Apple OS X operating system. Consequently, they could not switch over to Apple computers despite the latter's better design, superior operating system, and lower susceptibility to viruses.[57]

To increase your customers' switching costs and reduce their likelihood of changing suppliers, you can do the following:

- Make your products more attractive to customers than competitors' products by making the products more functional, by adding features, or by making peripheral components available.
- Sign long-term contracts with the producers of complementary products or services to ensure the availability of complements, which enhances the value to customers of the products that you provide.
- Make future generations of your product backward compatible with previous generations so that customers can easily upgrade their products (as Microsoft has done with multiple generations of Windows software).
- Make your products or services incompatible with the alternatives offered by your competitors.[58] For example, AOL refused to allow its instant messaging service to connect to those provided by other companies, which gave customers an incentive to remain in the AOL network.[59]

Technical Standards and Competition Between Systems

Technical standards often change the nature of competition from that between firms to that between systems. When more than one technical standard exists in an industry, some products are designed to be compatible with other products, and together they make up a technical system. Other products, designed not to be compatible with those products, but designed to be compatible with each other, make up a different technical system. The groups of companies that work together on the first system compete with the groups of companies that work together on the second system. For instance, not all MP3 players are compatible with all compression formats, creating several different MP3 systems. Thus, competition in this business occurs at the level of the system, with Apple's iPod/iTune system competing against other systems that provide music players and Web sites for downloading music.

When multiple different technical systems compete in an industry, you need to decide whether to make your products conform to all, or only some, of them. For example, until very recently, the Apple and Microsoft computer operating systems were not compatible, so the providers of applications software had to decide whether to offer two versions of their products or just align with one of the systems.

If you chose to conform to only one technical standard, then you need to decide which system is the best for your company to join. Moreover, you will have to design your products to be compatible with the other products in that system, while making them incompatible with products outside of the system.

Key Points

- Companies often battle to control technical standards by gaining the support of complementary products, having a killer application, making their products backward compatible, and managing customer and competitor expectations.
- If you lose a standards battle, you can exit the market, conform to the standard and compete on another dimension, or focus on a niche and meet its needs without conforming to the standard.
- If your technology becomes the industry standard, then you need to defend it against the efforts of other firms to dislodge it by keeping customer switching costs high.
- Technical standards often lead to competition between systems of companies, rather than between individual organizations.

Open Versus Closed Standards

Companies need to decide whether to create an **open standard**—a standard for which specifications are known by other companies—or a closed standard—a system for which those specifications are not known.

There are many advantages to having an open standard. First, an open standard will make it easier for other companies to understand how your products work and, therefore, how to build complementary products.[60] For example, JVC's decision to establish an open standard encouraged movie studios to provide more video content on VHS than on Betamax, and led the video recorder market to tip to the VHS standard.[61]

In fact, by attracting the providers of complementary products, an open standard can create a positive feedback loop that benefits a company. The more complementary products that are available for a product, the more customers are interested in buying it; and the more customers that are interested in buying a product, the more companies are interested in providing complementary products for it.

Take, for example, the case of the open-source software provider Linux, which provides software that any user can license for free in return for licensing any additions that they make to the software at no charge. As Figure 12.5 shows, the more people that used Linux software, the more software applications were created for it; and the more software applications that were created for Linux, the more people used it.[62]

Second, open standards encourage other companies, which might develop superior alternatives, to adopt your technology, mitigating the threat of competition.[63] For example, Sun Microsystems brought out minicomputers that used the UNIX operating system and adopted an open approach to its software, publishing its specifications and interface protocols.[64] This open system made it possible for other companies to write software for Sun's platform, which reduced their incentive to develop their own alternatives.[65]

Third, open standards allow you to generate a large installed base very quickly by attracting more users than a single company can support.[66] For example, Microsoft generated a much larger installed base for its DOS operating system than Apple did for

FIGURE 12.5
Positive Feedback in the Development of Linux

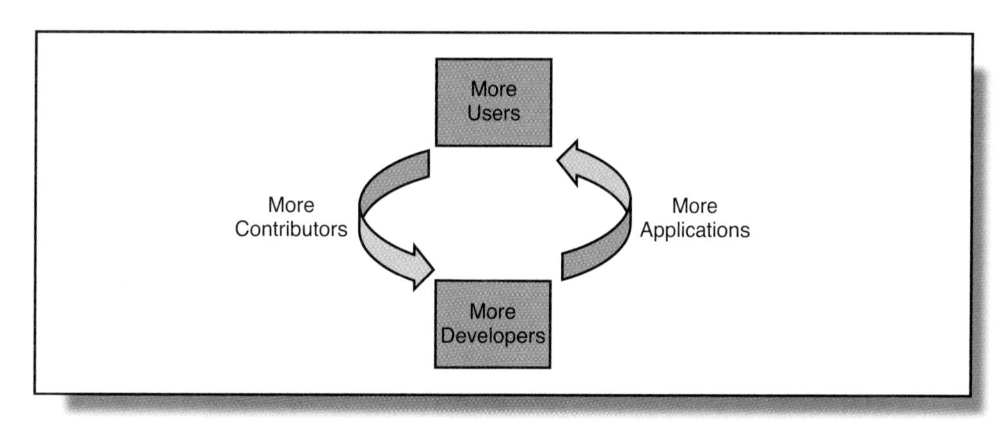

A positive feedback loop exists between developers and users of a computer operating system: The more users of the operating system there are, the more attractive the system is for developers; and the more developers there are, the more attractive the system is for users.

its Macintosh operating system because Apple Computer adopted a closed standard that did not allow anyone else to manufacture computers using its system, while Microsoft allowed anyone who wanted to make personal computers to use its system. Because Apple could not come close to producing the same number of computers as all other computer manufacturers combined, Microsoft DOS quickly developed a much larger installed base than Apple Macintosh.[67]

While an open standard benefits companies, it also has risks. By opening up your technology to other companies, you run the risk of losing control over your technology and, consequently, sales.[68] For example, Palm used open-source coding for its PDAs to increase adoption and to encourage the development of a technical standard. However, this approach made Palm vulnerable to competitors who could produce the PDA hardware at a lower cost than Palm.

By using an open standard, you also incur the risk that licensees might change your technology in a way that makes it unnecessary for them to pay you royalties. Thus, you might make less money off of your technology than if you had created a closed standard and never licensed it to others. For example, AMD was able to get out of paying royalties on Intel's K5 microprocessors after having made modifications to Intel's initial technology, undermining Intel's revenue stream from that technology.[69]

Furthermore, an open standard demonstrates to your competitors how your technology works, making it easier for them to imitate it. Because your competitors will be more likely to leapfrog your technology if they can imitate it, having an open standard could undermine your competitive advantage.

Two different types of open standards exist, each of which creates different opportunities and problems for managers and entrepreneurs. Some open standards are proprietary, like Microsoft Windows. The standard is open because Microsoft shares information about the operating system so that other companies can provide complementary products. However, the Windows standard is owned and controlled by Microsoft. Other open standards are nonproprietary, like the Linux operating system, which is not owned or controlled by a single company.

Proprietary and nonproprietary open standards both have their advantages and disadvantages. Nonproprietary open standards benefit companies by making their products better for customers. First, they are neutral, which means that customers don't have to accept the strategic direction of the owner of the standard, or its power over the development of their products. Second, they do not require customers to pay royalties, thus lowering costs. Third, they are easier to use and customize because more information about them is provided to users.[70]

However, nonproprietary open standards create several strategic problems for providers. First, they do not allow companies to control quality or the development of the standard in a way that benefits their products, as Microsoft has done with the Windows operating system.[71] Second, they do not provide strong incentives for companies to support their customers.[72] Third, they open up markets to competition, which can undermine a provider's competitive position, as occurred when computer operating systems switched from IBM to Unix.

GETTING DOWN TO BUSINESS

The Story of Open-Source Software[73]

Perhaps the best known story of open-source software is Linux, which begins in the mid-1980s, with Dr. Richard Stallman of MIT, the founder of the free software movement. Stallman's goal was to create an operating system that was not controlled by any company and was freely available to all users.[74]

Any operating system needs a "kernel" or set of commands that allow it to control key functions, like displaying information on the screen or downloading information from a Web site. In the early 1990s, Linus Torvalds, a computer science student at the University of Helsinki, created the "kernel" for the GNU software, which led to the creation of Linux, a free software operating system.[75]

An open-source software operating system differs from a proprietary system, like Microsoft Windows or Windows NT, in two important ways. First, open-source software is distributed as source code, which programmers can read, rather than as machine-readable binary code, which is difficult to adapt.[76] Second, open-source software licenses give users the right to alter the software as long as they make any of their changes freely licensable without royalties to other users.[77]

Unlike proprietary software, which is maintained by the company that owns it, the Linux operating system is maintained by a group of volunteer programmers around the world. Some for-profit companies, like Red Hat, offer contracts to companies to support their use of the Linux operating system in return for fees. However, the incentives to support the Linux system are less than that of a proprietary system, and there is far less organization behind the effort.[78]

In the late 1990s and early 2000s, Linux software became increasingly popular amongst users. Attracted by lower fees, greater flexibility to adapt the software, and superior robustness, many companies shifted from proprietary operating systems to the Linux system. Large companies, like Credit Suisse First Boston and Goldman Sachs, found the software to be very effective for running complex transactions on their servers, and small companies found it to be more cost effective and flexible than proprietary alternatives. As a result, from 2001 to 2006, Linux's installed base in the $50 billion server market grew by 25 percent, as compared to the growth of the overall installed base of 9 percent. Moreover, during this time, the installed base of Unix-based operating systems declined 5 percent, and that of Novell's operating system decreased by 15 percent.[79] Because of this rise in installed base, companies like IBM, HP, and Oracle began to write applications to work with the Linux operating system, providing a wide range of complementary software applications for the Linux system.[80]

This, of course, was a major threat to the companies, like Microsoft, that controlled the proprietary operating systems. If their customers switched to the Linux standard, they would lose a great deal of money. As a result, they responded by trying to show the advantages of having a vendor-supported operating system: someone responsible for fixing bugs, which reduced the need to conduct maintenance and development activities in-house or through contracts with third-party providers.

Key Points

- Open systems are valuable because they facilitate the creation of complementary products, create a positive feedback effect, permit the rapid creation of a large installed base, and discourage other companies from competing against your technology.
- Open systems risk the loss of control over technology, which could lead to a decline in sales.
- Two types of open systems exist; some, like the Microsoft operating system, are proprietary, while others, like the Linux operating system, are nonproprietary.
- Nonproprietary open standards have the advantage of being more attractive to customers; they are neutral, don't require royalty payments, and are easier for customers to use.
- Proprietary open standards help companies by reducing competition, by giving them control over the development of technology, and by providing a strong incentive to support the system.

Discussion Questions

1. How do technical standards affect firm strategy and competition? What are the trade-offs faced by managers and entrepreneurs in markets in which technical standards are present?
2. Why do some markets "tip" to a single standard, but others do not?
3. Why doesn't the best technology always emerge as the standard?
4. Should you fight a standards battle? Why or why not?
5. What can you do to win a battle over technical standards? Why are these things effective?
6. Should you use product preannouncements? Why or why not?
7. What should you do if you lose a standards battle? Explain your answer.
8. Should you seek an open or closed technical standard? What factors should you consider in your decision? Why?

Key Terms

Backward Compatibility: Making a product or service work with previous generations of products or services.

Installed Base: The number of users of a product at a point in time.

Open Standard: A standard whose specifications are known by other companies.

Preannouncements: The announcements of product release dates before the actual release occurs.

Technical Standards: The specifications that are necessary to make sure that the components of a technical system are compatible with each other.

Putting Ideas into Practice

1. **Identifying Technical Standards** The purpose of this exercise is to identify all of the technical standards that affect a laptop computer. Begin by making a list of all of the components of a laptop computer with wireless capability. Then, identify if there is a technical standard that governs each of these components. (If you know the name of the specific technical standard, that's great. Write it down. If you don't know the name of the standard, don't worry about it, just describe the standard.) When you are done, count up the number of components and the number of standards on your list. Do you have a lot or a few? Depending on what you found, explain why you think there are either a lot or only a few technical standards that affect a laptop computer with wireless capability.

2. **How Standards Emerge** Identify three technical standards. Investigate how those standards came into existence. (They don't all have to have come into existence the same way.) Explain why the mechanisms that you identified accounted for the emergence of the standards.

3. **Standards Battles** Identify a standards battle that was fought by two companies (other than the DVD versus DIVX battle described in the opening vignette). Investigate the tactics used by the winner to fight the standards battle. Did the choice of tactics affect that company's victory? Why or why not?

NOTES

1. Adapted from Dranove, D., and N. Gandal. 2003. The DVD-vs.-DIVX standard war: Empirical evidence of network effects and preannouncement effects. *Journal of Economics and Management Strategy*, 12(3): 363–386.
2. Greenstein, S., and V. Stango. Forthcoming. The economics of standards and standardization. In S. Shane (ed.), *Blackwell Handbook on Technology and Innovation Management*. Oxford: Blackwell.
3. Garud, R., and A. Kumaraswamy. 1993. Changing competitive dynamics in network industries: An exploration of Sun Microsystems' open systems strategy. *Strategic Management Journal*, 14(5): 351–369.
4. Utterback, J. 1994. *Mastering the Dynamics of Innovation*. Boston, MA: Harvard Business School Press.
5. Afuah, A. 2003. *Innovation Management*. New York: Oxford University Press.
6. Utterback, *Mastering the Dynamics of Innovation*.
7. Tushman, M., and L. Rosenkopf. 1992. Organizational determinants of technological change: Toward a sociology of technological evolution. *Research in Organizational Behavior*, 14: 311–347.
8. Besen, S., and G. Saloner. 1989. The economics of telecommunications standards. In R. Crandall and K. Flamm (eds.), *Changing the Rules: Technological Change, International Competition, and Regulation in Communication*. Washington, DC: Brookings Institution, 177–220.
9. Suarez, F., and J. Utterback. 1995. Dominant designs and the survival of firms. *Strategic Management Journal*, 16(6): 415–430.
10. Farrell, J., and C. Shapiro. 1992. Standard setting in high definition television. *Brookings Papers on Microeconomics*, Washington, DC: Brookings Institution, 52–58.
11. Dhebar, A. 1995. The introduction of FM radio (A), (B), and (C), *Harvard Business School Teaching Note*, Number 5-594-072.
12. Suzuki, O. 1998. The PC-based desktop video conferencing systems industry in 1998. *Stanford University Case*, Number SM 51.
13. Dvorak, P., N. Wingfled, and S. McBride. 2004. Technology titans battle over format of DVD successor. *Wall Street Journal*, March 15: A1, A8.
14. Axelrod, R., W. Mitchell, R. Thomas, D. Bennett, and E. Bruderer. 1995. Coalition formation in standard-setting alliances. *Management Science* 41(9): 1493–1508.
15. Hill, C. 1997. Establishing a standard: Competitive strategy and technology standards in winner-take-all industries. *Academy of Management Executive*, 11(2): 7–20.
16. Von Burg, U., and M. Kenney. 2003. Sponsors, communities, and standards: Ethernet vs. Token Ring in the local area networking business. *Industry and Innovation*, 10(4): 351–375.
17. Choi, J. 1994. Irreversible choice of uncertain technologies with network externalities. *RAND Journal of Economics*, 25: 382–401.
18. Arthur, B. 1989. Competing technologies, increasing returns, and lock-in by historic events. *Economic Journal*, 99: 116–131.
19. Tushman and Rosenkopf, Organizational determinants of technological change.
20. Kirsch, D. 2000. *Electric Vehicles and the Burden of History*. New Brunswick, NJ: Rutgers University Press.
21. Wade, J. Dynamics of organizational communities and technological bandwagons: An empirical investigation of community evolution in the microprocessor market. *Strategic Management Journal*, 16: 111–133.
22. David, P. 1985. Clio and the economics of QWERTY. *American Economic Review*, 75: 332–337.
23. De Vries, H., and G. Hendrikse. 2001. The Dutch banking chipcard game. *International Studies of Management and Organization*, 31(1): 106–125.
24. Sathe, V. 2003. *Corporate Entrepreneurship*. Cambridge, UK: Cambridge University Press.
25. Ibid.
26. Greenstein and Stango, The economics of standards and standardization.
27. Shapiro and Varian, The art of standards wars.
28. Yoffie, D., and M. Slind. 2006. DVD war. *Harvard Business School Case*, Number 9-706-504.
29. Ibid.
30. Ibid.
31. Dranove and Gandal, The DVD-vs.-DIVX standard war.

32. Lee, J., J. Lee, and H. Lee, 2003. Exploration and exploitation in the presence of network externalities. *Management Science*, 49(4): 553–570.
33. Shapiro and Varian, The art of standards wars.
34. Brandenburger, A. 1995. Power play (B): Sega in 16-bit video games. *Harvard Business School Case*, Number 9-795-103.
35. Arthur, W. 1996. Increasing returns and the new world of business. *Harvard Business Review*, July–August: 100–109.
36. Schilling, M. 2003. Technological leapfrogging: Lessons from the U.S. video game console industry. *California Management Review*, 45(3): 6–35.
37. Lee, Y., and G. O'Conner. 2003. New product launch strategy for network effects products. *Academy of Marketing Science Journal*, 31(3): 241–255.
38. Eliashberg, J., and T. Robertson. 1988. New product preannouncing behavior: A market signaling study. *Journal of Marketing Research*, 25(3): 282–292.
39. Dranove and Gandal, The DVD-vs.-DIVX standard war.
40. Eliashberg and Robertson, New product preannouncing behavior.
41. Riggins, R., C. Kriebel, and T. Mukhopadhyay. 1994. The growth of interorganizational systems in the presence of network externalities. *Management Science*, 40(8): 984–998.
42. Schilling, Technological leapfrogging.
43. Hoxmeier, J. 2000. Software preannouncements and their impact on customers' perceptions and vendor reputation. *Journal of Management Information Systems*, 17(1): 115–139.
44. Eliashberg and Robertson, New product preannouncing behavior.
45. Hoxmeier, Software preannouncements and their impact on customers' perceptions and vendor reputation.
46. Bayus, B., S. Jain, and A. Rao. 2001. Truth or consequences: An analysis of vaporware and new product announcements. *Journal of Marketing Research*, 38: 3–13.
47. Greenstein and Stango, The economics of standards and standardization.
48. The ideas in this section were provided to me by Shane Greenstein in an e-mail. I am deeply indebted to him for providing his insights as they are the basis of my own thinking on this issue.
49. Axelrod et al. Coalition formation in standard-setting alliances.
50. Schilling, M. 2005. *Strategic Management of Technological Innovation*. New York: McGraw-Hill.
51. Lee, Lee, and Lee, Exploration and exploitation in the presence of network externalities.
52. Greenstein and Stango, The economics of standards and standardization.
53. Lee, Lee, and Lee, Exploration and exploitation in the presence of network externalities.
54. Yoffie, D. 2006. Apple Computer 2006. *Harvard Business School Teaching Note*, Number 5-706-513.
55. Tam, P. 2003. Cloud over Sun Microsystems: Plummeting computer prices. *Wall Street Journal*, 242(76): A1, A16.
56. Lee, Lee, and Lee, Exploration and exploitation in the presence of network externalities.
57. Mossberg, W. 2006. Apple opens the Mac to Windows. *Wall Street Journal*, April 6: B1, B4.
58. Schilling, Technological leapfrogging.
59. Denison, D. 2000. The battle for instant messaging: AOL fending off challenges in a market that is growing into a global network. *Boston Globe*, December 11.
60. Garud and Kumaraswamy, Changing competitive dynamics in network industries.
61. Cusumano, M., Y. Mylonadis, and R. Rosenbloom, 1992. Strategic maneuvering and mass market dynamics: The triumph of VHS over Beta. *Business History Review*, 66: 51–94.
62. O'Mahoney, S., and C. Baldwin. 2000. IBM and Linux (A). *Harvard Business School Teaching Note*, Number 5-906-016.
63. Hill, Establishing a standard.
64. Rosenberg, M., and B. Silverman. 2001. Sun Microsystems Inc: Solaris Strategy. *Harvard Business School Case*, Number 9-701-058.
65. Ibid.
66. Brynjolfsson, E., and C. Kemerer. 1996. Network externalities in microcomputer software: An econometric analysis of the spreadsheet market. *Management Science*, 42(12): 1627–1647.
67. Arthur, Increasing returns and the new world of business.
68. Schilling, *Strategic Management of Technological Innovation*.
69. Hill, Establishing a standard.
70. O'Mahoney and Baldwin, IBM and Linux (A).
71. Ghemawat, P., B. Subirana, and C. Pham. 2005. Linux in 2004. *Harvard Business School Case*, Number 9-705-407.
72. O'Mahoney and Baldwin, IBM and Linux (A).
73. Adapted from C. Baldwin, S. O'Mahony, and J. Quinn, 2003. IBM and Linux (A). *Harvard Business School Case*, Number 9-903-083.
74. Ibid.
75. Ibid.
76. Ibid.
77. Ibid.
78. Ibid.
79. Ghemawat, Subirana, and Pham, Linux in 2004.
80. Ibid.

Chapter 13

Strategy in Networked Industries

Learning Objectives

After reading this chapter, you should be able to:

1. Define *increasing* and *decreasing returns*, identify industries subject to the two different models, and explain why some industries and not others are subject to increasing returns.
2. Define *network externalities*, explain the difference between direct and indirect network effects, and describe the relationship between network externalities and increasing returns.
3. Explain how firms can use network effects to their strategic advantage.
4. Identify when multiple networks can coexist in an industry, and explain how the presence of one or multiple networks affects firm strategy in networked industries.
5. Describe the firm strategies that are most effective in increasing returns businesses, and explain why these strategies need to differ from those that are most effective in decreasing returns businesses.

6. Explain why starting large is important in a business based on increasing returns.
7. Explain why building your installed base quickly is crucial in a business based on increasing returns.
8. Understand the tactics necessary to build your installed base rapidly.
9. Define *customer lock in*, describe how to get customers to ignore it, and explain why this is important in businesses based on increasing returns.
10. Explain why being the first mover is important in businesses based on increasing returns, and how to get to market quickly in these industries.

Network Effects: A Vignette[1]

Can anyone catch up to Apple Computer in the digital music business? Apple's iTunes music site controls 72 percent of the market for music online, a whopping 62 percent higher than the next largest competitor, eMusic. Apple controls a commensurate portion of the music player business, selling 75 percent of the MP3 players purchased.[2]

Apple's strategy in this business has been to create a closed network. Its iPod players are the only devices that can play songs downloaded from its iTunes Web site, and its iTunes site is the only place where its iPods can get songs. (However, Apple's iPods can play music in the MP3, AAC, and WAV formats that other companies use.)

Central to the iPod/iTunes strategy is a "digital rights management" system that protects the intellectual property of the music labels. Because users are limited to making only five copies of downloaded songs and can only play them on their iPods, Apple's system is more attractive to the record labels than many of the other digital music sites (see Figure 13.1).

Moreover, Apple pays heavy royalties to the music labels, which get two-thirds of the revenue from iTunes music downloads. As a result, Apple only makes money on the sales of iPods, and earns nothing from operating iTunes.[3]

So far this closed network has been successful. But Apple has tried this before. In personal computers, Apple used a closed network; its operating system only worked with its computers. Some experts believe that this closed approach caused Apple to miss out on the tremendous success of Microsoft and Intel in that business.

Apple is also facing a lot of pressure from European governments to open up its system, given the view of regulators that the closed system is unfair to consumers.[4] So it is unclear if Apple will be able to maintain its current strategy, especially in Europe.

Perhaps because of its experience with computers, because of regulatory pressure, or because it wants to reduce the power of the music labels, Apple is proposing a change to its system. In February 2007, Steve Jobs, CEO of Apple, posted a letter outlining his views on the digital rights management system. Under the new model, users of the iTunes site would no longer be limited to making only five copies of downloaded songs and to playing music only on iPods. They would be able to make as many copies as they wanted and to play them on any MP3 player.

Do you think Apple will be able to remain the dominant player in this industry? If so, what do you think Apple needs to do to remain dominant? Do you think that changing to a completely open system will help?

FIGURE 13.1 Different Digital Music Business Models

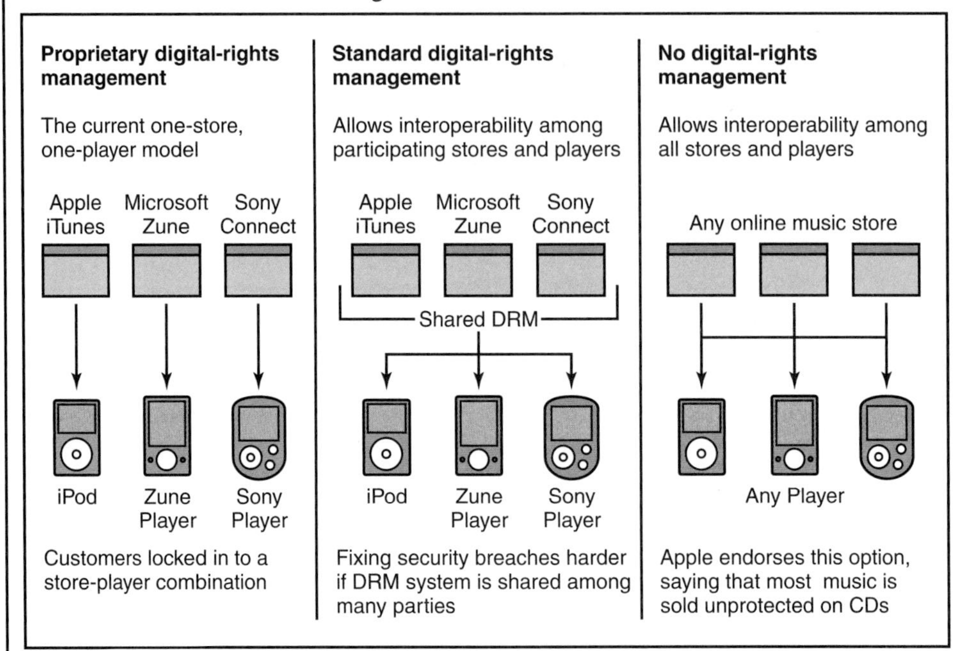

Apple is considering moving from the proprietary digital-rights management business model to the no-digital-rights management business model to maintain its control over the digital music business.

Source: Wall Street Journal, February 7, 2007: A11.

INTRODUCTION

Effective technology strategy depends on whether you operate in an industry based on **decreasing** or **increasing returns**. In industries based on decreasing returns, what you earn on the marginal product produced declines as the volume of output increases, but in industries based on increasing returns, what you earn on that marginal product increases with the volume of output. Therefore, as a technology manager or entrepreneur, you need to figure out how the two types of industries are different, and what effect these differences have on firm strategy. By identifying which type of industry you compete in, you can figure out what strategic actions your company should take to succeed.

This chapter discusses technology strategy in industries subject to increasing returns. The first section explains how increasing returns businesses differ from decreasing returns businesses, and the sources of increasing returns. The second section identifies specific strategic actions that you can take to enhance your company's performance in industries based on increasing returns.

Increasing Returns Businesses

Many businesses display decreasing returns (see Figure 13.2). Take coal mining, for example. At first, you get a high return on the marginal product produced because the veins that you tap first are the least costly ones to mine. However, as you produce more and more coal, you use up the easy-to-access veins and are forced to incur higher and higher costs to get the remaining deposits. As a result, your costs increase with the volume you produce, leading to decreasing marginal returns.

However, some businesses show increasing returns. Perhaps the best example of an increasing returns business is an Internet auction house. At first, this business has a low marginal return because the cost of attracting buyers and sellers to the site is very high. The absence of many sellers makes buyers wary, and the absence of many buyers makes sellers concerned. So you need to spend a lot of money on advertising, price discounting, and sales efforts to attract both buyers and sellers to the site. However, as the site sells more and more goods, buyers and sellers become more favorably disposed to it, reducing the costs of attracting them. As a result, costs decrease with the volume of transactions processed, while prices remain constant, generating increasing marginal returns.[5]

We can see this pattern of increasing returns if we look at real data for eBay, which is shown in Figure 13.3. By plotting the number of registered users of the auction house against the company's net income annually from 1998 to 2005, the figure shows the pattern of increasing returns. The more registered users eBay has, the more money the company makes per user.

Why Industries Display Increasing Returns

So knowing if a business displays increasing or decreasing returns is important to formulating an effective technology strategy. High-technology industries—pharmaceuticals, computers, and telecommunications to name a few—are more likely than low-technology industries to show increasing returns.[6] Why is this the case?

FIGURE 13.2
Increasing Versus Decreasing Returns

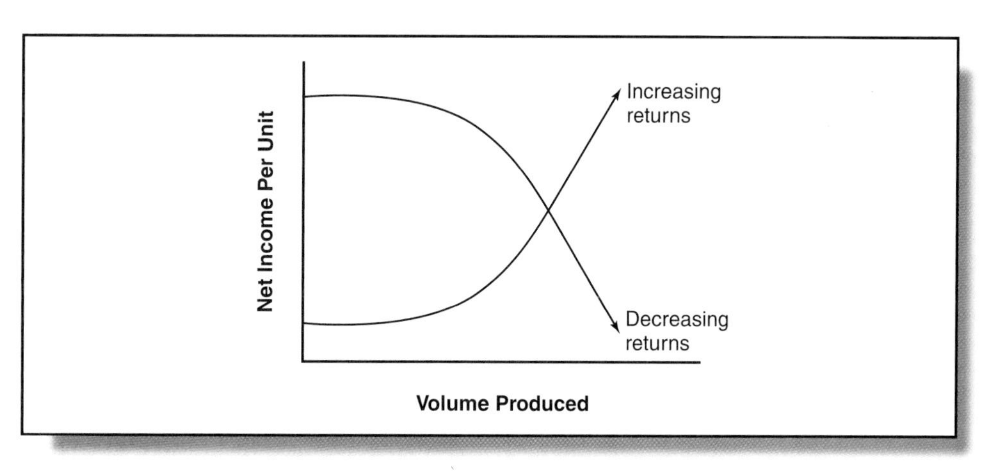

Net income per unit goes up with the volume produced for increasing returns businesses but goes down for decreasing returns businesses.

FIGURE 13.3
Increasing Returns at eBay

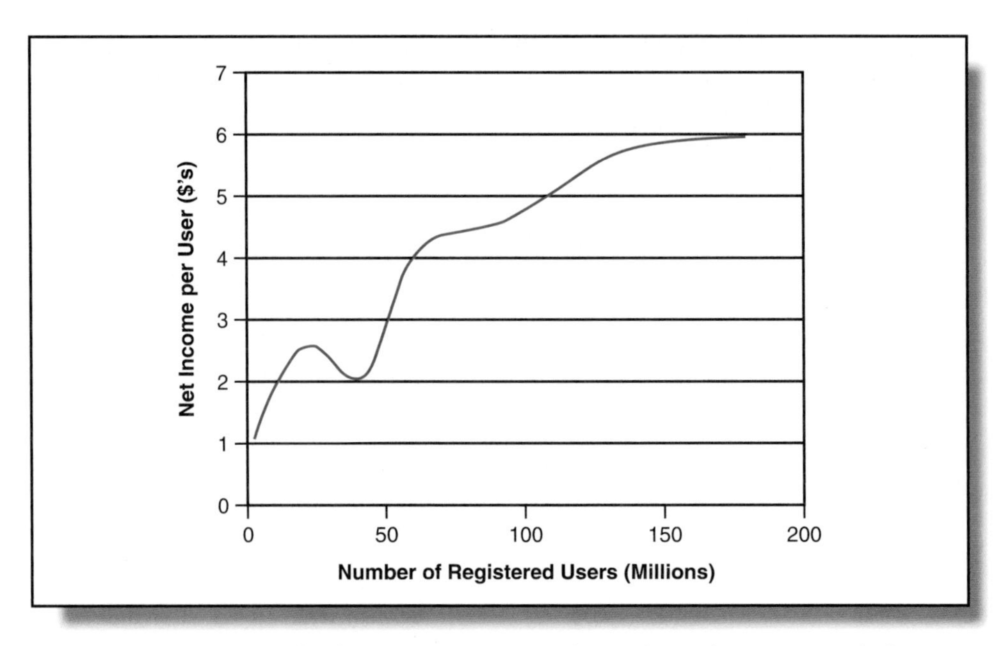

EBay's net income per user has been increasing as its volume of users has gone up, which is evidence of increasing returns to scale.

Source: Created by the author from data contained in eBay's annual reports.

The first reason was suggested previously. When up-front costs are high and marginal costs are low, unit costs drop dramatically as volume increases. Thus, when most costs are incurred up front and prices remain the same or increase with volume, profit margins rise as volume goes up. Take, for example, drug production. It costs hundreds of millions of dollars to research and test a new drug. However, once that drug has received FDA approval, most of the costs have been incurred. The cost of producing each capsule or vial of medication is very small. So if the prices at which the drugs are sold are kept constant, then the profit margins generated on the additional units sold are higher than the profit margin on the first unit. (This is why many people are upset about how pharmaceutical companies price drugs, but that is an issue for another course and another book.)

Now you might think that increasing returns look just like economies of scale. And they do, up to a point. With economies of scale, the cost per unit declines—and per unit profit margin goes up—as volume increases, just like with increasing returns, because the fixed costs are amortized over more units. However, with economies of scale, unlike with increasing returns, the per unit cost eventually starts going up again. Take a steel mill, for example. At a certain volume of production, diseconomies of scale set in because it takes a huge amount of communication and coordination cost to operate a giant steel mill with hundreds of thousands of workers. This eventual increase in cost, as volume increases, is not present in industries, like software, that are subject to increasing returns. After Microsoft sells a single master disk of its operating system to an original equipment manufacturer, which reproduces it, its marginal cost per unit starts to go down. But unlike the case with economies of scale, the marginal cost keeps going down forever.[7]

A second reason why an industry might display increasing returns is producer learning. When learning curves are steep, firms increase their efficiency at producing and selling products as they make and market more units. As a result, the more that a firm produces, the lower its costs become. Again, given constant prices, this declining marginal cost leads to increasing marginal profit.

A third reason why an industry might display increasing returns lies in **network externalities**. Network externalities exist when the purchase of a product by a new customer creates additional value for existing customers. For instance, the more people that you know that use the social networking Web site Facebook, the more valuable it is for you to be in Facebook.[8]

Networked products display increasing returns because they allow companies to raise prices as they sell more units. Because network externalities increase a product's value to existing customers as additional customers adopt it, the price that customers are willing to pay for the product increases with the volume sold. Therefore, companies can charge later-to-adopt customers more for networked products than they can charge earlier-to-adopt customers.

Moreover, when customers pay for a product on an ongoing basis—as with cell phone service—then companies have the opportunity to raise prices on additional units of networked products sold to existing customers. For example, over time, eBay and PayPal have both increased the prices that customers pay to use their services. Because companies can increase prices on networked products as volume increases, while holding their costs constant, the marginal profit on networked products go up with the quantity sold.

Types of Network Effects

Network externalities come in two varieties: direct effects and indirect effects. **Direct network effects** are network externalities that come from the direct interaction of users. As Figure 13.4 shows, the classic example is the telephone. When only a handful of people had telephones, people could not assume that telephones could be used to communicate, reducing the value of the tool. As the number of people who had telephones grew, the ability to rely on phones for communication increased, enhancing their value to users.

Robert Metcalfe, the inventor of the Ethernet, came up with a formula to quantify direct network effects. He said that the value of a telecommunications network is proportional to the square of the number of devices on that network, a relationship now known as **Metcalfe's law**.[9]

One of the main implications of Metcalfe's law is that communications products, like e-mail, fax machines, telephones, and social networking Web sites, show greater than linear growth. Another implication is that their value grows faster than their cost, leading to higher profit margins as the networks get bigger. For example, if you have a network of 10 telephone users and you add another one, you have increased your network by 10 percent. However, you have increased the value to the users—the number of possible connections between them—from 10^2 (100) to 11^2 (121), an increase of 21 percent.

Indirect network effects are network externalities that develop when the presence of **complementary products**—products that are used along with the focal product—increases a product's value.[10] As Figure 13.5 shows, complementary products increase a focal product's value because the product is more useful if the complement is present.[11] For example, the value of a video game console increases with the number of

FIGURE 13.4
Direct Network Effects

Telephones display strong network effects; the more people who purchase telephones, the more valuable telephones are to the people who already have them.

Source: http://en.wikipedia.org/wiki/Network_effects.

games that you can play on it, and the value of a DVD player increases with the number of movies that you can watch on it.[12] These network effects are indirect because the complementary product does not affect the focal product's functionality, but does make it more desirable.

Indirect networks are often complex because network effects can influence suppliers as well as customers. For instance, Xbox has game developers, who provide the content for customers to play, as well as customers who buy the consoles. As the size of the network increases, the value of participating in the network increases for both the customers and the game developers. Moreover, these network effects influence each other, with the number of console owners affecting the value of the network to game developers and vice versa.[13]

FIGURE 13.5
Indirect Network Effects

Cable television displays indirect network effects because there is little good programming provided when the number of customers is few.

Source: Universal Press Syndicate.

Strategic Issues in Networked Industries

Networked industries, such as video games, fuel cell vehicles, instant messaging, Internet search, and DVDs, are becoming a very important part of the world economy. In fact, one researcher estimates that they account for the majority of the revenue of 60 of the top 100 companies.[14] Therefore, technology strategists need to understand the strategic issues raised by network effects.

One important issue concerns maintenance of the network. Networks can be controlled by a single firm, like Microsoft in the case of the Xbox video game or Apple in the case of the Macintosh computer, or they can be shared by more than one firm, as in the case of the VISA credit card system.[15] When a network is controlled by a single firm, that firm is responsible for the maintenance of the network's platform. For instance, eBay is based on a platform in which users share a common architecture; make use of common components, such as Web pages and bid management software; and follow common rules, such as those for registration and bidding.[16] Ebay is responsible for setting the architecture and rules for this network.

When a network is shared, maintaining the platform is more difficult because multiple parties need to work together to manage it. The easiest case occurs when the companies that share the platform are working together as strategic partners. Under those circumstances, the companies can simply negotiate the network maintenance tasks among themselves.

However, sometimes, the companies that use a network are rivals, as is the case with credit card companies who share the Visa card platform.[17] In this case, the companies sharing the platform need to have mechanisms in place to ensure that they cooperate on its maintenance. If they allow their competition to spill over into platform management, they may allow a different platform to succeed in competition against theirs.

Another important strategic issue for networked industries concerns the possibility of sustaining more than one network. In some industries, competing networks can exist, such as is the case in video games like Microsoft Xbox and Sony Playstation.[18] However, most of the time, multiple networks are temporary and are the result of a competitive process between businesses that has not yet played itself out.

TABLE 13.1
Are Multiple Networks Possible?
In May 2005, MySpace.com was not the most used social networking site, but by May 2006, it was almost four times as large as the next most used site, suggesting that the social networking Web site business might converge on only one network.

Site	U.S. Visitors in May 2005 (thousands)	U.S. Visitors in November 2005 (thousands)	U.S. Visitors in May 2006 (thousands)
Myspace.com	15,578	26,684	51,411
Classmates.com	16,968	15,158	14,792
Facebook.com	6,964	11,065	14,069
MSM Spaces	3,140	4,191	9,566
Xanga.com	8,397	7,901	7,146
Flikr.com	923	2,336	5,163
LiveJournal.com	7,394	4,843	3,904
Hi5.com	2,959	3,355	3,013
Tagged.com	456	1,373	2,078
Bebo.com	2,013	1,500	1,752
Friendster.com	1,269	1,468	1,083
Linkedin.com	196	354	519
Orkut.com	113	83	210

Source: Adapted from http://www.imediaconnection.com/content/10623.asp.

In the long run, industries can sustain multiple networks only if customers use those networks for different purposes. For example, as Table 13.1 shows, MySpace and Facebook are competing networks in the online social networking industry. Both companies will survive over time only if they are useful to customers for different purposes. If the two Web sites are perfect substitutes for each other, then customers will see an advantage in switching to the Web site with more users. Why? Because they will derive more value from being part of the larger network. Moreover, this effect will accelerate over time. As people migrate from the smaller network to the larger one, the value advantage of being part of the larger network will grow, further exacerbating the incentive for customers to shift from the smaller one to the larger one.

A third issue concerns the use of strategic action to exploit network effects. Companies can, and do, take action to turn both direct and indirect network effects to their advantage. Take the example of Palm Computing. Palm enhanced direct network effects by creating a beaming capability that allowed information to be transferred between Palm PDAs. And it increased indirect network effects by working hard to get developers to write software for its operating system.[19]

A final strategic issue concerns the competition that occurs from outside the network. Even when an industry has converged on a single network, the company that controls the network still faces the potential for competition from players in another network that want to expand into a new market. For example, Microsoft overtook RealNetworks's lead in the streaming media market by bundling its media player with its PC operating system and its media server software with its Windows NT server operating system. Because it did not charge extra for the added software in this bundle, many content providers switched to Windows Media Player from RealNetworks's alternative.[20]

GETTING DOWN TO BUSINESS
Starting a New Company in a Networked Industry[21]

In September 2005, eBay purchased Skype, a VOIP telephone company founded in 2003, for $1.5 billion. Skype provides a free service that permits computer-to-computer telephone calls of up to 5 people at a time, text messaging of up to 50 people at a time, and free video messaging. Users simply have to have a broadband connection and download free software from the company's Web site that permits computers running Windows, Macintosh OS X, Microsoft Pocket PC, and Linux to use the Skype system.[22]

To transfer the voice, text, and video messages, Skype uses a proprietary signaling protocol that does not work with the VOIP signaling protocol that many VOIP companies, like Vonage and AT&T Call Vantage, use.[23]

At the time of the acquisition, Skype was one of the larger VOIP companies, with 54 million registered users. However, it made money on very few of its customers. It only charged a fee to the 5 percent of its customers that used SkypeOut and SkypeIn, which were higher level services with more features. Its customers for base-level services paid nothing.[24]

The company also faced competition from a variety of sources including VOIP providers, like Net2Phone; other PC-to-PC providers, like Yahoo!; Cable companies, like Comcast, that were bundling cable services, VOIP, and high-speed Internet access; and the telephone companies, like AT&T, that were cannibalizing their existing telecommunications businesses by offering VOIP.

Despite the heavy competition faced by Skype, eBay's board thought that the start-up was attractive as an acquisition target because its business complemented eBay's business. It facilitated communications between buyers and sellers on an auction site, allowed eBay to offer pay-per-call services, and allowed cross-selling of products to the 99 percent of Skype customers who were not eBay customers.[25]

Whether or not eBay's board was right and Skype would provide the benefits that it expected, the acquisition of a two-year-old company for $1.5 billion shows that it was a successful start-up. Clearly, its founders adopted the right approach to creating value as a new company in an industry based on increasing returns.

Key Points

- Some industries, particularly knowledge-intensive ones, are subject to increasing returns, which means that profit margins increase with the volume of production.
- Industries display increasing returns when they have high up-front costs and low marginal costs, when producer learning is high, and when network effects are present.
- Network effects can be direct and come from the interaction of users, or indirect and come from the presence of complementary goods.
- Metcalfe's law explains why there are increasing returns in industries that have direct network effects; the value of a network is proportional to the square of the number of devices in it.
- Industries subject to network effects face important strategic issues that do not confront other industries, including the maintenance of the network, the possibility of sustaining more than one network, the use of strategic action to exploit network effects, and competition from other networks.

STRATEGY FOR INCREASING RETURNS

To be successful, your company needs to adopt a different strategy in industries based on increasing returns than in industries based on decreasing returns. In increasing returns industries, you need to establish large-scale operations, build your installed base quickly, get customers to ignore lock in, and be a first mover.

Start Large

You should start your business on a large scale if you operate in an industry based on increasing returns. Although limits to the supply of, or access to, raw materials and customer preferences usually preclude you from taking control of the entire market in industries based on decreasing returns, you need to compete for control of the market in industries based on increasing returns.

These industries tend toward natural monopolies (the minimum efficient scale of the industry is larger than the size of the market in maturity) or are subject to strong network effects in which the cost to users of having more than one platform are very high and the benefits to having them are very low.[26] As a result, these industries are winner-take-all.[27] Successful firms tend to become de facto technical standards, while unsuccessful businesses tend to fail. For example, in the computer software industry, the Windows operating system became the de facto technical standard and gave Microsoft market power.

To establish operations on a large scale, you need to make large bets.[28] Because profit margins increase with the volume of production, you need to attract customers first and make profits second, which generates deeply negative cash flow in the early years of operation. Moreover, the development of products in these industries is often very expensive, as is the case when a company develops a video game console based on a custom design for a semiconductor.[29]

Increasing returns businesses are, thus, not for the faint of heart. Only the most dominant businesses in each segment will survive, and the large investments that you need to compete drive up the magnitude of the downside loss from failure. For example, the Internet grocery delivery start-up Webvan lost several hundred million dollars of its investors' money when it failed. And PayPal spent $150 million before it reached a breakeven level of sales.[30]

Moreover, starting large and targeting a large market creates huge core rigidities for a company. This makes it very difficult for the company to be flexible and adapt to changes that unfold over time, which is often important with uncertain new technology. Thus, delaying action, preserving options, and remaining flexible and adaptive, are often not possible, increasing the risk of the bet.

For start-ups, increasing returns mean raising large amounts of money to build the business; the time-honored method of financing a business—bootstrapping (building a business using internal cash flow)—is not very effective. Given the cyclicality of venture capital markets, this also means that start-ups have to pay a lot of attention to when they raise money and how much they raise when they get it.[31] Furthermore, businesses based on increasing returns, like biotechnology, show the kinds of ownership patterns displayed in Figure 13.6; their need to raise larger and larger rounds of money leads their founders to have smaller and smaller ownership shares, resulting in a very small average level of founder ownership at the time that the companies go public.

FIGURE 13.6
Amount Raised and Ownership in Biotechnology

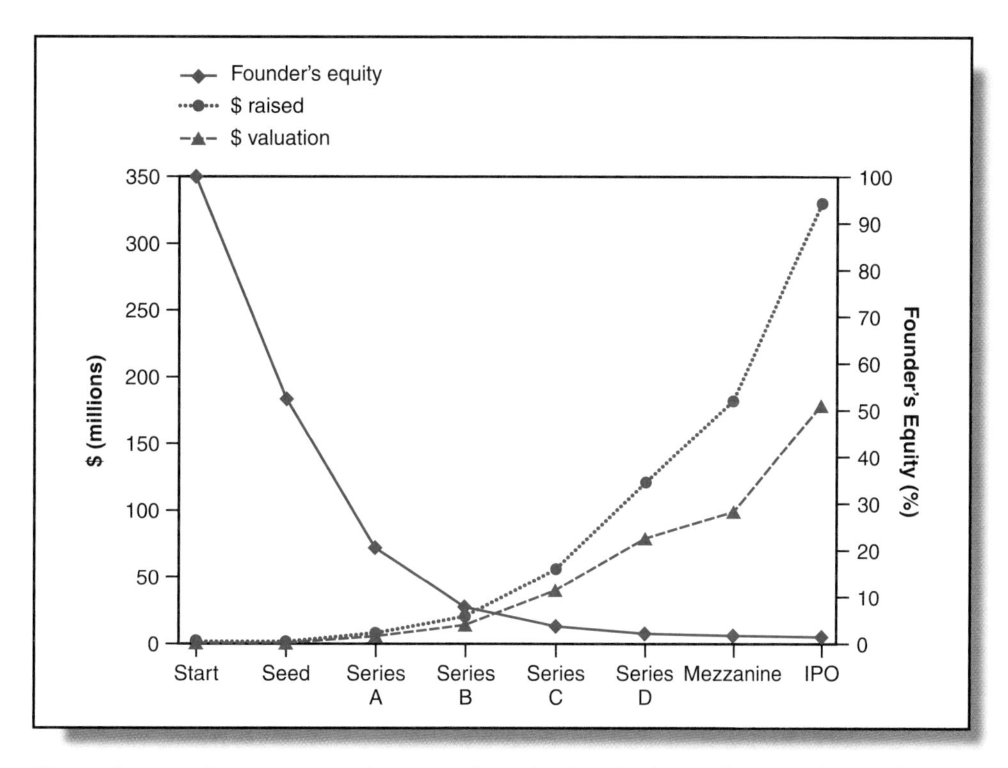

The need to raise large amounts of money in biotechnology leads founders to end up with a very small share of ownership of their companies at the time of an initial public offering.

Source: http://www.nature.com/nbt/journal/v21/n6/fig_tab/nbt0603–613_F4.html.

Build a Large Installed Base Quickly

You need to build a large **installed base**—the number of current users of a product or service—quickly in industries based on increasing returns.[32] First, rapid growth of the installed base helps you to make your product the industry standard, which gives you long-lasting leverage over other firms in your value chain. Networked industries are more likely to converge on a high-volume product than on a low-volume product, as the standard, because customers benefit the most from adopting the product that has the highest installed base.[33] For example, Mirabilis, an Israeli company in the instant messaging (IM) business, developed the most popular IM system. Potential customers preferred Mirabilis's product because the company had more users than any other provider of instant messaging.

Second, a large installed base makes your product more attractive to the providers of complementary products. These companies face less risk if they make their new products and services compatible with dominant technologies than if they make their products compatible with other technologies because customers prefer the products with the largest installed base. For example, DoMoCo chose to go with Compact HTML instead of the Wireless Application Protocol for its i-mode phone to ensure that its product was compatible with the technology that had the largest installed base, even though Wireless Application Protocol was technically superior to Compact HTML.[34]

The availability of complementary products increases the value of your product to customers, which, in turn, makes it easier for you to sell more of it. For example, when Microsoft was getting started, Bill Gates built up the installed base for MS-DOS by striking a deal with IBM to put MS-DOS in its personal computers. This effort boosted Microsoft's installed base and motivated other software companies to create MS-DOS compatible products. The availability of compatible products, in turn, increased the adoption of MS-DOS by other computer manufacturers.[35]

Third, a large installed base pushes your company down the learning curve. As you may remember from Chapter 10, the learning curve makes companies better at supplying products as they provide more of them. Therefore, if your company moves down the learning curve faster than its competitors, your product will become more attractive to customers than other alternatives, which will spur its adoption.[36]

Fourth, a large installed base also improves the economics of developing new technologies. The larger the installed base, the faster that you can recoup the cost of developing new versions of your product. For example, because it has a much larger installed base, Microsoft can break even on the cost of developing a new version of its operating system software in a matter of weeks, whereas it takes Apple several years.[37]

Finally, building your installed base quickly will keep out competition by creating a positive feedback loop that is hard for competitors to break. Take eBay, for example. Figure 13.7 compares the number of registered users of eBay against the number of auctions occurring quarterly from 1998 to 2006. It demonstrates the positive feedback between users and sellers; the number of users leads to an increase in the number of auctions, while an increase in the number of auctions leads to an increase in the number of users.

FIGURE 13.7
Positive Feedback Effects in Networked Industries

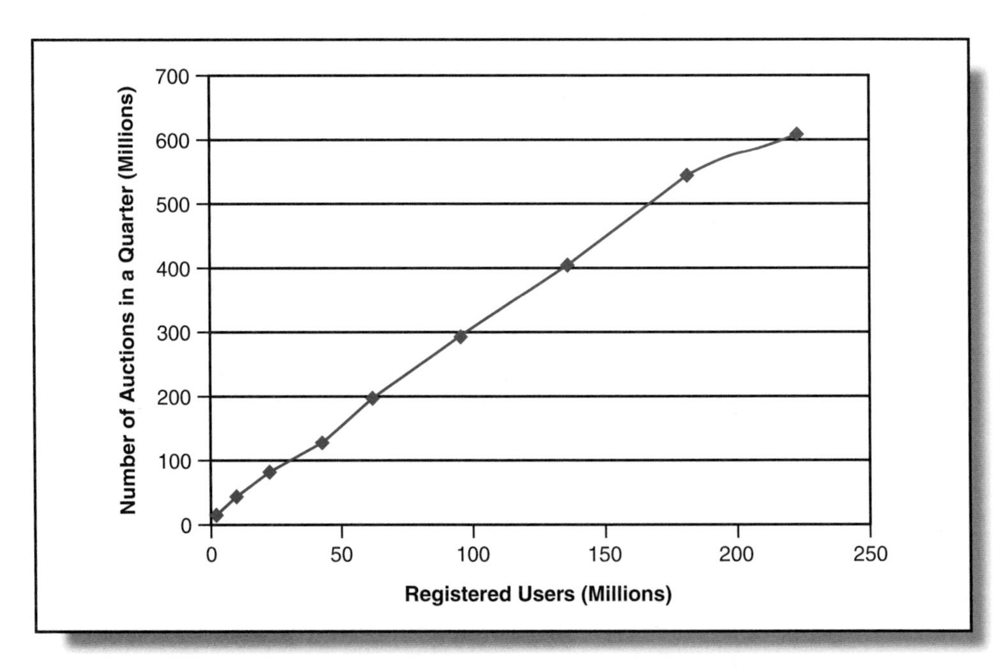

Plotting the number of registered users of eBay against the number of auctions occurring in a quarter shows the positive feedback between users of eBay and sellers on eBay: The more users there are, the more auctions there are; and the more auctions there are, the more users there are.
Source: Created from data contained in eBay's annual reports.

So, clearly, if you operate in a business based on increasing returns, you want to build your installed base quickly. So what can you do to increase the size of your installed base? Basically, you can do three things. First, you can use penetration pricing.[38] By charging a low price, you will attract more customers, which will build your installed base.[39] Charging a low price might mean giving away your product to customers, as America Online did initially,[40] or it might even mean paying customers to adopt your product, as did PayPal, which put $10 in the account of each new user.[41]

Of course, paying for adoption, or even giving away your product, is a high-risk way to build your installed base. If other companies do the same thing, then you won't be able to build your installed base faster than them, and you will be left swimming in red ink. Moreover, even if competitors don't copy your strategy, you need to make sure you have some mechanism for generating a profit later. Otherwise, you will forever lose money on the sale of your product.

Second, you can build your installed base quickly by bundling your product with other products that are already popular with customers. For example, Microsoft and Netscape both bundled their products with personal computers. As a result, customers who were buying computers anyway received these products.

The use of bundling is particularly helpful in building your installed base if the bundled products are complementary. By bundling complementary products, you can create a positive feedback loop between the two products, which further helps to build your installed base.[42] For example, Microsoft sells the Xbox console and games like *Halo* together in its video game system, which increases the positive feedback between sales of the console and sales of the games that play on it.[43]

Third, you can target the mass market from the start,[44] rather than following the approach to product adoption discussed in Chapter 3 (targeting the innovators and early adopters first, and then transitioning to the early and late majority of the market). Because the early and late majority is larger than the innovators and early adopters, this approach will allow you to go after, and hopefully capture, a larger number of customers right away. If you capture more customers right off the bat, then you will build your installed base faster.

In short, by pricing low, bundling your products, and targeting the mass market, you can build your installed base quickly. As Figure 13.8 shows, this can give you an

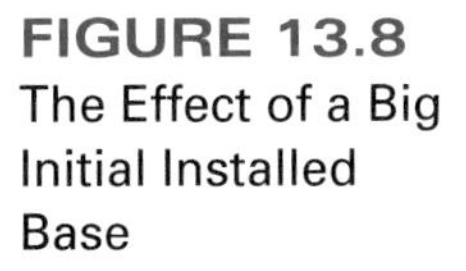

FIGURE 13.8
The Effect of a Big Initial Installed Base

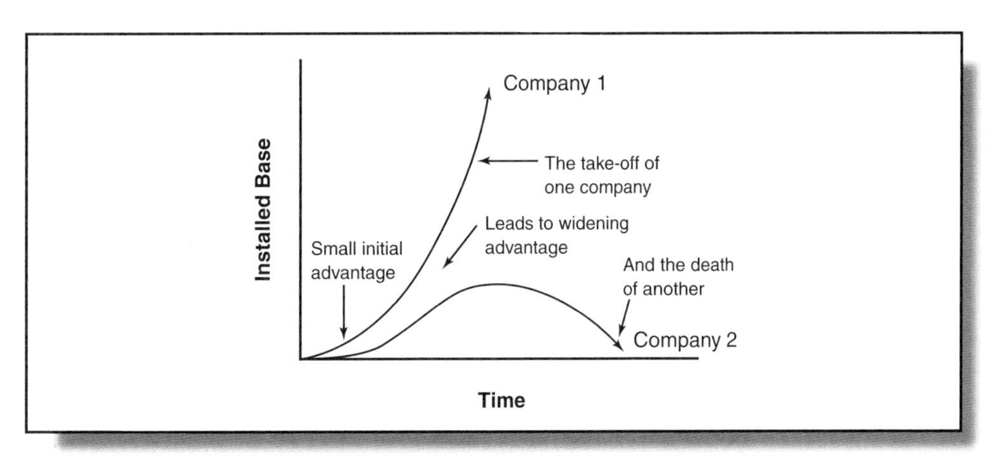

A small initial difference in an installed base can lead one company to get an advantage that leads to the demise of its competitors.

advantage over competitors, which increases over time and eventually causes them to drop out of the market.

Get Customers to Ignore Lock In

Customer lock in occurs when customers view the costs of switching suppliers as too high to justify doing so. For example, the airlines are "locked in" to specific suppliers of aircraft because it is expensive for their employees to learn to operate and maintain airplanes.[45] If an airline's pilots and mechanics have learned to use Boeing products, then the airline doesn't want to incur the cost of retraining them to fly and maintain Airbus planes.[46] Similarly, people with large digital music collections might be locked in to the iPod/iTunes system because switching players will leave them unable to play their music collections.

Customers are wary of purchasing new products that have high switching costs because they are afraid of being locked into a product that is not going to be around in the future, which will cause them to incur high costs to switch products later on.[47] Therefore, in increasing returns industries, you need to get customers to ignore the potential for lock in and adopt your product.

So how do you do this when there is still uncertainty about whether or not your product will be around in the future? One way is to offer customers their money back if they are unhappy with your product or service. A money-back guarantee will reduce customers' switching costs, thereby making them more likely to adopt your new product or service.[48]

Another way is to use a "razor–razor blade model." With this model, you create a unique system, composed of a base product, like a video game console, and additional components, like game cartridges, that are not compatible with your competitor's base product. You sell the base product at cost and the components at a high margin. (This approach was named for the early safety-razor companies that sold their razors at cost to attract customers and made their profits on the sales of replacement blades, which were uniquely designed for their razors.)

Customers will buy into your system because the low up-front price gets them to try your product to see if it has value. Because most customers are myopic, they will underestimate how many units they will purchase over time, particularly if there is uncertainty about the value of the product. As a result, they will be attracted by the low initial cost and discount the high cost of additional purchases.

Once customers have purchased your system, they are better off buying your components, rather than changing systems, even if your components cost more than competitors'. Switching systems requires your customers to incur the up-front cost of buying into another system, as well as the training necessary to get up to speed on that system. As long as the difference between the cost of your components and those of your competitors is less than the sum of the up-front cost of your competitor's system plus the cost of getting up to speed on it (amortized over the customer's projected horizon for the use of the system), then your customer has an incentive to stay with your product.

The razor–razor blade model can be seen in a variety of industries. For instance, Hewlett-Packard sells its printers at a loss to get customers to purchase them, and makes all of its profits on the sale of ink cartridges.[49] Similarly, Microsoft sells its Xbox consoles at below cost and makes its profits on game software.[50]

Customer lock in poses an additional problem for new companies. Because these companies are seeking customers that established companies already serve, they need to get customers to break lock in and leave their established suppliers. For instance, cellular telephone service providers often sell telephones that only work on their systems and require customers to sign multiyear contracts. As a result, new entrants to this business cannot attract customers unless they can find ways for customers to switch suppliers.[51]

Be a First Mover

In industries based on increasing returns, you should be a first mover. In these businesses, a product is likely to become dominant if its installed base becomes larger than that of competing products—even if this outcome is just the result of blind luck. By moving early, you can build your installed base ahead of your competitors, making it hard for them to catch up. That is because products take off when their installed base begins to grow.

Take eBay, for example. Because eBay was one of the first Internet auction houses, it built its installed base ahead of its competitors. Its larger customer base made it easier for the company to attract new customers (it had the most eyeballs looking at it). As a result, eBay remained ahead of its competitors. Similarly, in online payment systems, PayPal defeated many competitors, including Yahoo!, Citibank, eBay, BidPay, CheckFree, and Paying Fast, because it moved quickly to create such a large installed base that competitors could not catch up.[52]

To be a first mover, two things are crucial. First, you need to get your product to market quickly. As a result, you might not want to perfect it, or add features to it, before entering the market. If your efforts to make your product better delay its launch, they will actually *decrease* your chances of attracting customers. In increasing returns businesses, customers often do not find it worthwhile to switch from an inferior product to a superior one that comes out later. Consequently, your best strategy is to get to market fast with a simple product that has few features but can be mass-produced easily.[53] Over time, you can improve your product in the process of operating your business. If your customers face high switching costs, they will stick with your product despite its imperfections.

Second, you need to use contractual mechanisms to create the business's value chain, as occurs when you sign agreements with other companies to manufacture your products for you.[54] You can get your products to market more quickly by contracting with other firms because your contractors already have their operations in place. This, of course, is why so many new companies in industries based on increasing returns, like software, are virtual companies. They do not establish production, but, instead, license to, or form strategic alliances with, other firms to make their products for them.

Licensing your technology to other companies also helps you to encourage the adoption of your products. This is true even if you can't charge others a licensing fee, or when your licensee is a competitor, because licensing encourages product adoption, which generates additional sales,[55] and reduces the incentive for other companies to develop competing products.[56] Therefore, many companies in increasing returns industries rely heavily on licensing. For example, when Sun developed the SARC microprocessor, it licensed its components to several manufacturers, which facilitated the rapid adoption of the Sun processor.[57]

Limitations to a Strategy Based on Increasing Returns

While you need to adopt a different strategy in industries based on increasing returns than in industries based on decreasing returns, doing so is not always easy. Industries do not always show increasing returns just because theory says that they should. Therefore, you have to evaluate whether increasing returns are really present in your business before you set your strategy. It may turn out that your industry is based on decreasing returns after all, or that switching costs are lower than theory says that they should be. Or late movers' technologies may turn out to be much better than those of first movers, leading customers to switch suppliers rather than being "locked in" to the first mover's product. In some cases, network effects may only apply below a certain "critical mass" of customers. Once this number has been reached, increases in network size may no longer create additional value for customers, so returns might not increase with network size.

Key Points

- Because increasing returns businesses are winner-take-all, firms in industries based on increasing returns need to make large bets.
- Firms in increasing returns industries need to build a large installed base quickly because customer adoption decisions are affected by the size of the installed base.
- Penetration pricing helps to build a large installed base quickly, but it comes at the risk of lack of profitability.
- Bundling and targeting the mass market help companies to build a large installed base rapidly.
- Firms in increasing returns industries need to minimize customer lock in to attract customers; using the razor–razor blade model is a good way to do this.
- First mover advantages are large in industries subject to increasing returns because the likelihood of new customer adoption increases with the size of the installed base.
- Getting to market quickly is more important than having superior technology in industries subject to increasing returns because customers are not likely to switch to the superior technology if the inferior technology has already been adopted by a large installed base.
- Contracting, rather than owning the different parts of the value chain, facilitates getting to market quickly and is a valuable approach in industries based on increasing returns.
- You have to evaluate whether increasing returns are really present in your business before you set your strategy because industries do not always show increasing returns when theory says that they should.

Discussion Questions

1. Think of an industry with which you are familiar. Is it based on increasing or decreasing returns? Why do you make that argument?
2. Of the different reasons why industries display increasing returns, which do you think is most important? Why?
3. What should your strategy be if your technology is subject to direct rather than indirect network effects? Should it be the same or different? Why?
4. Will network effects always lead to convergence on a single network? Why or why not?
5. What strategic actions should you follow in an increasing returns business? How are those strategic actions different from those that you should follow in a decreasing returns business? What will happen if you follow a strategy designed for a decreasing returns business in one based on increasing returns?
6. If you are not the first to market in an industry based on increasing returns, should you try to catch up to the first mover in terms of installed base? Why or why not? If your answer is yes, what can you do

to catch up to the first mover in terms of installed base? Why will those actions help?

7. Apple Computer makes a lot of money selling iPods but doesn't make any money selling iTunes. It pays about $0.65 per downloaded song to the music label that produced it, and $0.22 to credit card processors, leaving it about $0.12 per song to pay for the cost of operating its Web site. This makes Apple's business a reverse "razor–razor blade" business, in which the item that is purchased repeatedly is sold at cost and the item that is purchased once is sold at a high margin.[58] Why do you think that Steve Jobs set up the business this way?
8. What additional strategic issues do networked industries raise for start-up companies? Why are start-ups confronted by these additional issues?

Key Terms

Complementary Products: The products that are used along with the focal product.

Customer Lock In: A situation in which customers will not shift to competitor products even though the alternative provided by the competitor is better.

Decreasing Returns: A situation in which the returns that you receive on the marginal product produced decrease with the volume produced.

Direct Network Effects: The network externalities that come from the direct interaction of users.

Increasing Returns: A situation in which the returns that you receive on the marginal product produced increase with the volume produced.

Indirect Network Effects: The network externalities that result from the availability of complementary goods.

Installed Base: The number of users of a product at a point in time.

Metcalfe's Law: A principle that the value of a telecommunications network is proportional to the square of the number of devices on that network.

Network Externalities: A situation in which the purchase of a product by a new customer increases its value to existing customers.

Putting Ideas into Practice

1. **eBay's Installed Base** The purpose of this exercise is to understand the role of installed base in networked industries. The data in Table 13.2 are figures for the number of registered users on the eBay online auction site at the end of December in each year indicated. Plot the number of registered users against the year. Is the number of users flat or growing? If it is growing, what is the pattern of growth (e.g., linear, accelerating, decelerating)? Why does the pattern in the number of users over time take this shape? Why doesn't it take another shape?
2. **Increasing or Decreasing Returns** The purpose of this exercise is to help you understand increasing and decreasing returns to scale. Table 13.3 shows the net income per unit for two different products. Use these data to calculate a scatter plot of the number

TABLE 13.2 Data for Exercise 13.1

The number of registered users of eBay has grown dramatically each year since 1997.

Year	Registered Users (Millions)
1997	0.0
1998	2.2
1999	10.0
2000	22.5
2001	42.4
2002	61.7
2003	94.9
2004	135.5
2005	180.6
2006	222.0

Source: From data contained on eBay's Web site, http://www.ebay.com.

TABLE 13.3
Data for Exercise 13.2
This table shows the net income per unit produced for two products.

	Product 1	Product 2
Customers Served	Net Income/Unit	Net Income/Unit
0	0	0
5	$1.00	$7.00
10	$1.80	$6.00
20	$2.10	$5.10
40	$2.50	$4.30
80	$3.00	$3.60
160	$3.60	$3.00
320	$4.30	$2.50
640	$5.10	$2.10
1280	$6.00	$1.80

of customers against the net income per unit for both products. Do these products display constant returns to scale? If so, explain why. If not, what pattern do they display? Explain why the product displays the pattern that you have identified. Suggest some products that might follow the pattern shown for Product 1. Explain why those products might display the returns to scale shown for Product 1. Suggest some products that might follow the pattern shown for Product 2. Explain why those products might display the returns to scale shown for that product.

3. **Understanding Network Externalities** Think about the following products: automatic teller machines, VCRs, diesel-powered cars, coal-fired power plants, VOIP phones, and surgical instruments. Are they subject to network externalities? If not, then explain why not.

   If they do, then explain what kind of network effects they display. Also explain why there are network externalities. What causes them? Do they continue to increase with the number of customers, or do they "saturate" once you reach a critical mass of customers? Explain why the pattern occurs.

Notes

1. Adapted from Smith, E. 2006. Can anybody catch iTunes? *Wall Street Journal*, November 27, R1, R4.
2. Ibid.
3. Yoffie, D., and M. Slind. 2006. Apple Computer, 2006. *Harvard Business School Case*, Number 9-0706-496.
4. Wingfield, N., and E. Smith. 2007. Job's new tune raises pressure on music firms. *Wall Street Journal*, February 7: A1, A11.
5. Bettis, R., and M. Hitt. 1995. The new competitive landscape. *Strategic Management Journal*, 16: 7–19.
6. Arthur, B. 1996. Increasing returns and the new world of business. *Harvard Business Review*, July–August: 100–109.
7. Yoffie, D., D. Mehta, and R. Seseri. 2006. Microsoft in 2005. *Harvard Business School Case*, Number 9-705-505.
8. Harkey, M., and W. Barnett. 2006. Teaching note for Facebook. *Stanford Graduate School of Business Teaching Note*, Number E-220TN.
9. Metcalfe's law, http://En.wikipedia.org/wiki/Metcalfe's law.
10. Lee, Y., and G. O'Conner. 2003. New product launch strategy for network effects products. *Academy of Marketing Science Journal*, 31(3): 241–255.
11. Hill, C. 1997. Establishing a standard: Competitive strategy and technology standards in winner-take-all industries. *Academy of Management Executive*, 11(2): 7–20.
12. Greenstein, S., and V. Stango. Forthcoming. The economics of standards and standardization. In S. Shane (ed.), *Blackwell Handbook on Technology and Innovation Management*. Oxford: Blackwell.
13. Eisenmann, T. Managing networked businesses: Course overview for educators. *Harvard Business School Note*, 5-807-104.
14. Ibid.
15. Ibid.
16. Ibid.
17. Ibid.
18. Ibid.
19. Corts, K. The rise and fall (?) of Palm Computing in handheld operating systems. *Harvard Business School Teaching Note*, Number 5-703-520.

20. Eisenmann, Managing networked businesses.
21. Adapted from Eisenmann, T. 2006. Skype. *Harvard Business School Case*, Number 9-806-165.
22. Ibid.
23. Ibid.
24. Ibid.
25. Ibid.
26. Eisenmann, Managing networked businesses.
27. Arthur, Increasing returns and the new world of business.
28. Ibid.
29. Eisenmann, Managing networked businesses.
30. Tam, P., and R. Buckman. 2007. Tech start-ups have money to burn, but choose thrift. *Wall Street Journal*, January 18: B1.
31. Eisenmann, Managing networked businesses.
32. Hill, Establishing a standard.
33. Katz, M., and C. Shapiro. 1986. Technology adoption in the presence of network externalities. *Journal of Political Economy*, 94(4): 822–841.
34. Moon, Y. 2002. NTT DoCoMO: Marketing i-mode. *Harvard Business School Teaching Note*, Number 5-503-097.
35. Arthur, Increasing returns and the new world of business.
36. Arthur, B. 1987. Competing technologies: An overview. In G. Dosi (ed.), *Technical Change and Economic Theory*. New York: Columbia University Press, 590–607.
37. Yoffie, D., and M. Kwak. 1999. Apple computer 1999. *Harvard Business School Teaching Note,* Number 5-799-150.
38. Lee, Y and O'Conner, New product launch strategy for network effects products.
39. Brynjolfsson, E., and C. Kemerer. 1996. Network externalities in microcomputer software: An econometric analysis of the spreadsheet market. *Management Science*, 42(12): 1627–1647.
40. Lee, and O'Conner, New product launch strategy for network effects products.
41. Eisenmann, T., and L. Barley. 2006. PayPal Merchant Services. *Harvard Business School Case*, Number 9-806-188.
42. Lee and O'Conner, New product launch strategy for network effects products.
43. Eisenmann, Managing networked businesses.
44. Shapiro, C., and H. Varian. 1999. The art of standards wars. *California Management Review*, 41(2): 8–33.
45. Afuah, A. 2003. *Innovation Management*. New York: Oxford University Press.
46. Arthur, Increasing returns and the new world of business.
47. Van Hove, L. 1999. Electronic money and the network externalities theory: Lessons for real life. *Netnomics*, 1: 137–171.
48. Schilling, M. 2003. Technological leapfrogging: Lessons from the U.S. video game console industry. *California Management Review*, 45(3): 6–35.
49. Tam, P. 2004. As cameras go digital, a race to shape habits of consumers? *Wall Street Journal*, November 19: A1, A10.
50. Yoffie, Mehta, and Seseri, Microsoft in 2005.
51. Ali, S. 2006. Cellphone users are unshackled by ruling. *Wall Street Journal*, December 7: B4.
52. Eisenmann and Barley, PayPal Merchant Services.
53. Arthur, Increasing returns and the new world of business.
54. Ibid.
55. Sun, B., J. Xie, and H. Cao. 2004. Product strategy for innovators in markets with network effects. *Marketing Science*, 23(2): 243–254.
56. Brynjolfsson and Kemerer, Network externalities in microcomputer software.
57. Rosenberg, M., and B. Silverman. 2001. Sun Microsystems Inc: Solaris Strategy. *Harvard Business School Case,* Number 9-701-058.
58. Yoffie and Slind, Apple Computer, 2006.

Chapter 14

Collaboration Strategies

Learning Objectives

After reading this chapter, you should be able to:

1. Identify different modes of doing business, and explain why some are considered contractual and others are considered vertically integrated.
2. Explain why certain types of technologies are better suited than others to exploitation through contractual modes of doing business.
3. Understand how the different contractual modes of doing business work.
4. Describe strategic alliances and understand their pros and cons.
5. Identify the different types of licensing that firms can undertake, and explain how companies can use them to their advantage.

6. Explain why strategic alliances and licensing agreements are more common in some industries than in others.
7. Describe contract manufacturing and outsourcing and understand their pros and cons.
8. Describe joint ventures and explain the advantages and disadvantages that they offer to companies.
9. Explain how firms can improve their performance at using contractual modes of doing business.
10. Understand the difficulties that arise when large, established firms contract with small, start-up firms.

Vertical Integration: A Vignette[1]

In 1978, Genentech, one of the pioneers in the then-emerging biotechnology industry, invented a way to synthesize the gene for human insulin. Genentech contracted with Eli Lilly, the dominant producer of pig and cow insulin, to manufacture and market human insulin using its patented technology.[2] In 1980, Genentech figured out a way to make interferons (proteins that the immune system produces in response to viruses or other foreign agents) using genetic engineering techniques.[3] The company then sold the rights to market its interferons to Hoffmann-La Roche.[4]

Genentech's initial business model, in which it focused on research and commercialization and licensed its discoveries to established pharmaceutical companies that manufactured and marketed products from them, has proved to be the dominant approach in biotechnology. Today, half of all new products approved by the U.S. Food and Drug Administration are based on genetic engineering, many of them developed by biotechnology companies, but almost all manufactured and marketed by pharmaceutical firms that were originally founded long before the biotechnology revolution.

During the same period that biotechnology came into existence, the personal computer industry emerged, with Apple Computer as one of its pioneers. Unlike Genentech and the early biotechnology companies, Apple did not collaborate with an established computer company to manufacture and distribute computers based on its technology but, rather, produced and sold its own computers. And today, far less contracting occurs between start-ups and established firms in the computer industry than in the pharmaceutical industry.

Why do new biotechnology firms tend to invent new technologies and then license them to established firms that manufacture and market them, but new computer hardware firms do not? This chapter explores the characteristics of industry environment, technology, and firms that determine when companies will use contractual (like biotechnology) versus vertically integrated (like computer hardware) modes of exploitation of new technologies.

INTRODUCTION

Figuring out how to configure the value chain for your company is an important technology strategy issue. Your decision about your company's governance structure—whether to own or contract with other firms to provide manufacturing, marketing, and distribution, and whether or not to vertically integrate into the supply of raw materials and other inputs—has major strategic implications for your company. Which approach is better depends on your industry, the nature of the technology that underlies your products and processes, and your company's strategy and structure.

This chapter explores different modes of doing business. The first section describes the difference between contractual and vertically integrated modes, and explains how the nature of technology affects the choice between the two. The second section describes several common contractual forms, including licensing, joint ventures, strategic alliances, outsourcing, and contract manufacturing, and discusses the advantages and disadvantages of each. The third section offers insight into when your company should work alone and when it should collaborate with other firms. The final section explains how to set up effective collaborative organizational arrangements.

CONTRACTUAL AND VERTICALLY INTEGRATED MODES OF DOING BUSINESS

As a technology entrepreneur or manager, you need to think about whether your company's value chain will be vertically integrated or contractual. **Vertically integrated** means that your company owns all of the parts of the value chain (e.g., supply, manufacturing, marketing, distribution, etc. . .), while **contractual** means that your company signs agreements with independent firms to provide part of the value chain, as occurs with licensing, joint ventures, strategic alliances, contract manufacturing, and outsourcing.

Vertically integrated and contractual modes of doing business each have advantages and disadvantages. While vertical integration gives you greater control, it also imposes greater coordination costs and demands the smooth utilization of assets across stages of the value chain.[5] Contracting reduces costs but exposes you to the risks that come from having limited control over the activities of your contracting partners.

When choosing whether to use a vertically integrated or contractual mode of business, you need to look at the advantages and disadvantages of each. For example, you might compare the costs and benefits of licensing technology from another firm with acquiring that company to obtain its technology.

Although the balance between vertical integration and contracting varies by industry, companies in almost all industries use both modes for the different parts of the value chain, from product development to production to distribution. As a result, competitors in the same industry often choose very different modes of doing business, making the way that a company configures its value chain an important aspect of technology strategy. For instance, Intel conducts most of its research outside the organization by funding university researchers; IBM does most of its research inhouse. This difference leads the two companies to compete, at least in part, on the basis of which approach to research is more effective in making different kinds of semiconductors.[6]

So how should you determine what the governance structure of your business should be? It should depend, in large part, on two dimensions of your core technology: discreteness and tacitness. Contractual modes of exploitation are more effective when technologies are discrete (like a drug), than when technologies are part of a system (like computer software) because discrete technologies can be used alone, while systemic technologies need to be used with complementary technologies (see Figure 14.1). When technologies are part of a system, the development of products and services requires coordination of the different components of the system so that complementary changes can be made to them. Vertical integration facilitates this coordination because it allows a single firm to decide how to make these changes, rather than requiring different companies to come to an agreement that satisfies all sides.

If the value from the joint sale of the different components that make up a systemic product is evenly distributed across the components, you sometimes can exploit a systemic technology through a contractual approach. However, if the value generated from the joint sale of the components is unevenly distributed, you usually cannot. The companies that own the components from which most of the value is created will have a strong incentive to exploit the technology, while the companies that own the other components will not. For example, when Microsoft initially sought to enter the video game business, it tried to convince personal computer manufacturers, like Dell, to make consoles to run its software. However, because video game consoles are typically sold at cost, while profit is made from selling the games themselves, Dell and other computer makers had little interest in working with Microsoft to develop video games.[7]

Vertical integration is particularly important if a technology is systemic and technical standards have not been established.[8] For example, 30 years ago, the companies that produced applications software also produced systems software because technical standards ensuring the compatibility of applications and system software did not

FIGURE 14.1
Discrete and Systemic Technologies

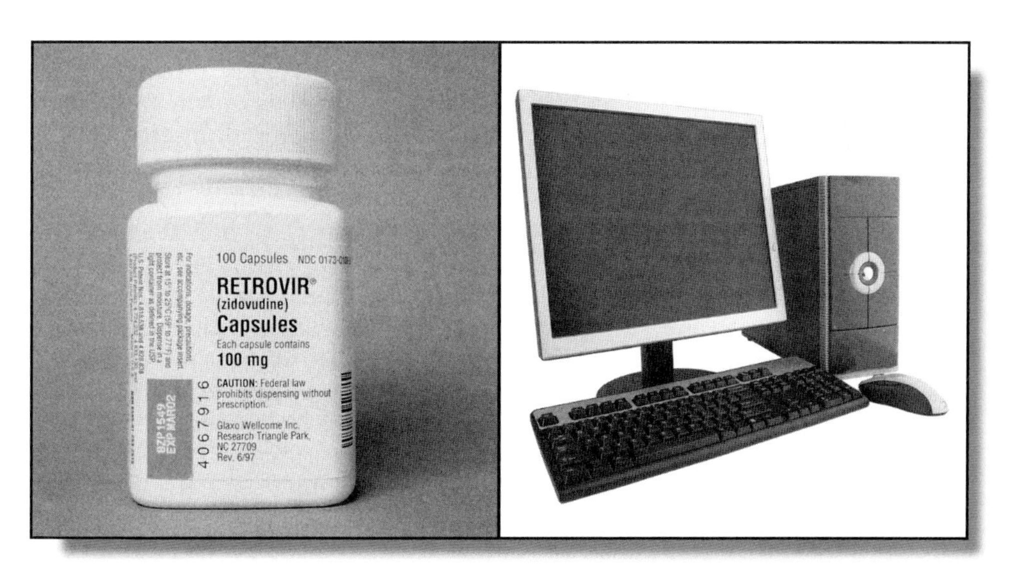

Contractual modes of business are more effective with discrete technologies, like drugs, than with systemic ones, like computers.

Source: (a) PhotoEdit Inc. and (b) Getty Images Inc. RF.

yet exist. Now that these standards are in place, however, different companies produce the two types of software.[9]

A single company will also need to make all of the components to a systemic technology if all of them need to be developed before any value can be generated.[10] If different companies are responsible for different components, but no value can be created until all of them have completed development, a chicken-and-egg problem develops, in which none of the component producers wants to develop its component until the others have done so.[11]

Take, for example, the history of radio. When radios were first invented, RCA developed both the production of the hardware (radios) and the production of the software (broadcasting) because other companies were unwilling to enter the broadcasting business until customers owned radios and could listen to the broadcasts, and were unwilling to enter the device manufacturing business until there were broadcasts that customers could listen to on the radio.[12]

Established companies have an advantage over new companies in the development of innovations that are based on systemic technologies. This advantage results from the relative importance of existing technology to the creation of a new systemic product or service.[13] With a systemic technology, the value generated by the new product can be reaped only if a firm has in place the other technology with which the new product will operate. Because it is already in operation, an established firm is likely to have this technology in place. However, a new firm will have to create it. The cost of creating the rest of the system will lead a new firm to gain less than an established firm from a new product based on systemic technology.[14] In contrast, a discrete technology does not require the creation of a system of other technologies to be deployed. Therefore, established firms have no advantage over new firms in the development of products based on these technologies.

Contractual modes of doing business are more effective when the knowledge underlying a technology is easily codified. To have a contract, you need to be able to write enforceable agreements that indicate both parties' rights and obligations, specify how the fulfillment of those obligations will be measured, and provide for clear legal remedies in the event of a dispute.[15] This specification is more difficult if the knowledge underlying a technology is tacit and cannot be written down clearly in words, blueprints, designs, or drawings. Take, for example, a technology to produce a new composite material. Even if you know how to make the material, you will have a hard time licensing the technology if you cannot write down the process to do it. Your inability to specify the process will make it difficult to identify each party's obligations or how the fulfillment of those obligations will be measured.

Moreover, when knowledge is tacit, people learn how to do something by doing it, and know many things that they cannot articulate. To transfer tacit knowledge, people need to show others what they know, rather than just giving them written documents. Because the movement of people is much easier within organizations than between them, contracting between independent organizations tends to work poorly when knowledge is tacit.[16]

Take, for example, the case of Infinera, a start-up company that invented a photonic integrated circuit. Because the knowledge underlying the creation of this integrated circuit was largely tacit, the company had no way to transfer the knowledge of how to make the chip to semiconductor manufacturers, and had to build a fabrication plant to manufacture it.[17]

Key Points

- Vertical integration occurs when a company owns all parts of the value chain; contracting occurs when a company signs an agreement with an independent firm to provide part of the value chain.
- Contractual forms of doing business are composed of strategic alliances, technology licensing, outsourcing, contract manufacturing, and joint ventures.
- The nature of technology affects the choice between vertical integration and contracting because systemic technologies and tacit knowledge are best exploited through vertically integrated companies.

Alliances, Licensing, Joint Ventures, Contract Manufacturing, and Outsourcing

If you decide to use a contractual form of governance, then you have a number of more specific choices to make: You have to decide whether to use strategic alliances, licensing, joint ventures, contract manufacturing, or outsourcing.

Strategic Alliances

A **strategic alliance** is a relationship between two firms to work together to achieve a common business goal. It can be formal and specified in a written contract or informal and the result of an unwritten agreement. When alliances are made between firms at different stages of the value chain, they are called **vertical alliances**. For example, Millennium Pharmaceuticals has a strategic alliance with Bayer Pharmaceuticals in which Millennium develops promising drug targets and Bayer obtains FDA approval for them, and manufactures and markets the drugs.[18] When alliances are made between firms at the same stage in the value chain, they are called **horizontal alliances**. The partnership between Kodak and Hewlett-Packard to develop photographic printers is an example of a horizontal alliance.[19]

The use of both vertical and horizontal alliances to govern technology businesses has been growing in recent years, particularly between companies from different countries. As Figure 14.2 shows, in 2003 there were more than three times as many international technology alliances as there were in 1980.

Advantages and Disadvantages of Strategic Alliances

So what makes strategic alliances so popular? They offer eight advantages to technology firms:

- First, strategic alliances permit companies to specialize in the capabilities in which they have a comparative advantage. For example, the skills needed to develop drugs are very different from those needed to manufacture them because research and development, which seeks to show the efficacy of drugs, is quasi-academic, and manufacturing, which seeks to produce replications that meet regulatory standards, is production-oriented. To manufacture drugs, companies need to hire people with different skills and attitudes from the ones that they need to develop those drugs. Consequently, many biotechnology firms form strategic alliances with pharmaceutical firms and focus on their comparative advantage in R&D, while the

FIGURE 14.2
International Technology Alliances

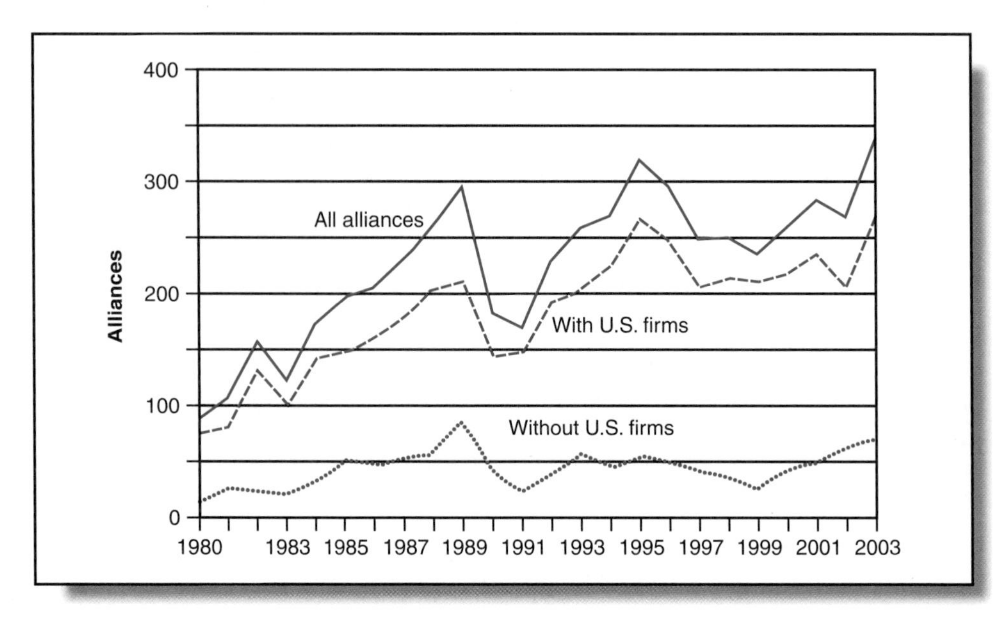

The number of international technology alliances has been rising over the past two decades, and a consistent portion of those alliances involve U.S. companies.

Source: Science and Engineering Indicators, 2006, http://www.nsf.gov/statistics/seind06.

pharmaceutical firms concentrate on their comparative advantage in obtaining regulatory approval, manufacturing, and marketing.[20]

- Second, strategic alliances reduce the expense of building the business's value chain by making it possible to use another company's assets at marginal cost. This cost reduction is particularly important to start-up firms, which lack cash flow from existing activities and so face a high cost of capital. (Because lending money to, or buying shares in, new businesses is riskier than lending money to, or buying shares in, established companies, new companies must pay a risk premium—a higher interest rate on a loan or more shares in an equity investment—to obtain the same amount of money from financial markets as established firms.[21]) For instance, Imax Inc., a small company in Ohio that is developing fuel cells for the army, has formed an alliance with W. L. Gore & Associates to avoid incurring the cost of creating its own the fuel cell membrane from scratch.[22]
- Third, strategic alliances allow companies to develop new competitive advantages, such as economies of scale from increased production volume, lead time from the use of other companies' assets, and improvement on the learning curve by sourcing knowledge from other firms.[23] For example, Sony partnered with Samsung to make flat-panel television sets to move up the learning curve in liquid crystal display technology,[24] while Cisco partnered with Microsoft to develop a new networking product in 18 months, instead of the four years it would have taken to do it alone.[25]
- Fourth, strategic alliances allow companies to learn capabilities outside of their current areas of technical and market expertise.[26] For example, many of the pharmaceutical firms formed alliances with biotechnology companies after the genetic engineering revolution so that their R&D units would learn how to conduct drug development using genetic engineering and genomics.[27]

- Fifth, strategic alliances allow companies to investigate new technologies and markets at relatively low cost because the magnitude of the investment needed to investigate a new technology or market through a strategic alliance is less than the magnitude of the investment needed to do so independently.[28]
- Sixth, strategic alliances provide companies with real options that allow them to investigate possible acquisition targets.[29] By forming strategic alliances with other companies, you can make small investments that allow you to gather information about their cultures and capabilities, which permits you to make more informed selection of acquisition targets, to more accurately value those targets, and to better estimate the costs of merging them into your organization.[30]
- Seventh, strategic alliances allow companies to increase their profits by reducing price competition that dissipates the economic rents from providing a product or service.[31] Of course, you have to be very careful if you do this in the United States. Regulators see the formation of strategic alliances to reinforce market power as anticompetitive behavior that is prohibited by antitrust laws.
- Eighth, strategic alliances are useful in setting standards in an industry, which encourages customer adoption of products and the provision of complementary products by other firms. For example, General Motors and Toyota have formed an alliance to set standards for fuel cell and electric vehicles in the hopes that it will encourage customer adoption of such vehicles.[32]

However, strategic alliances also have three significant drawbacks:

- First, alliances to manufacture, market, and distribute products keep companies from developing capabilities in these areas.[33] For example, Microsoft lacks capabilities in procurement, manufacturing, and distribution because the company has traditionally concentrated on the development of computer software, and has formed strategic alliances with other companies to produce the hardware on which its software runs. Because these capabilities are necessary to offer devices that integrate computer hardware and software, Microsoft has had trouble entering the video game business and lost billions of dollars in its initial foray into the production of the Xbox.[34]
- Second, strategic alliances expose companies to the risk of loss of proprietary knowledge. Because you must transfer knowledge to your alliance partners to make your alliances work, you expose your company to the risk that your counterparts will take knowledge that you did not intend to transfer, and increase the likelihood that your proprietary knowledge will diffuse to other firms.[35]
- Third, when two large companies decide to work together, it raises concerns that the alliance will reduce competition, raise prices, and hurt customers, leading to regulatory scrutiny that can increase costs significantly.[36] (See Table 14.1 for a summary of the pros and cons of strategic alliances.)

Joint Ventures

A **joint venture** is legal partnership between two firms to own equity in a third firm formed by the two partners. For example, 3Com and the Chinese firm Huawei have formed a joint venture company named H-3C to produce Ethernet switches and IP routers.[37]

TABLE 14.1 Advantages and Disadvantages of Strategic Alliances
Strategic alliances have advantages and disadvantages that you need to balance when deciding whether or not to use them.

ADVANTAGES OF ALLIANCES	DISADVANTAGES OF ALLIANCES
Help develop new competitive advantages	Hinder the development of capabilities
Allow specialization of capabilities	Risk loss of proprietary knowledge
Facilitate learning about new markets and technologies	Create antitrust problems
Allow investigation of acquisition targets	
Reinforce market power	
Reduce the cost of creating the value chain	
Establish technical standards	

Like other contractual modes of business, joint ventures have several advantages:

- First, joint ventures help companies to learn new capabilities that enhance their performance or are necessary to exploit new technologies or to enter new markets. For instance, Samsung formed a joint venture with Sony to make flat-panel televisions to learn how to make better displays than the versions it used in cell phones and computer monitors.[38]
- Second, joint ventures help companies to improve their efficiency by allowing companies to concentrate on common activities in a joint venture, minimizing the duplication of activities, and increasing scale economies. They also permit companies to exploit synergies in their operations, as occurs when the manufacturing operations of one company are combined with the marketing activities of another.[39]
- Third, joint ventures allow companies to pool resources to undertake activities that they cannot afford to undertake alone.[40] For example, Advanced Micro Devices established a joint venture with United Microelectronics Corporation to create a 300 mm wafer semiconductor fabrication facility because the high cost of establishing a semiconductor foundry (now averaging $2.5 billion) makes producing microprocessors in a dedicated facility too costly for companies with less than several billion dollars of annual revenue.[41]

However, joint ventures also have some important drawbacks:

- First, the creation of joint ventures exposes you to counterparts that use the joint venture as a window on your technology or capabilities, or as a place to test and cherry-pick talent or to dump poor-performing employees.
- Second, the success of joint ventures depends on the willingness of both parties to make large investments in mechanisms to coordinate the organizational structures, cultures, and strategies of the parent organizations, which raises the cost of operating joint ventures.

An important special case of joint ventures in technology-intensive industries is the **joint research organization**, or a joint venture established for the purposes of conducting research. For example, the Semiconductor Research Corporation, a joint effort of the major semiconductor firms, finances precommercialization research on semiconductors and related technology, largely at academic institutions.[42]

FIGURE 14.3
Joint Ventures

As this historical cartoon depicting the partnership between Thomas Edison, the inventor of the electric light, and Charles Brush, inventor of the arc lamp, shows, joint ventures between technology companies have long been part of business strategy.
Source: Getty Images Inc.—Hulton Archive Photos

Joint research organizations are becoming increasingly common because rising technical complexity requires companies to work with counterparts with complementary expertise to develop new technologies.[43] In the United States, interest in these organizations has risen since the passage of the National Cooperative Research Act of 1984, which exempts joint R&D efforts from antitrust barriers as long as that activity is not in restraint of trade.

Like other kinds of joint ventures, joint research organizations are not all successful. The most effective ones focus on research that extends the technical capabilities of the firms supporting them, and where controls are in place to keep the parent companies from appropriating their counterparts' proprietary knowledge.[44]

Licensing

Licensing is an arrangement under which one organization permits another to use its technology in return for the payment of a fee. Sometimes companies are the licensors of technology to other businesses, as was the case when Intel licensed its instruction set for producing microprocessors in personal computers to AMD.[45] Other times, universities, government laboratories, and nonprofit organizations are the licensors to companies, as was the case when Michigan State University licensed its patent to shikimic acid, a basic ingredient in the flu drug Tamiflu, to Roche Pharmaceuticals.[46]

Licensing is an important part of technology strategy, particularly to the start-up companies that have based their entire business model on it. For example, the initial business model of biotechnology firm Abgenix involved licensing the company's genetically engineered mouse to pharmaceutical and biotechnology companies that had identified specific disease targets to treat with antibodies.[47]

Even some very large companies generate a significant portion of their revenue from technology licensing. For example, at IBM, royalties from licensing account for 20 percent of net income,[48] while at Texas Instruments they account for more than 55 percent.[49] As a result, many high-technology companies, like Microsoft, have established business units to license their technology to other firms.[50]

Often firms license to others technology that they are longer using, or that is no longer central to their business. For example, Lockheed Martin converted idle patents on flight simulators into an equity investment in a start-up video game company, and Microsoft licensed some of its unused patents to Toshiba.[51] Boeing licensed laser technology and factory automation tools that do not apply to the aerospace business to firms in other industries,[52] while General Electric licensed out a microorganism that digests oil spills to the major oil companies,[53] and Monsanto licensed genes that it has developed to other companies to produce crop seeds.[54]

Companies also have used technology licensing to grow their businesses. By licensing in technology from other organizations, these companies have moved into more lucrative lines of business. For example, tiny Haloid Corporation licensed Chester Carlsson's xerography patents and transformed a small printing business into the multinational corporation Xerox; and Suhas Patil licensed a single MIT semiconductor gate array patent to create the billion-dollar semiconductor company Cirrus Logic.

Advantages and Disadvantages of Licensing

Licensing is a valuable tool of technology strategy. It can provide a mechanism to earn a financial return on a technology that your company has developed, even if your company lacks the capabilities or resources to develop or manufacture new products or services that use the technology, or wants to avoid the costly and difficult process of creating manufacturing and marketing capabilities.

The returns that you can earn from licensing your technology can be quite high because licensing allows you to transfer technology to other firms and avoid creating duplicative assets that lower the returns of all companies in the industry. Combined with the much lower financial or human resource expenditures that it demands, this means that licensing can generate very high profit margins.[55] For example, IBM estimates that it would have to have an extra $20 billion in sales every year to earn the same net income as the $1 billion it earns from royalties on licensing its technology.[56]

Licensing also facilitates rapid market penetration. By transferring technology to other companies that already have manufacturing or marketing assets in place, you can get your technology to market more quickly. For instance, Sony licensed its compact disk technology to 30 manufacturers to get that technology into products quickly.[57] (As we saw in Chapter 13, the ability to achieve rapid market penetration is very important to firms in industries subject to increasing returns.)

Licensing allows you to manage risk in technology development by treating new technologies as real options. By obtaining the right, but not the obligation, to make use of a technology invented by a university, supplier, or customer only if that entity achieves a key milestone in technology development, you can mitigate the risk of advancing the technology.[58]

Licensing helps you to avoid intellectual property disputes. When other firms hold key patents that limit your freedom to operate in a particular technology area, you can mitigate the threat of patent litigation by licensing the other firms' patents. You can also obtain additional protection of your own intellectual property by licensing it to other firms, particularly those with deep pockets that can enforce the patents against infringement, because your licensees can be made responsible for defending your intellectual property.

On the other hand, licensing has drawbacks that you need to understand. Much like strategic alliances, licensing hinders capability development because other firms develop, manufacture, market, and distribute products using your technology, denying you the opportunity to learn how to do these things. If these activities add a lot of value to the base technology, or if they permit the creation of strong competitive advantages through trade secrecy, patenting, specialized manufacturing, or learning curves, you may be worse off than if you created these capabilities. In fact, many firms have chosen to develop, manufacture, and market their own products even when they initially lacked the capabilities to do so, because the long-term benefits of capability development exceeded the short-term benefits of greater revenue from licensing. For instance, Abgenix, a biotechnology company that had originally developed and licensed the genetic mouse, ultimately expanded into the production of its own drugs to ensure that the company generated capabilities in late-stage drug development and regulatory approval.

In addition, you will likely generate lower total profits by licensing a technology than from producing a product or service based on the technology because most of the value generated by new products comes from the development, manufacture, and marketing of the products than from the technology itself.[59] For example, Xerox licensed the technology for the Ethernet to Bob Metcalfe who founded 3Com, earning a small licensing fee, rather than the billions of dollars that 3Com made off of the invention.

Furthermore, by licensing your technology to another company, you might create a competitor that you otherwise would not have. For example, Intel licensed its microprocessor technology patents to AMD in the early 1980s to win a contract with IBM as the supplier of microprocessors for the IBM personal computer because IBM insisted on having a second source supplier.[60] This licensing agreement strengthened AMD, which was, at the time, a young semiconductor company, and helped it to grow into a large and formidable competitor.

TABLE 14.2
Dos and Don'ts of Technology Licensing
By doing certain things and avoiding others, you can make your company more effective at technology licensing.

Do . . .	Don't . . .
Understand your negotiating partner's interests, desires, and pressure points	Assume that you have all of the information
Know what you have to out-license or what you want to in-license	Do a license just to record a deal; make sure your partner is capable
Look for comparable transactions	Try to create a zero-sum game
Involve legal counsel in legal issues	Be inflexible on deal terms
Include a best effort clause	Forget the issue of fundamental fairness
Establish performance milestones	Give rights to future discoveries
Include more than just patents (e.g., know-how, trademarks)	Consider the agreement to be the end of the relationship

Exclusive and Nonexclusive Licenses

As Figure 14.4 shows, licensing agreements have a number of different terms. One of the most important of these is whether the license is granted on an **exclusive** or a **nonexclusive** basis. An exclusive license gives the right to use a technology to only one company; a nonexclusive license grants that right to many companies.[61]

FIGURE 14.4 A License Agreement

This agreement is made between U.S. Computer CORP (USCC) and Software, INC. (SI). USCC and SI agree as follows:

1. LICENSES: The license from SI to USCC covers: (a) the technology that SI has relating to videoconferencing in a standalone format; (b) any source and object code related to videoconferencing SI has that will work in the VC; (c) any technology SI has developed for direct connection to the personal computer; (d) all development tools applicable to the "Licensed Technology."

The Licensed Technology is as it exists on the Effective Date in written and electronic documents, including schematics, data base tapes, software, source and object code for delivery to USCC, and except for the foregoing, does not include delivery of any physical products; provided, however, future modifications and enhancements to the Licensed Technology pursuant to this Agreement shall become part of the "Licensed Technology."

SI will immediately deliver to USCC the Licensed Technology and hereby grants to USCC an exclusive nonassignable world-wide license to use the Licensed Technology and to make, have made, use, market and sell products containing or embodying such Licensed Technology to sell direct computer products for itself or to Original Equipment Manufacturers under their brand names.

← This clause indicates what is licensed by the licensor to the licensee

2. ROYALTY: On any system that USR sells that incorporates any Licensed Technology USCC agrees to pay SI two percent (2%) of gross sales.

← The royalty is what the licensee has to pay to the licensor for the technology

3. VERIFICATION: SI is entitled to audit the records of USR through SI auditor, provided that (a) such audit shall occur no more than once per year, and (b) such auditor (i) shall be acceptable to USCC and (ii) shall have executed an appropriate nondisclosure agreement. If such an audit discloses a deficiency in the royalty paid of greater than five percent (5%), then USCC will pay the reasonable cost of such audit plus interest on the deficiency from the time due until paid of twelve percent (12%) simple interest per annum.

← The licensor can check the licensee's records to make sure that they aren't cheating

4. IMPROVEMENTS: USCC will provide SI technical and other confidential and proprietary information that USCC determines is necessary or useful for SI to improve the modem and speakerphone functionality for the video communicator. SI shall use the USCC information only for its own branded system products, and shall not sublicense or otherwise disclose the USC information to third parties for use in their products or for any other reason. Each party agrees to license to the other party any enhancements it make to the Licensed Technology.

← Improvements to the technology are part of the license

5. TECHNICAL SUPPORT: SI will provide engineering support to USCC, for all Licensed Technology, including all such technology initially delivered to USCC and all enhancements. Such engineering support shall be sufficient to enable USCC to achieve its objectives of volume shipments as soon as possible.

← Technical help by the licensor is part of the deal

6. EXCLUSIVITY: SI agrees that it will not grant to any third party any licenses to use the Licensed Technology, or any portion thereof, or to make, have made, use, market or sell products with such Licensed Technology, or any portion thereof.

← No one else can license the technology

7. WARRANTIES: SI warrants that SI owns all rights in and to all other portions of the Licensed Technology, free of any liens, claims, encumbrances or other restrictions that would impair USCC rights under this Agreement. The foregoing warranties exclude any warranty that the Licensed Technology does not infringe the intellectual property rights of any third party.

← The licensor guarantees that is the owner of the technology

8. CONFIDENTIAL INFORMATION: The parties will keep confidential any information provided to it by the other party that is proprietary to the other party and marked confidential; provided such information shall not be considered proprietary once it is in the public domain by no fault of the other party.

← Both sides will keep the other's information confidential

Licensing agreements have several standard clauses.

Source: Adapted from http://contracts.onecle.com/8x8/usrobotics.lic.1997.05.05.shtml

Exclusive licenses have the advantage of providing licensees with a greater incentive to expend effort and money to get a product or service based on the licensed technology to market. In contrast, nonexclusive licensees have an incentive to **free ride** because they can benefit from the efforts of other licensees to develop the technology without incurring the costs.

However, exclusive licensing is riskier than nonexclusive licensing. With exclusive licensing, you will earn no return if you are poor at choosing licensees and select one that is unable to get the technology to market; with nonexclusive licensing, you are diversified across the talents of different licensees. The risk of a licensee **shelving a technology** (licensing a technology without intention to develop it just to keep others from licensing it) is also greater with exclusive licensing than with nonexclusive licensing because a single nonexclusive licensee cannot shelve a technology by itself.

Cross–Licensing

When two companies each have patents to technologies necessary to develop a product, neither party will invest in the development of that product for fear of a patent infringement lawsuit. This stalemate can be resolved by **cross-licensing**, an agreement between two parties to license reciprocally.

To minimize transactions costs, cross-licensing agreements rarely specify individual patents. Rather, the firms agree to license patents to all of their existing and future inventions in a particular field of use for a specified period of time.[62] The royalty payment on a cross-licensing agreement is typically calculated by estimating the value of the patent portfolios of both firms and subtracting one from the other.[63] For example, when Sun Microsystems and Microsoft considered cross-licensing as part of an effort to make their technologies compatible, the two organizations each valued their own and each other's patent portfolios to determine which company would have to pay net royalties to the other.[64]

Cross-licensing is particularly important in industries like semiconductors and computers, in which the cost of closing down a factory for even a few days (because of an injunction in a patent litigation case) can drive a firm into bankruptcy,[65] and in which technologies are cumulative and systemic, and so require the use of patents held by many different companies.[66]

When cross-licensing is important, firms often develop large patent portfolios to motivate competitors to cross-license with them[67] because companies prefer to cross-license with counterparts that have equal-sized patent portfolios to balance the amount of technology each party needs to give up.[68] Moreover, having a lot of patents makes it easier to threaten a patent infringement countersuit, which encourages competitors to cross-license. Therefore, start-up companies, which usually have a small number of patents, are often at a competitive disadvantage in industries in which cross-licensing is important because they lack a patent portfolio large enough to get other firms to cross-license with them.

Outsourcing and Contract Manufacturing

A third contractual mode of doing business is **outsourcing**, or assigning, some aspect of operations to another company under contract. All parts of the value chain can be outsourced. For example, the large pharmaceutical firms, like Merck, outsource research to universities and nonprofit research institutes,[69] and the major auto companies, like Honda and Toyota, outsource engineering design work to

FIGURE 14.5
Outsourcing R&D

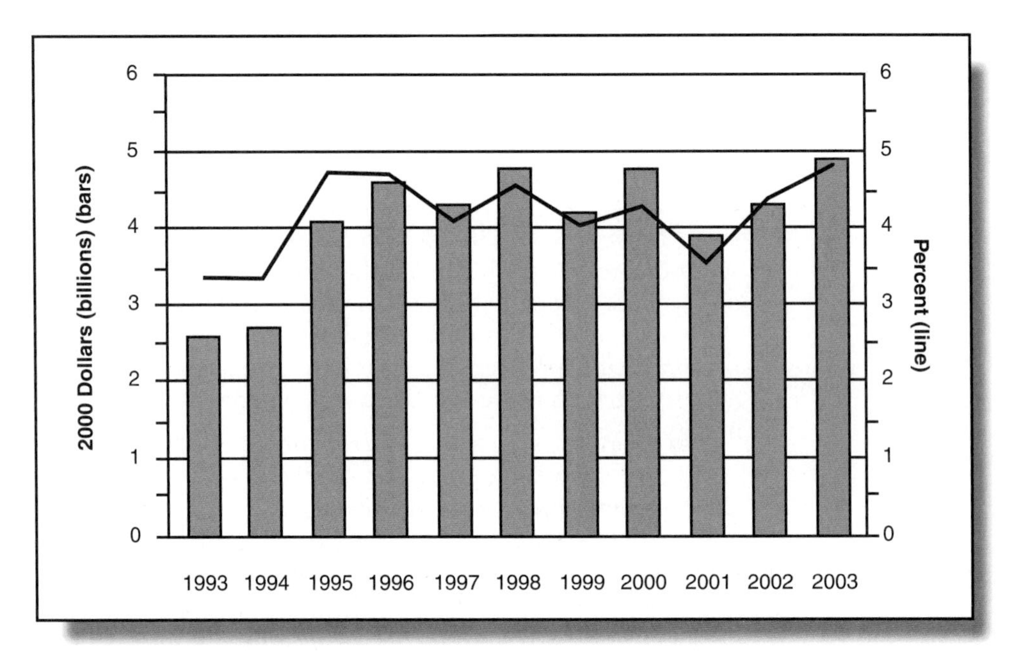

The amount and proportion of R&D that is outsourced still remains fairly small.
Source: Science and Engineering Indicators, 2006, http://www.nsf.gov/statistics/seind06.

other companies.[70] However, as Figure 14.5 shows, outsourcing of R&D remains fairly uncommon.

When companies outsource the manufacturing of their products to another company, outsourcing is called **contract manufacturing**.[71] For example, Microsoft uses contract manufacturers to produce its Xbox 360 video game device.[72]

Outsourcing provides several benefits:

- One, it allows companies to specialize in the parts of the value chain where their core capabilities lie,[73] permitting them to move up the learning curve,[74] and increasing their production volume, thereby reaping the benefits of economies of scale. For instance, many semiconductor firms concentrate on designing semiconductors and outsource manufacturing to specialist facilities, allowing them to make smaller and faster semiconductors.[75]
- Two, outsourcing allows companies to expand their operations without investing in additional plant and equipment or increasing their labor force, which reduces the risk they have to bear and enhances their flexibility.[76] If demand shrinks and they need to reduce their operations, they can do so more easily than if they produced the product themselves because they have made no capital investment in increased operations and need only change the terms of their contracts when they come up for renewal. (This is particularly efficient when demand is highly variable.[77]) Moreover, the supplier of the outsourced service is better able to bear the risk of demand shrinkage because its capital investment is diversified across multiple customers, limiting its risk to industry shrinkage, and not to decline in demand for a particular company's products.[78]
- Three, outsourcing allows a company to accelerate the pace at which it begins operations because the supplier of the outsourced service—be that

product design, manufacturing, or marketing—is already in operation and does not need to scale up.[79] For example, research has shown that automakers are able to develop new car models a third of a year faster if they outsource engineering design to other firms than if they conduct the same work internally.[80]

While outsourcing provides the benefits described previously, those benefits come at the expense of several potential downsides:

- First, outsourcing risks the loss of technology to the firm providing the product because that company needs to have detailed information about your technology and could use that information to compete with you. Many consumer electronics companies have confronted this problem because they have engaged in contract manufacturing with Chinese manufacturers, who have learned to create competing products from the information provided by American and Japanese electronics companies.
- Second, outsourcing will keep companies from developing capabilities at the activities that they outsource, and the absence of those capabilities could hinder the company's performance.[81] For instance, 3M discovered that it could not understand its customers' needs when it outsourced market research because doing so created too many intermediaries between product developers and customers, and kept the company from conducting proprietary market research.[82]
- Third, outsourcing involves high transaction costs because firms must write contracts to specify how the activity will be conducted, which they do not have to do if the activity is conducted in-house.[83] When the outsourced activity is complex, the cost of writing these contracts—in terms of incomplete and unenforceable agreements—often becomes so high as to deter companies from engaging in the practice.

The costs and benefits of outsourcing suggest several rules of thumb for you to follow when considering it. You should contract out only those activities at which significant cost savings can be achieved, and which are not central to the mission of your company.[84] Thus, outsourcing support services, such as employee benefit services and accounting, or low value-added activities for which specialists exist, like product assembly, makes sense;[85] outsourcing activities directly linked to the creation of product features that customers value, that involve technology with which you could develop a competitive advantage, or that require significant investment over time to build the skills to develop your products, does not.[86]

For this reason, many observers believe that Boeing is taking a major risk by outsourcing the production of sections of the tail, the fuselage, the landing gear, and even the wings, of its new Dreamliner airplane.[87] While outsourcing is saving Boeing billions of dollars, it is also reducing the company's ability to produce future generations of aircraft. Consequently, Boeing may be mortgaging its future by outsourcing.

Key Points

- Strategic alliances are relationships between two firms to work together to achieve a common goal; they can be vertical (between firms at different stages of the value chain) or horizontal (between firms at the same stage of the value chain).

GETTING DOWN TO BUSINESS

Contract Manufacturing in the Medical Device and Equipment Industry[88]

MedSource Technologies is a company founded in 1999 to provide contract manufacturing of components to the medical device industry. Medical devices generally contain more than 20 parts—things like plastic, wire, and machined metal—which original equipment manufacturers (OEMs) typically obtain from their suppliers. MedSource Technologies's founders believed that OEMs would want to focus on designing and selling medical devices and would see value in purchasing parts from them.[89]

Contract manufacturing is popular in the medical device supply business because economies of scale in production are large. A supplier like MedSource Technologies could produce many more parts than any OEM could make alone. The tooling and manufacturing processes for many components in medical devices demand a very large up-front investment.[90] However, many OEMs are small, and many of the products that they produce are new and only needed in low volume. Therefore, a supplier can spread production across many more units than the OEM itself, thus reducing the cost of the parts used in the medical devices.[91]

In addition, there is a strong learning curve in the production of parts for medical devices. As the volume produced goes up, the makers of those parts are able to improve quality a great deal. Thus, the suppliers can move further up the learning curve than the OEMs because they produce more of each part.[92] This is very important in the medical device market, given the devastating effects of product failure.[93]

To make this strategy work, MedSource Technologies had to develop standardized processes. If its processes weren't standardized, the OEMs would face a significant risk. The contract manufacturers, like MedSource, could "hold them up." That is, they could force the OEMs to renegotiate their contracts to more favorable terms for the contract manufacturer. But if the manufacturing processes were standardized, this problem would be mitigated because the OEMs could simply switch to another contract manufacturer if the supplier tried to renegotiate.[94]

- Strategic alliances permit the specialization of capabilities, the development of new capabilities, and the investigation of potential acquisition targets; create market power; set standards; and reduce costs; but can hinder the creation of capabilities; raise antitrust issues; and risk the loss of proprietary knowledge.
- Joint ventures, which are contractual agreements between two firms to create and hold equity in a third firm, help companies to learn new capabilities, improve efficiency, and pool resources, but can lead to opportunistic action by parent companies and demand costly efforts to blend organizational structure, culture, and strategy to be effective.
- Licensing is an arrangement in which a firm permits another to use its intellectual property in return for a fee; it helps firms to enhance financial returns, achieve rapid market penetration, manage risk, and reduce intellectual property disputes, but can hinder the development of capabilities, lower profits, and create competitors.
- Licensing can be exclusive (to only one licensee) or nonexclusive (to multiple licensees); exclusive licensing gives licensees stronger incentives to develop technology, but raises the risk of choosing the wrong licensee.
- Cross-licensing, or bilateral licensing, is important in industries in which patents belonging to many firms are needed to develop a single product.
- Outsourcing is the assignment of a company's operations to another firm; when manufacturing is outsourced, it is called contract manufacturing.

FIGURE 14.6
Outsourcing Is Ubiquitous

Outsourcing is an increasingly popular form of business.
Source: http://www.cartoonstock.com/directory/o/outsourcing.asp.

- Outsourcing allows firms to capitalize on their core capabilities, expand their operations at lower risk, and increase their pace of activity, but hinders capability development, risks the loss of proprietary knowledge, and involves high transaction costs.
- Outsourcing is more successful if firms focus on noncore activities at which large savings can be garnered.

Work Alone or Collaborate

Given the advantages and disadvantages of strategic alliances, licensing, joint ventures, outsourcing, and contract manufacturing, how do you know if you should collaborate or go it alone? As Figure 14.7 shows, there are four dimensions that you need to consider in making the decision: the strategic factors that create the advantages and disadvantages of collaborative arrangements, the nature of your core technology, the type of firm that you run, and the industry in which you operate.

First, you need to identify the trade-offs between the different modes of doing business across the different dimensions discussed earlier in this chapter. Are any of the factors particularly important for your business? If so, those factors might be so

FIGURE 14.7
Forces to Consider

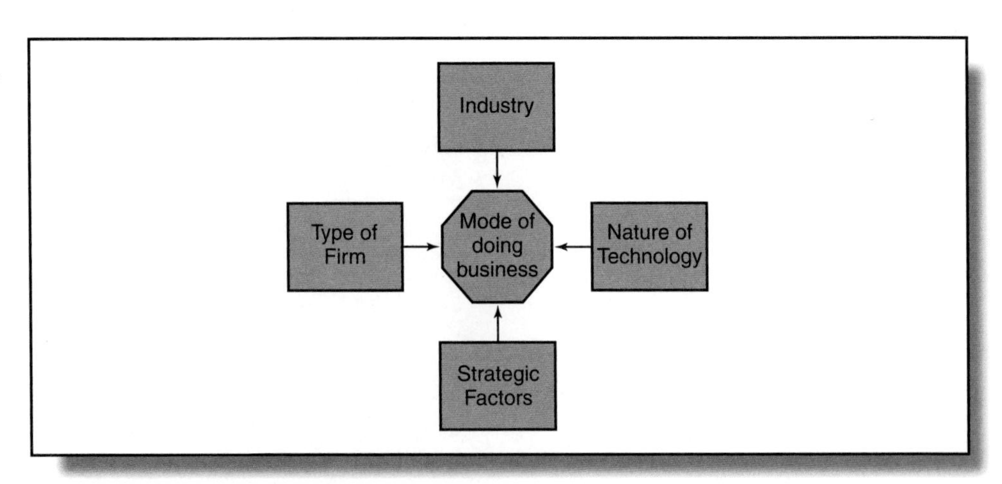

The four key forces that you should consider in determining your mode of doing business are the industry you are in, the type of firm you are running, the nature of your technology, and the key strategic factors discussed in this chapter.

great as to swamp the others and lead you toward one mode of doing business over another for strategic reasons.

Second, you need to consider the nature of your technology. Based on the dimensions discussed earlier in the chapter, is your technology amenable to contractual forms of doing business? If it is, is there one contractual mode that is better than another for that technology?

Third, you should consider whether you are operating a small, start-up firm or a large, established company. The resource and capabilities constraints of new firms often lead them to make different decisions about business mode than established firms that are exploiting the same business opportunities. Because start-ups often lack cash, they need to weigh the demand for money against other factors in determining whether or not to use contractual forms of business, and cash needs often win out. For example, SmartShopper Electronics, a start-up that offers a device that allows people to record and print out lists of groceries and errands, uses contractors to produce the voice-recognition and printing software that goes on the inside of the product; to design the product and logo; to make, test, and box the electronics; and to receive the shipments, warehouse the inventory, and fulfill orders.[95] An established company pursuing the same opportunity might do many of these things in house.

Finally, you need to think about your industry because industry characteristics have a large influence on the effectiveness of contractual modes of doing business. Contractual forms of business are more important in industries subject to network effects because first mover advantages are very strong in these industries.[96] As Chapter 13 explained, licensing permits faster market penetration than in-house production and distribution, thus allowing a rapid increase in the installed base, which is so important in networked industries. For example, Microsoft was able to compete with Netscape, which had a year's lead time on the launch of its browser, by contracting with online provider AOL to distribute its browser. By licensing out its Web browser, Microsoft was able to catch up to Netscape before the latter company had an insurmountable lead in installed base.[97]

Contractual forms of doing business are more valuable in industries in which opportunities are short-lived because licensing allows you to move quickly. Take, for example, the source of an opportunity for a new electronic voting machine that resulted from the problems with "hanging chads" in the 2000 U.S. presidential election. The frequency of elections means that you have to get your new product developed and sold within a relatively short period. Creating a voting machine value chain from scratch would likely take too long for you to meet the window of opportunity in this business, but establishing it through contracting would permit you to meet the window of opportunity.

Markets for Knowledge

Researchers have found that markets for technology face two major problems, which make contractual modes of doing business more effective in some industries than in others: disclosure and hold-up. As Nobel Prize–winning economist Kenneth Arrow explained, markets for technology are undermined by the **disclosure paradox**: To get buyers to pay for your new technology, you need to provide them with evidence that it is worth something. However, the act of showing prospective buyers why a new technology is valuable often demonstrates how the technology works. And once the buyers know how the technology works, they have no reason to license it—you just gave it away for free![98]

Having an effective patent allows a firm to extricate itself from the disclosure paradox because the patent protects an innovator against imitation after disclosure. Thus, as Figure 14.8 shows, in biotechnology, where patents are very effective, contractual modes of business, like strategic alliances and licensing, are very common. At the same time, the cost and difficulty of drug discovery, regulatory approval, drug manufacture, and pharmaceutical marketing make the value of specialization very high,[99] and avoiding the expense of duplicating these things valuable.[100]

Markets for technology are also undermined by **hold-up** problems. Hold-up occurs when one party to a contract demands that the agreement be renegotiated in its favor, and the other party acquiesces because the termination of the agreement would be worse for it than the new terms of the agreement.

Hold-up is more likely to occur if there are only a few counterparts with whom a company could contract and if the assets that are needed to use the technology are specialized. (Remember the discussion in Chapter 10 about specialized assets? They are assets made specific to a transaction, like a piece of manufacturing equipment

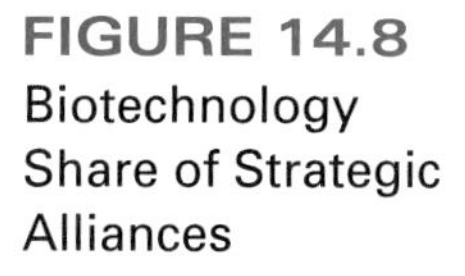
FIGURE 14.8 Biotechnology Share of Strategic Alliances

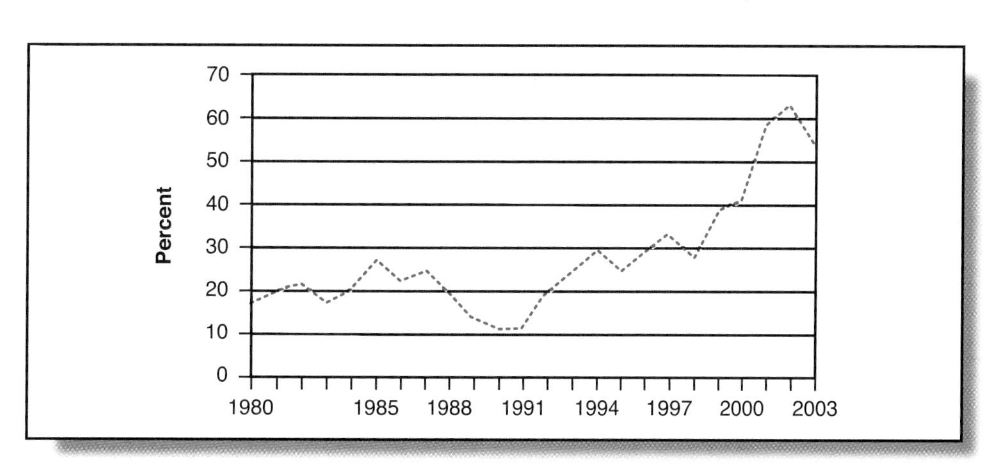

Strategic alliances are most common in the biotechnology industry, which now accounts for over half of all strategic alliances created.

Source: Adapted from *Science and Engineering Indicators*, 2006, http://www.nsf.gov/statistics/seind06/c0/c0s2.htm.

that has to be built and can only be used to produce the licensor's product.) If there are small numbers of potential counterparts, then you cannot easily switch partners if problems arise, giving your counterpart leverage to hold you up.[101]

If the assets that you need to use the technology are specialized, then you are vulnerable to your counterpart's efforts to renegotiate the terms of the agreement in his or her favor because the financial returns that can be gained from the second best use of the asset are much less than those that can be gained from the best use. Consequently, it is rational to accede to your counterpart's demands to renegotiate the terms of the agreement,[102] as long as the new, worse, terms would still allow you to generate higher returns than those earned from the second best use of the asset.[103] Thus, companies often will not contract for technology if doing so will require them to invest in specialized assets for fear of being taken advantage of by opportunistic counterparts, leading markets for these technologies to fail,[104] and requiring firms to own their own manufacturing, marketing, and distribution assets.

Key Points

- To decide whether you should use contractual or vertically integrated modes of doing business, you need to consider the industry in which you operate, the nature of your core technology, the type of firm that you run, and the strategic factors that create the advantages and disadvantages of collaborative arrangements.
- The nature of intellectual property protection, competitive dynamics, network effects, the nature of returns, and the rate of change in your industry affect whether you should use contractual or vertically integrated modes of doing business.
- The limited resources and capabilities of new firms often lead them to make different decisions about business mode than established firms exploiting the same business opportunities.
- Markets to sell technology often fail because of disclosure and hold-up problems.

TABLE 14.3 Characteristics of Different Modes of Business
This table compares the five different modes of business discussed in this chapter on the dimensions of speed to market, cost, control over IP, the risk/return relationship, specialization of capabilities, development of new capabilities, antitrust problems, the real options approach, and help with standards setting.

Mode of Business	Speed to Market	Cost	Control over IP	Risk/Return	Specialization of Capabilities	Development of New Capabilities	Antitrust Problems	Real Options Approach	Standards Setting
Integration	Slow	High	High	High/High	Low	High	Low	Low	Low
Strategic alliances	Moderate	Moderate	Low	Moderate/Moderate	High	Moderate	High	High	High
Licensing	Fast	Low	Low	Low/Low	High	Low	Moderate	High	High
Joint ventures	Moderate	Moderate	Low	Moderate/Moderate	Moderate	Moderate	High	Moderate	Moderate
Outsourcing	Fast	Low	Moderate	Low/Low	High	Low	Moderate	Low	Low

Making Contractual Arrangements Work

Research shows that half of all contractual business arrangements fail, often at significant cost to the parties involved.[105] Take, for example, General Motors's strategic alliance with Fiat to develop small cars, which had to be undone at the cost of several billion dollars, when the joint efforts of the companies failed to save either company any money.[106]

Designing Effective Contractual Arrangements

To design effective contractual organizational arrangements, you should do the following:

1. Know your own firm's strategy and capabilities at different parts of the value chain and have a clear picture of how the contractual arrangement that you are considering will help with that strategy.
2. Assess the match between your business and your prospective counterparts on such dimensions as technology, skills, resources, strategies, and corporate culture to ensure that they are complementary.[107]
3. Conduct **due diligence** (the verification of the credentials) on your transaction partners to ensure that they will behave honestly and fairly, and have appropriate assets and capabilities.
4. Keep your partner from acting self-interestedly at your expense by keeping them away from any information or technology that you do not want to disclose, creating a team with the responsibility and authority for managing your relationship with your counterpart, setting up policies and procedures—such as audits and site visits—to verify that your partners are adhering to the terms of the contract, establishing routines and structures to facilitate communication between the two organizations,[108] and making reciprocal financial investments[109] that provide "hostages" to deter opportunistic action.[110]

Contracting Between Large and Small Firms

Many contractual organizational arrangements are created between large, established firms and small, start-ups because new firms are very good at developing new products and finding the initial market niches in which to exploit them, while established firms are very good at using manufacturing, marketing, and distribution to capture the mass market.[111] However, these contractual arrangements raise additional management issues. Because large, established companies are often hierarchical bureaucracies, start-up companies are often stymied by the inaction of their counterparts and the difficulty of knowing whom to work with to solve problems.

Moreover, the relationships between the two types of firms are asymmetrical because a single large, established company will often contract with many small, start-ups, while a small start-up company will typically contract with one large firm. As a result, the large companies are more diversified than the small companies, and so are more willing to terminate poor performing relationships. Take, for example, the reaction of Amylin Pharmaceuticals, a small biotechnology company, and its partner, Johnson & Johnson, to poor results from a study testing the drug Symlin that the companies were jointly developing. Amylin, which was developing no other drugs at the time, wanted to persist with the project despite the poor results, while Johnson & Johnson, which had many better performing alternatives, wanted to terminate the alliance.[112]

The differences in the portfolios of products also lead to conflict between new and established firms over decisions to bring new products to market. The established firms, which have multiple new product development projects, often want to select the one with the greatest prospects for success, while the new firms, which usually have only one project, want theirs to be selected, even if it is not the best. For example, the small biotechnology firm Tanox ended up in a dispute with its strategic alliance partners, Novartis and Genentech, over TNX-901, a drug to prevent allergic reactions to foods like peanuts. Tanox wanted to introduce this drug to the market after researchers had achieved favorable results in clinical studies. However, Novartis and Genentech decided to move forward with a different drug in their portfolios, Xolair, and focus it on asthma and hay fever treatments because Xolair showed more promise than TNX-901, and because asthma and hay fever are more common than food allergies.[113]

Finally, established and start-up firm partners also disagree about when to bring new products to market because established companies often want to delay new product introduction to reduce cannibalization of their existing products, while new firms, which have no products in the market yet, want to enter immediately. For instance, the biotechnology company Amylin and its pharmaceutical partner, Eli Lilly, disagreed about when to introduce the diabetes drug Exenitide because the drug would postpone the point when diabetics need to take insulin, a product that Eli Lilly makes significant profits selling.[114]

Key Points

- Contractual modes of business often fail at a high cost.
- Success at contracting is enhanced by knowing your company's strategy, formulating a plan, assessing your company's fit with its collaboration partner, conducting due diligence, and managing the relationship once the contract has been signed.
- Contractual relationships between large and small firms face the problems of large firm bureaucracy stymieing action, and asymmetry between the two parties, which leads them to make different decisions about project selection and termination, and the timing of market entry.

Discussion Questions

1. Why do firms collaborate with other firms? What are the risks of collaboration? How might the risks be mitigated?
2. Information technology and biotechnology are both high-tech industries. Why are contractual modes of business more common in one than in the other?
3. When should an established company license its technology to others? What about a start-up? Why are those conditions the best for licensing?
4. What are the pros and cons of outsourcing parts of the innovation process? Are there technological competencies that should not be outsourced? Why or why not?
5. What can you do to better manage a contractual organizational arrangement? Why do these things help?
6. Why do you think that contractual organizational arrangements are becoming increasingly common?
7. Which do you think is the most important in determining whether or not to use contractual modes of business: strategic issues, industry characteristics, the nature of the technology, or the type of firm? Why do you choose that dimension?

KEY TERMS

Contract Manufacturing: The outsourcing of manufacturing to a company that specializes in manufacturing on behalf of others.

Contractual: A mode of business in which a company signs an agreement with an independent organization to provide part of the value chain.

Cross-licensing: An agreement between two parties to license reciprocally.

Disclosure Paradox: The inability to get others to contract for knowledge unless the knowledge is disclosed, combined with the unwillingness for others to pay for knowledge once it has been disclosed.

Due Diligence: The verification of the credentials of another party.

Exclusive License: A license that gives only one party the right to use a technology.

Free Ride: The tendency of one party to let others do the work necessary to receive a benefit.

Hold-up: The tendency of one party to a contractual agreement to take advantage of the other's vulnerabilities to renegotiate the terms of the agreement.

Horizontal Alliance: A partnership between firms at the same stage of the value chain to work together to achieve a common goal.

Joint Research Organization: A joint venture established to conduct research.

Joint Venture: A legal partnership between two firms to own equity in a third firm formed by the two partners.

Licensing: A form of business, under which one firm permits another to use its technology in return for the payment of a fee.

Nonexclusive License: A license that grants many parties the right to use a technology.

Outsourcing: A mode of business in which a company assigns some aspect of its operations to another company under contract.

Shelving a Technology: The act of licensing a technology with no intention of developing it simply to keep others from licensing it.

Strategic Alliance: A partnership between two firms to work together to achieve a common goal.

Vertical Alliance: A partnership between firms at the different stages of the value chain to work together to achieve a common goal.

Vertically Integrated: A mode of doing business in which the company owns more than one part of the value chain.

PUTTING IDEAS INTO PRACTICE

1. **Deciding Whether or Not to License a Technology** Go to your university's technology transfer office Web site. (You can look on your university's homepage or go to Google and type in "technology transfer office" to get to it. If your university does not have a technology transfer office, or you cannot find it, you can pick another university and look for its technology transfer office.) On the technology transfer office Web page, find the place where the university lists inventions available for license. Read through those inventions and select one for investigation. The goal of your investigation should be to determine whether the university should invest its time and money in trying to license the technology.

   Once you have chosen the invention that you are going to evaluate, identify some potential licensees for the technology. (To do this, think of what the technology does and what companies would have a need for that feature or features.) Now list the advantages and disadvantages of using licensing as a mechanism for exploiting the technology. Evaluate the pros and cons of licensing this invention, and determine whether this technology is a good candidate for licensing.

2. **Evaluating a Strategic Alliance** The purpose of this exercise is to analyze a strategic alliance between a venture capital–backed biotechnology firm by the name of GlycoFi and pharmaceutical giant Merck. The purpose of the alliance is to use GlycoFi's yeast-based glycoprotein optimization platform to develop new drug candidates, such as oncology drugs and vaccines against infectious diseases. GlycoFi is a small private firm, focused on the development of a technology platform that optimizes the growth of a protein in yeast, which can improve the effects of a drug.[115] The company has approximately 60 patents on this technology. The company does not manufacture or market any products. Merck is a large, vertically integrated pharmaceutical firm that develops, manufactures, and markets drugs worldwide.

   Is this strategic alliance a good idea for both parties? Why or why not? To answer this question, address: (1) Why each party wants to form the alliance, (2) the strengths and weaknesses that each of the parties brings to the alliance, (3) the potential sources of conflict between the two firms, and (4) the risks that each party incurs in forming the alliance.

3. **Evaluating Contract Manufacturing** This exercise is designed to help you evaluate contract manufacturing. Assume that you have started a company to exploit a new semiconductor design. Should you manufacture your own chip or contract with a foundry to make it for you? Why do you make this recommendation? (To answer these questions, you should consider the advantages and disadvantages of using a foundry).

   Now assume that you have decided to use a foundry and are considering choosing IBM. Go to www-03.ibm.com/chips/asics/foundry to learn about IBM's semiconductor foundry. Should you select IBM? What are the advantages and disadvantages of choosing IBM as your foundry?

NOTES

1. Adapted from Gans, J., and S. Stern. Forthcoming. Managing ideas: Commercialization strategies for biotechnology. *Kellogg Biotechnology Review.*
2. Ibid.
3. http://en.wikipedia.org/wiki/Interferon
4. Robbins-Roth, C. 2000. *From Alchemy to IPO.* Cambridge, MA: Perseus Press.
5. Miles, R., and C. Snow. 1992. Causes of failure in network organizations. *California Management Review*, 34(4): 53–72.
6. Chesbrough, H. 1999. Intel Labs (A) Photolightography strategy in crisis. *Harvard Business School Teaching Note*, Number 5-601-120.
7. Guth, R. 2005. Microsoft's Xbox reflects new focus on hardware. *Wall Street Journal*, November 22: A1, A6.
8. Chesbrough, H., and D. Teece. 2002. When is virtual virtuous? *Harvard Business Review*, August: 5–11.
9. Ibid.
10. Schilling, M. 2003. Technological leapfrogging: Lessons from the U.S. video game console industry. *California Management Review*, 45(3): 6–35.
11. Van Hove, L. 1999. Electronic money and the network externalities theory: Lessons for real life. *Netnomics*, 1: 137–171.
12. Dhebar, A. 1995. The introduction of FM radio (A), (B), and (C). *Harvard Business School Teaching Note*, Number 5-594-072.
13. Winter, S. 1984. Schumpeterian competition in alternative technological regimes. *Journal of Economic Behavior and Organization*, 287–320.
14. Ibid.
15. Sanchez, R. 1996. Strategic product creation: Managing new interactions of technology, markets, and organizations. *European Management Journal*, 14(2): 121–138.
16. Chesbrough and Teece, When is virtual virtuous?
17. Thurm, Scott. 2005. Start-up claims quantum leap in fiber optics for telecom gear. *Wall Street Journal*, May 26: B4.
18. http://biotech.about.com/gi/dynamic/offsite.htm?site=http%3A%2F%2Fwww.fool.com%2Fportfolios%2Frulebreaker%2F2001%2Frulebreaker010509.htm
19. Allen. K. 2003. *Bringing New Technologies to Market.* Upper Saddle River, NJ: Prentice Hall.
20. Rothaermel, F. 2001. Incumbent's advantage through exploiting complementary assets via interfirm cooperation. *Strategic Management Journal*, 22(6/7): 687–699.
21. Evans, D., and L. Leighton. 1989. Some empirical aspects of entrepreneurship. *American Economic Review*, 79: 519–535.
22. Vanac, M. 2005. Painesville project is offering promise. *The Plain Dealer*, November 25: C1, C4.
23. Yeheskel, O., O. Shenkar, A. Fiegenbaum, and E. Cohen. 2001. Cooperative wealth creation: Strategic alliances in Israeli medical-technology ventures. *Academy of Management Executive*, 15(1): 16–23.
24. Dvorak, P., and E. Ramstad. 2006. TV marriage. *Wall Street Journal*, January 3: A1, A6.
25. Cisco Systems Inc.: Acquisition integration for manufacturing (A). *Harvard Business School Case*, Number 9-600-015.
26. Yeheskel, Shenkar, Fiegenbaum, and Cohen, Cooperative wealth creation.
27. Galambos, L., and J. Sturchio. 1998. Pharmaceutical firms and the transition to biotechnology: A study in strategic innovation. *Business History Review*, 72: 250–278.
28. Schilling, M., and K. Steensma. 2001. The use of modular organizational forms: An industry-level analysis. *Academy of Management Journal*, 44(6): 1149–1168.
29. Folta, T., and K. Miller. 2002. Real options in equity partnerships. *Strategic Management Journal*, 23(1): 77–88.
30. Pisano, G. 1990. The R&D boundaries of the firm: An empirical analysis. *Administrative Science Quarterly*, 35: 153–176.

31. Gans, J., and S. Stern. 2003. The product market and the market for "ideas": Commercialization strategies for technology entrepreneurs. *Research Policy*, 32: 333–350.
32. Mohr, J., S. Sengupta, and S. Slater. 2005. *Marketing of High Technology Products and Innovations* (2nd edition). Upper Saddle River, NJ: Prentice Hall.
33. Schilling, M., and C. Hill. 1998. Managing the new product development process: Strategic imperatives. *Academy of Management Executive*, 12(3): 67–81.
34. Guth, Microsoft's Xbox reflects new focus on hardware.
35. Das, T., and B. Teng. 1999. Managing risks in strategic alliances. *Academy of Management Executive*, 13(4): 50–62.
36. Mohr, Sengupta, and Slater, *Marketing of High Technology Products and Innovations.*
37. http://news.zdnet.com/2100-1035_22-5926305. html
38. Dvorak and Ramstad, TV marriage.
39. Caloghirou, Y., G. Hondroyiannis, and N. Vonortas. 2003. The performance of research partnerships. *Managerial and Decision Economics*, 24: 85–99.
40. Yeheskel, Shenkar, Fiegenbaum, and Cohen, Cooperative wealth creation.
41. Burgelman, R., and P. Meza. 2003. Intel beyond 2003: Looking for its third act. *Stanford Business School Case*, Number SM-106.
42. Schilling, M. 2005. *Strategic Management of Technological Innovation*. New York: McGraw-Hill.
43. Mowery, D. 1995. The boundaries of the U.S. firm in R&D. In N. Lamoreaux and D. Raff (eds.), *Coordination and Information: Historical Perspectives on the Organization of Enterprise*. Chicago: University of Chicago Press.
44. Caloghirou, Hondroyiannis, and Vonortas, The performance of research partnerships.
45. http://news.com.com/2100-1040-257059.html?legacy=cnet
46. Pollack, A. 2005. Is a bird flu drug really so vexing? *New York Times*, November 5: B1, B13.
47. Gourville, J. 2002. Abgenix and the Xenomouse. *Harvard Business School Teaching Note*, Number 5-503-046.
48. Rivette, K., and D. Kline. 2000. Discovering new value in intellectual property. *Harvard Business Review*, 78(1): 2–12.
49. Jaffe, A., and J. Lerner. 2004. *Innovation and Its Discontents*. Princeton, NJ: Princeton University Press.
50. Guth, R. 2005. Microsoft plans licensing venture to boost revenue. *Wall Street Journal*, May 5: B4.
51. Zaun, T. 2005. 2 Giants agree to cross-license some patents. *New York Times*, May 14: B2.
52. Arora, A., A. Fosfuri, and A. Gambardella. 2001. *Markets for Technology: The Economics of Innovation and Corporate Strategy*. Cambridge, MA: MIT Press.
53. Narayanan, V. 2001. *Managing Technology and Innovation for Competitive Advantage*. Upper Saddle River, NJ: Prentice Hall.
54. Kilman, S. 2006. DuPont, Syngenta set venture to license crop-biotech traits. *Wall Street Journal*, April 11: A10.
55. Tao, J., J. Daniele, E. Hummel, D. Goldheim, and G. Slowinski. 2005. Developing an effective strategy for managing intellectual assets. *Research Technology Management*, 48(1): 50–58.
56. Rivette and Kline, Discovering new value in intellectual property.
57. Allen, *Bringing New Technologies to Market.*
58. Schilling, *Strategic Management of Technological Innovation.*
59. Gourville, Abgenix and the Xenomouse.
60. http://www.amd.com/us-en/Weblets/0,,7832_12670_12686,00.html
61. Kesan, J. 2000. Intellectual property protection and agricultural biotechnology. *American Behavioral Scientist*, 44(3): 464–503.
62. Grindley, P., and D. Teece. 1997. Managing intellectual capital: Licensing and cross-licensing in semiconductors and electronics. *California Management Review*, 39(2): 92–111.
63. Jaffe and Lerner, *Innovation and Its Discontents.*
64. Guth, R., and D. Clark. 2004. Behind secret settlement talks: New power of tech customers. *Wall Street Journal*, April 5: A1, A14.
65. Reitzig, M. 2004. Strategic management of intellectual property. *Sloan Management Review*, 45(3): 35–40.
66. Ziedonis, R. 2004. Don't fence me in: Fragmented markets for technology and the patent acquisition strategies of firms. *Management Science* 50(6): 804–820.
67. Tao et al. Developing an effective strategy for managing intellectual assets.
68. Grindley and Teece, Managing intellectual capital.
69. Quinn, J. 2000. Outsourcing innovation: The new engine of growth. *Sloan Management Review*, 41(4): 13–27.
70. Clark, K. 1989. Project scope and project performance: The effects of parts strategy and supplier involvement on product development. *Management Science*, 35(10): 1247–1263.
71. Schilling, *Strategic Management of Technological Innovation.*
72. Guth, Microsoft's Xbox reflects new focus on hardware.

73. Venkatesan, R. 1992. Strategic sourcing: To make or not to make. *Harvard Business Review*, November–December: 1–11.
74. Quinn, Outsourcing innovation.
75. Schilling, *Strategic Management of Technological Innovation*.
76. Quinn, Outsourcing innovation.
77. Schilling and Steensma, The use of modular organizational forms.
78. Quinn, J. 1999. Strategic outsourcing: Leveraging knowledge capabilities. *Sloan Management Review*, 40(4): 9–20.
79. Quinn, Outsourcing innovation.
80. Clark, Project scope and project performance.
81. Schilling and Steensma, The use of modular organizational forms.
82. Thomke, S. 2002. Innovation at 3M (A). *Harvard Business School Case,* Number 9-699-012.
83. Schilling and Steensma, The use of modular organizational forms.
84. Sakol, M. 1994. Supplier relationships and innovation. In M. Dodgson and R. Rothwell (eds.), *The Handbook for Industrial Innovation*. Aldershot, UK: Edward Elgar, 268–274.
85. Quinn, Strategic outsourcing.
86. Venkatesan, Strategic sourcing.
87. Wayne, L. 2006. Boeing bets the house. *New York Times*, May 7, Section 3: 1, 7.
88. Adapted from Huckman, R. 2005. MedSource technologies. *Harvard Business School Teaching Note*, Number 5-605-014.
89. Ibid.
90. Huckman, MedSource technologies. *Harvard Business School Teaching Case*.
91. Huckman, MedSource technologies. *Harvard Business School Teaching Note*.
92. Ibid.
93. Huckman, MedSource technologies. *Harvard Business School Teaching Case*.
94. Huckman, MedSource technologies. *Harvard Business School Teaching Note*.
95. Vanac, M. 2007. Entrepreneurs' device listens, then prints out your grocery list. *The Plain Dealer*, January 3: C1, C3.
96. Baumol, W. 1999. Licensing proprietary technology is a profit opportunity, not a threat. *Research Technology Management*, 42(6): 10–11.
97. Schilling and Hill, Managing the new product development process.
98. Arrow, K. 1962. Economic welfare and the allocation of resources for inventions. In R. Nelson (ed.), *The Rate and Direction of Inventive Activity*. Princeton, NJ: Princeton University Press.
99. Galambos and Sturchio, Pharmaceutical firms and the transition to biotechnology.
100. Gans, J., D. Hsu, and S. Stern. 2002. When does start-up innovation spur the gale of creative destruction? *RAND Journal of Economics*, 33(4): 571–586.
101. Pisano, The R&D boundaries of the firm: An empirical analysis.
102. Peteraf, M. 1993. The cornerstones of competitive advantage: A resource-based view. *Strategic Management Journal*, 14: 179–191.
103. Azoulay, P., and S. Shane. 2001. Entrepreneurs, contracts and the failure of young firms. *Management Science*, 47(3): 337–358.
104. Chesbrough and Teece, When is virtual virtuous?
105. http://www.accenture.com/xd/xd.asp?it=enweb&xd=ideas%5Coutlook%5C7.2001%5Calliances.xml
106. Boudette, N., N. Shirouzu, and S. Power. 2006. GM-Renault-Nissan wouldn't be easy, past auto pacts show. *Wall Street Journal*, July 3: B1, B2.
107. Kanter, R. 1994. Collaborative advantage: The art of alliances. *Harvard Business Review*, July–August: 96–108.
108. Das and Teng, Managing risks in strategic alliances.
109. Kanter, Collaborative advantage.
110. Das and Teng, Managing risks in strategic alliances.
111. Markides, C., and P. Geroski. 2005. *Fast Second: How Smart Companies Bypass Radical Innovation to Enter and Dominate Markets*. San Francisco: Jossey-Bass.
112. Abboud, L. 2005. How Eli Lilly's monster deal faced extinction—but survived. *Wall Street Journal*, April 27: A1, A9.
113. Hamilton, D. 2005. How Genentech, Novartis stifled a promising drug. *Wall Street Journal*, March 5: A1, A10.
114. Abboud, How Eli Lilly's monster deal faced extinction—but survived.
115. http://www.findarticles.com/p/articles/mi_m0EIN/is_2005_Dec_12/ai_n15929740

Chapter 15

Strategic Human Resource Management of Technical Professionals

Learning Objectives

After reading this chapter, you should be able to:

1. Explain how corporate culture influences innovation.
2. Understand how to manage R&D personnel.
3. Explain how to organize technical communications.
4. Design the right organizational incentives for innovation.
5. Understand the role of external ties and social networks in innovation.
6. Explain how to increase the creativity of a company's employees.

7. Identify the different types of product development teams, understand the advantages and disadvantages of each, and explain when to deploy the different types of teams.
8. Describe virtual teams, and explain how they affect product development.

Strategic Human Resource Management: A Vignette[1]

Bill Gates, the founder of Microsoft, has always believed that human resources are the most important asset that Microsoft has. For this reason, the company has always made hiring, motivating, and keeping talented employees a key part of the company's approach to the creation of competitive advantage.[2]

Microsoft has operated on the philosophy that to write software a person needs to be very intelligent. The company targets elite schools in the United States and elsewhere looking for the most talented graduates.[3] It uses "strike teams" of recruiters to pick up talented employees when other companies have layoffs.[4] And its hiring standards are very high, so few applicants are selected for jobs.[5]

One of the drawbacks of this hiring strategy is that Microsoft programmers are not very good at working together. Moreover, people can advance far in the organization as individual contributors, rather than as team players. This reinforcement of individual achievement has led people in different units to talk to each other very rarely and has created a reluctance to share software code.[6] And, it has created a chronic lack of people who can lead others in the complex task of software development.[7]

The company has an elaborate strategy to motivate employees. It convinces employees that they are on a mission to "change the world." Employees are also given assignments that stretch their abilities very early in their careers.[8]

The company campus is a lot like a college campus, which encourages employees to stay and work long hours.[9] In fact, many employees actually have futon-like couches in their offices that they can sleep on.[10]

Standards are very high, and performance evaluation is very competitive. A forced curve is used so that one-quarter of employees receive below expectation performance appraisals.[11]

Employees are rewarded with equity, with grants of stock (in the early days) and stock options (later on) being very common.[12] When many employees left the company thinking that they could make more money and be closer to the cutting edge at dot-com companies, senior executives were given additional stock options, with the options being reissued to ensure that they were "in the money."[13]

To ensure that employees feel like they are operating in a small company, Microsoft has continuously been restructured into groups of no more than a couple of hundred people, each with authority and responsibility for a particular product or program.[14]

The result of these policies has been high job satisfaction among employees. Satisfaction rates have always exceeded those of other information technology companies, and Microsoft has had lower rates of employee attrition than other companies in its sector.[15]

Introduction

While some portion of success at innovation is the result of luck, effective management also helps. Therefore, as a technology entrepreneur or manager, you need to understand how to manage innovation. Researchers have identified many things that you can do to make your company better at it: establishing an innovative culture, understanding the motivations of your R&D personnel, creating effective communications within your organization and with important external stakeholders, designing the right incentives for employees to create new products, linking basic research and applied development, and designing the right organizational structure.

This chapter examines the management of human resources for effective innovation. The first section discusses the role of corporate culture in enhancing technological innovation. The second section describes effective human resource management techniques for R&D personnel. The third section explains how to have more creative employees. The final section identifies different types of product development teams and explains how you should manage them.

Corporate Culture

As a technology entrepreneur or manager, you are responsible for establishing your **organization's culture** (the set of beliefs, norms, values, and attitudes common to the members of an organization). Because your employees' willingness to innovate, as well as their performance at it, depends on your corporate culture, having the right culture is important to managing innovation successfully (see Figure 15.1). Fortunately, it is not difficult to learn what kind of culture will make your company innovative because researchers have identified several dimensions of organizational culture that enhance innovation.

FIGURE 15.1
Obstacles to Innovation

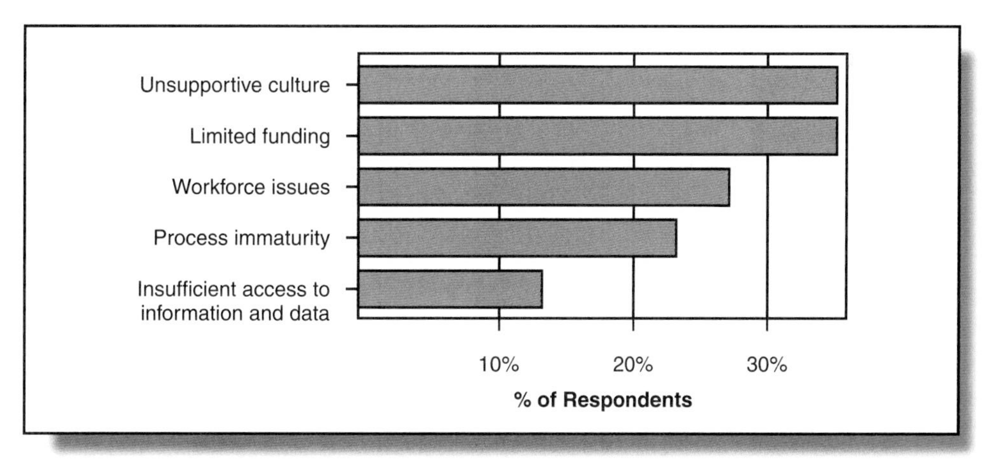

According to research by IBM Consulting Services, managers at many companies believe that having an unsupportive culture is one of the biggest obstacles to innovation.

Source: Data from IBM Business Consulting Services; http://imaginatik-marketing.smugmug.com/photos/68901840-S.gif.

First, welcoming the involvement of all organization members makes organizations more innovative by encouraging initiative and proactive problem solving by everyone in the company from the entry-level employees to the CEO.[16] Therefore, if you want your company to be innovative, you should make all employees aware of the value of innovation and the company's strategy to be innovative. You also should make innovation a shared responsibility of everyone in the company by involving all of them in discussions of innovation, relating it to every job in the company, and making everyone's performance appraisal and compensation dependent upon it.[17]

Second, creating a respect for knowledge and knowledge development makes organizations more innovative. Therefore, you should inculcate employees with the belief that staying ahead of competitors in knowledge development will help your company to succeed. You also should encourage employees to engage in self-examination and to challenge the core assumptions of the organization.[18] Because innovation demands the development of new knowledge and a willingness to challenge the status quo, organizations with a culture that supports such activities are more innovative.

Third, having a diversity of points of view makes companies more innovative. Diversity provides the variance in ideas necessary to come up with creative solutions to problems. Therefore, if you want your company to be innovative, you should create a climate in which your employees are comfortable expressing points of view that are different from those of others, and encourage them to work together despite differences in their ways of thinking. You should also allow people to deviate from organizational norms in both defining problems and in identifying solutions to them.[19]

Fourth, tolerating failure makes organizations more innovative.[20] People will not try to do new things if they know that they will be punished if they are unsuccessful. Moreover, people will try to blame others for failure if it is not tolerated.[21] Therefore, you should create systems to learn from your failures. For example, Eli Lilly developed an audit system to examine failed drug compounds to determine the reasons for their failure, and to make recommendations for improving the drug development process that led to the failures.[22]

You should also avoid punishing people who are "responsible" for failed innovative efforts by giving them poor performance evaluations. Rather, you should reward them for what they did right. For example, Eli Lilly's former chief scientific officer, W. Leigh Thompson, threw "failure parties" to celebrate the excellent work done by employees when that work resulted in failed drug development efforts simply because of bad luck.[23]

Fifth, encouraging constructive criticism makes organizations more innovative. Few innovations are developed exactly as they were originally envisioned, but, rather, are typically developed by testing the validity of ideas with others and improving them in response to criticism. Therefore, you should discourage the blanket acceptance of all new ideas as they are first articulated, as well as unconstructive criticism that offers little more than rejection of new ideas.

Key Points

- Organization culture is the set of beliefs, norms, values, and attitudes common to the members of an organization; research shows that it influences an organization's innovativeness.
- Organizations are more innovative if they involve all members in innovation, have a respect for knowledge and knowledge development, are open to a diversity of points of view, accept failure, and encourage constructive criticism.

Success at innovation depends on how you manage the people involved. This includes the human resource practices that you use with your R&D personnel. It also incorporates the way that you manage the flow of information within the organization and between your organization and its suppliers, customers, and competitors. Finally, it includes compensation schemes and other mechanisms to motivate employees.

Managing R&D Personnel

You need to manage R&D employees differently from employees in other parts of your company. R&D jobs are very different from other jobs. Technology development is uncertain and involves a great deal of change. It also requires the use of the scientific method rather than other approaches to problem solving. The time horizon for the work is long; R&D personnel do not focus on the day-to-day problems that many people in the organization face, and which have immediate and direct impact on company performance.[24]

Moreover, people who choose technical professions, like science and engineering have different personality traits, values, and beliefs than those who choose business-oriented professions, like accounting and finance. As a result, the attitudes and beliefs of R&D employees are different from those of people working in other parts of most companies.[25] Often R&D staff are less interested in the commercial outcomes of solving technical problems than in technical problem solving itself and are motivated by the work itself, rather than by extrinsic factors, like compensation.[26]

The differences in both the environment of R&D laboratories and the characteristics of people who work in technical professions mean that the human resource practices that you should put in place in the R&D units of your company are different from those that you should put in place in other parts of your organization. You should give your R&D personnel interesting and important work, and reward them for their technical accomplishments.[27]

You should also give your R&D personnel sufficient time and financial and human resources to accomplish the tasks that they are asked to do. For example, when Markem Corporation, a maker of printing equipment, needs to design a new piece of machinery, it frees the engineers assigned to the project from their other work responsibilities. As a result, they have the time to focus on designing the new machine, rather than splitting their time between designing a new machine and their day-to-day activities, which leads neither to be done well.[28]

Your R&D personnel will be more productive if you allow them to build strong links to, and exchange ideas with, technical personnel outside the organization[29] by attending scientific conferences, and by presenting and publishing technical papers.[30] In fact, research has shown that scientists at pharmaceutical firms are willing to earn less money if they are allowed to pursue their own scientific agendas and work hours, and are allowed to publish the results of their research.[31] Even outside of pharmaceuticals, high-tech companies, like SAS Institute, send their employees to industry conferences and encourage them to write books and technical papers.[32]

However, the need to manage R&D personnel differently from other employees is not without a cost. By allowing your scientists and engineers to publish their research, or even to exchange ideas with their counterparts at other companies, you will increase the likelihood that your company's technical knowledge will spill over to competitors. Therefore, you need to balance the benefits of greater productivity of R&D personnel against the loss of proprietary knowledge when you establish human resource practices for your R&D laboratory.

Managing Internal Communication

As a technology manager or entrepreneur, you also have to create strong communication within your organization.[33] Open communication among employees—particularly informal communication—enhances innovation by facilitating goal setting and work planning,[34] increasing information exchange, and reducing misunderstandings.[35] Therefore, many high-technology companies spend a lot of time and money on enhancing their internal communication channels. For instance, GlaxoSmithKline recently revamped its system for awarding employee bonuses to give more money to employees who communicated more frequently with their coworkers.[36]

As Figure 15.2 shows, you can enhance your company's internal communications in several ways. First, you can colocate personnel that need to communicate, such as marketing and engineering personnel.[37] Research has shown that communication is enhanced by physical proximity because people who are colocated communicate more often, more informally, and more in person (a richer form of communication than the telephone or e-mail) than people not located together. For example, when Motorola was developing its Razr phone, it took engineers, designers, marketing personnel, and even senior managers and put them together in one large room with four-foot cubicles. The structure of this space enhanced the level of communication among the people on the Razr phone project.[38]

Second, you can rotate people through jobs in different parts of your company.[39] Job rotation exposes employees to the roles that different functional areas

FIGURE 15.2
Ways to Encourage Technical Communication

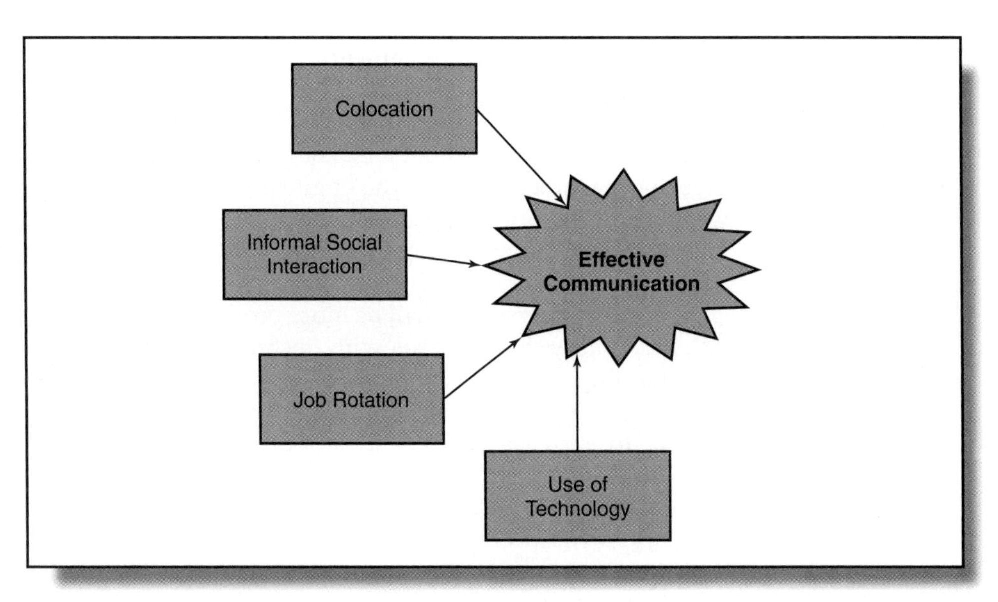

Companies have four main mechanisms for encouraging effective communication within the organization: colocation, informal social interaction, job rotation, and the use of technology.

play in developing new products and services. Moreover, it allows employees to form social ties to people in other parts of the organization, which provides an important source of informal communication among disparate parts of the company.[40]

Third, you can use meetings, expositions, and information technology to enhance communication within the organization. Many companies establish and encourage employees to use company-wide intranets, databases of expertise, and electronic archives about past innovation efforts.[41] Other companies sponsor internal events to encourage the exchange of information. For example, the software company SAS Institute has internal R&D expositions where developers show their new software to nontechnical employees.[42]

Social Networks and External Ties

If you want to enhance innovation in your organization, you also need to encourage strong external communication,[43] which enhances the ability of people to innovate in several ways.[44] At the most basic level, external communication provides access to information about sources of opportunity, which increases the likelihood that a person will identify an innovation. For instance, people with ties to university researchers often identify innovations that make use of university inventions because they are more likely than the average person to know about, and think about, these technologies.

Strong external communication also facilitates innovation because it improves decision making about potential innovations. For example, it is easier to evaluate the idea of developing a particular nanomaterial if you know that someone has made a breakthrough in the chemistry that would make that material possible, than if you do not.

Furthermore, strong external communication often gives the advantage of time because it provides the information necessary to develop the innovation before others compete for a needed resource, particularly when a rare source of supply is required. For example, suppose that there was a technological development that would allow dramatic shrinkage in the size of portable phones to the size that could fit on a tie clip. That technological change would generate the opportunity for you to establish a new type of cellular telephone service. Because providing cellular telephone service also requires access to some part of the radio spectrum, which is licensed by the government, identifying this opportunity requires you to access a part of the radio spectrum before others. If you do not, competition might bid up the price of the radio spectrum, and the opportunity would no longer exist.

To have strong links to external sources of information, you need to encourage your R&D and marketing employees to engage in boundary-spanning activities, such as attending conferences, visiting customers and suppliers, and participating in trade associations.[45] R&D employees have the training and experience to understand information about newly developed technologies that create opportunities for innovation, while marketing personnel have the training and experience to understand information about customers' preferences and unmet needs. Thus, enhancing the exposure of these personnel to external information about the sources of opportunity will help your organization to identify opportunities for innovation.

Social networks are crucial to the effective acquisition of external information. Research has shown that information tends to flow between people who are connected to each other because people often tell each other things that cannot be learned in other ways.[46] In fact, for many types of information, by the time it can be accessed in other ways, it is "old news."

Moreover, social networks enhance the transfer of information because people are more willing to believe information transferred from people who they know than from strangers or from nonhuman sources. When people obtain information from others in their social network, their knowledge of the person who provided the information helps them decide whether or not to believe the information. For instance, suppose that you hear that there is a new technological discovery—say cold fusion. You need to know whether that information is real or a hoax. If it is real, that information will help you to identify the opportunity to develop a technological innovation—a power plant built on cold fusion. However, if it is a hoax, then that information will not help you. By obtaining information about the discovery from someone in your social network, you will have a better sense of whether or not the information is real than if you obtain that information some other way. Thus, having employees with external social ties will help your organization to innovate.

Having employees with a central position in key social networks will help even more. As Figure 15.3 shows, people with a central position in a social network have better access to information than people with a noncentral position.[47] When people within a company are in a position in a social network that links together parts of the company, or links the company to other organizations, those people often serve as brokers, making sure that information is transferred between the different parties, which makes them better at innovation than people who are more distally located.[48]

Developing Incentives

Innovation is costly to people because it is risky, stressful, and prone to failure. Therefore, people need an incentive to innovate or they will not do it. Unfortunately, most companies do not compensate people in a way that provides an incentive to innovate. By paying people straight salaries (and perhaps small bonuses) regardless

FIGURE 15.3
Social Network of Researchers

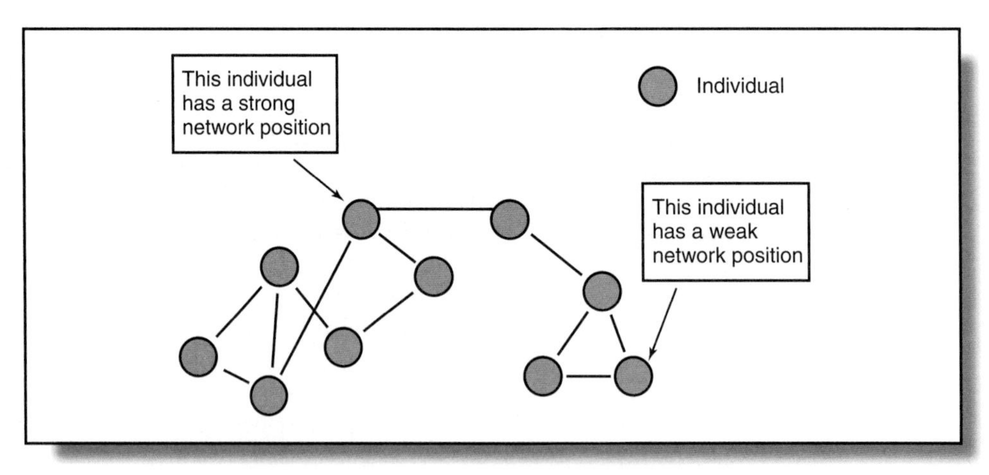

Some researchers have a better social network for gathering information useful for innovation because they are connected to more people than others, either directly or indirectly.

Source: http://en.wikipedia.org/wiki/Image:Social-network.png.

of what ideas that they come up with, most companies remove an important incentive to innovate—money. So it should not be surprising that people in most companies focus their attention on other aspects of their jobs and rarely innovate.[49]

Companies that are successful at innovation, however, reward innovators appropriately. These rewards involve additional variable compensation, such as payments that are related to the organization's performance at the innovation. For example, GlaxoSmithKline awards scientists bonuses if they contribute to the identification of potential new drugs.[50]

While providing bonuses is good, awarding equity is even better. Equity provides employees with the potential for upside rewards that compensate them for bearing the risk of failure and the adverse career consequences that innovation could entail.

Moreover, it helps companies to overcome the moral hazard problem that occurs when the innovator does not bear the cost of failure. When people gain from success but do not bear the cost of failure (as occurs when people get bonuses for successful innovation), they have an incentive to engage in excessively risky activity if that activity increases the upside potential. Why? If the positive outcome occurs, they will earn more money, but if the negative outcome prevails, they will be no worse off. In contrast, when people are compensated from the profits from innovation (get equity), then they have an incentive to take the course of action that has the largest expected value.

Providing equity also attracts the most talented product development and marketing personnel, who often prefer to work in companies where they have the potential for the large gains that equity ownership makes possible.[51] This is one reason why new companies are often more successful than established companies at innovation. The norms of equity (no pun intended!) in large organizations make it difficult to award some employees, and not others, significant equity holdings, limiting the use of equity as an incentive for innovation to only those innovations undertaken in separate, independent, units where these norms do not hold.[52] Moreover, the upside potential of the equity holdings in smaller firms is greater than that in larger firms because the smaller firms have more room for growth. As a result, Microsoft can no longer compete with the stock options that Google can provide to attract talent, leading its most talented software engineers to migrate to Google.[53]

Before you write off large companies as dinosaurs that cannot innovate, you should consider the fact that the use of equity as an incentive to encourage employees to innovate has its own drawbacks. If employees gain financially from successful innovation, they have an incentive to focus their attention on only that activity. This focus could lead them to fail to work on other projects, even when those projects are better for the firm.[54] Moreover, it leads them to persist with projects that do not have promising futures. Because they will gain financially if the innovations succeed, and have little to lose from continuing to pursue the innovations, employees with equity ownership have an incentive to persist with innovative efforts even if they have a low probability of success.

These incentives are so strong that some companies have sought to avoid offering equity to employees responsible for innovation. For instance, DEKA Research, the company responsible for the Segway Human Transporter and the IBOT Mobility System (an advanced wheelchair that can climb stairs), does not give equity or stock options to its engineers to avoid these incentive problems.

Key Points

- Managing R&D personnel is different from managing other employees because the work that they do is different, and because people with certain personality traits, values, and beliefs select technical professions.
- You should give your R&D personnel interesting and important work, reward them for technical accomplishment, and allow them to build strong links to technical personnel outside your organization.
- Innovation is enhanced by strong internal communication, which can be built by colocating employees, rotating people through different jobs, and making effective use of information technology, meetings, and events.
- External ties, particularly those that take the form of social networks, increase access to the information necessary to identify opportunities for innovation.
- Employees need an incentive to innovate, which is best accomplished by compensating them with equity; however, giving employees equity provides perverse incentives, which need to be managed.

ENHANCING CREATIVITY

Creativity, which is the ability to come up with novel solutions to problems, helps people to innovate by identifying patterns in disparate pieces of information about such things as technological change, market needs, and industry structure. It also helps people think of different ways to combine old information, which is useful because many innovations are nothing more than novel recombinations of existing information. For example, many innovations in nanomanufacturing are just recombinations of existing techniques that have long been used in mechanical engineering, chemistry, and semiconductor engineering.[55]

So what can entrepreneurs and managers do to encourage creativity in their companies? They can hire people with the right individual characteristics to be creative and put them in an environment that enhances creativity.[56]

Research has shown that a variety of individual attributes affect creativity, including personality, motivation, and knowledge.[57] Studies have shown that people with greater self-efficacy, more tolerance of risk, greater persistence, more internal locus of control, and greater tolerance of ambiguity are more creative than other people. So to have more creative employees, you can hire people with these personality traits.[58] You can also select people with **intrinsic motivation**—motivation that comes internally rather than from external rewards, like salary or bonuses—because the intrinsically motivated are more creative than the extrinsically motivated.[59] You can hire people who have a deep knowledge of their fields,[60] but who also are intellectually curious,[61] have generalist skills, and are willing to solve problems outside of their area of specialization because people with these characteristics are more creative than others.[62]

If you want creative employees, you should also pay careful attention to your organization's work environment. Having a flexible environment, where tasks are not standardized and are more varied,[63] where people are permitted to experiment with ideas, and where people have autonomy and the right to determine how to do their jobs, encourages creativity.[64] For instance, at 3M, a highly innovative company, employees are allowed to allocate a portion of their time to whatever ideas they see fit to pursue without asking permission of their supervisors. This freedom stimulates the creativity of 3M's employees.[65]

You should also create some slack in the system. Eliminating slack hinders creativity because people need time and money to be creative.[66] Thus, organizations whose budgets include slush funds, which employees can use to develop new products or processes, or to pursue new ideas, display more creativity than organizations without these funds.

You should create work groups composed of people from different functional backgrounds because having a diversity of approaches encourages creativity.[67] Moreover, groups that are too homogenous and cohesive drive out creative people[68] who do not want to act like everyone else or to conform to norms.[69]

Finally, you should teach your employees to reason by analogy. Because most new ideas are based on recombinations of old ideas, teaching people to look for analogies helps them to come up with creative ideas. For example, Georges de Mestral, a Swiss scientist, came up with the idea for Velcro after he saw burrs sticking to his dog when he took him for a walk in the Alps.

If creativity is so useful, then why don't all companies encourage all of their employees to be creative all of the time? The answer is that creativity has a downside, which makes efficiency better than novelty some of the time (see Figure 15.4). Creating novel ideas comes at the expense of spending time and money implementing old ideas. Moreover, high levels of novelty lead to limited compatibility between activities because novel ideas are, by definition, different from each other.

To understand these problems, think of an organization developing a new product where everyone is creative all of the time. All of the employees would be spending all of their time coming up with new product ideas instead of reproducing old ones. And these ideas could not be linked together easily because they would all be novel. As a result, the organization would be very inefficient at exploiting its ideas. In short, there is clearly an optimum level of creativity beyond which the downside of creativity exceeds its benefits.

Key Points

- Creativity, or the ability to create new and useful ideas, enhances innovation.
- Creativity depends on individual attributes as well as the characteristics of environments in which people work.

FIGURE 15.4
Creativity Versus Efficiency

Organizations don't want all of their employees' efforts devoted to the development of creative, cutting-edge ideas because such an approach makes them inefficient at exploiting old ideas.
Source: http://www.andertoons.com.

- To have more creative employees, you can hire people with greater self-efficacy, more tolerance of risk, greater persistence, more internal locus of control, greater tolerance of ambiguity, more intrinsic motivation, and deep knowledge of their fields.
- Creativity is enhanced by encouraging flexibility, building slack into the system, creating work groups composed of people from different functional backgrounds, and teaching employees to reason by analogy.
- Creativity has a downside; sometimes efficiency is better than novelty.

GETTING DOWN TO BUSINESS
Consistent Innovation at 3M[70]

With $17 billion in annual sales, 3M is a major multinational company headquartered in St. Paul, Minnesota. In operation for over 100 years, the company produces and sells over 60,000 different products—Scotch tape, medical devices, firefighting foam, and adhesives that bond panels in airplanes, among other things. The company has developed and maintained a strategy of generating 40 percent of its sales from products launched in the past four years.

The key to 3M's ability to consistently innovate in a wide variety of areas is found in the company's culture and management policies. The company spends over $1 billion per year on R&D, both internally and at universities and small businesses, which allows the company to develop new technologies and to create new products and services from its existing technology platforms.

Equally, if not more importantly, the company seeks to create a culture of innovation by employing a set of reinforcing human resource management policies. The company has a "15 percent rule," which encourages employees to spend up to 15 percent of their time on projects of their own choosing, as long as the projects have the potential to result in new products, new processes, or new ways of doing business. This policy spurs experimentation and innovation in areas not previously identified by senior management.

Another dimension of the company's culture is the "benevolent blind eye," which instructs managers to turn a "blind eye" to displays of independence by their subordinates. This policy encourages employees to act independently and has allowed 3M to develop a number of products that senior managers had not initially believed would make sense, and which likely would have been quashed in less tolerant organizations.

The company also offers genesis grants of up to $50,000 for employees to develop prototypes and market tests of products that they think are worth pursuing. These grants provide a mechanism through which innovative employees can develop new product ideas to the point where they can persuade others in the organization of their value.

Finally, 3M has made a huge investment in enhancing communication among its employees. It uses information technology to capture existing knowledge and transfer it to new employees. It organizes new product forums where employees from multiple divisions discuss their latest products, and technical forums where employees from around the company present papers and exchange new ideas.

PRODUCT DEVELOPMENT TEAMS

In most companies, teams of employees conduct product development. Teams can often develop new products more rapidly and more effectively than individuals because of the additional expertise and effort that they provide. In large corporations, product development teams are usually required because the magnitude of the effort needed to create a new product is so great that a single person does not have enough time to do it.

For example, the development of a new version of the Microsoft operating system software takes more person-hours than any single software developer has ever lived!

Because teams are an important part of the product development process, as a technology entrepreneur or manager, you need to learn how to manage them. This involves understanding a variety of things about these teams, including their size, duration, composition, structure, leadership, and location.

Let's start with team size. Product development teams are most effective if they are of moderate size. If teams are too small, then companies miss out on the benefits of joint effort and diversity of skills that teams provide. However, if teams are too large, then their performance is hindered by communication costs, a lack of identity, and free riding.[71]

At Google, for example, engineers work in three- to five-person teams.[72] But the optimal team size depends on the industry in which the company operates and the type of project that the team is undertaking. The best team size for modifying Internet search engines is probably too small for aircraft manufacturers designing a new passenger jet.

The optimal length of team tenure is neither too long nor too short. If teams do not work together for a long enough time—usually less than one year[73]—then the high cost of initial organizing, and the lack of information exchange among strangers, hinders the product development process.[74] For example, Friendster has had to fix a lot of bugs in its social networking Web site because its strategy of hiring software engineers only when they are needed has led to the creation of project teams with no cohesion or experience working together.[75]

However, if product development teams work together for too long—typically more than five years[76]—they become insular and unwilling to accept outside ideas, and their creativity declines.[77] Therefore, you should reconstitute your product development teams at the start and end of specific projects to break down the segmentation, functional mentality, and lack of holistic understanding that develops in more permanent teams.[78]

Functional Versus Cross-Functional Teams

Functional teams are made up of people from a single discipline, like engineering; **cross-functional teams** are composed of people brought together from different disciplines, like engineering and marketing.[79]

The choice of functional or cross-functional teams is important for you, as a technology entrepreneur or manager. Which one you should select will depend on the balance between the advantages and disadvantages of both types of teams. Functional teams provide managers with clear expertise in the team's domain and well-defined authority and responsibility. This clarity helps team managers to supervise and evaluate team members effectively.[80]

Functional team members can develop more specialized skills in their functional areas than cross-functional team members.[81] (For example, a chemical engineer will develop a deeper understanding of chemical engineering if he or she always works in teams of chemical engineers.[82]) Because specialization keeps your employees current in their fields, and may even make them experts at their jobs,[83] functional teams allow you to exploit the benefits of comparative advantage.[84]

Functional teams are also easier and less expensive to coordinate than cross-functional teams. Because functional team members are similar in outlook and background, they communicate better and work more closely than cross-functional team members.[85] However, cross-functional teams have two advantages over functional teams.[86] First, they are better at integrative activities, like product development, that involve joint problem solving by marketing, product design, and manufacturing personnel.[87] The varied skills, beliefs, thought processes, ideas,[88] and approaches that different functional areas bring to a collaborative effort facilitate this type of joint activity.[89]

Moreover, functional teams are often poor at solving problems that involve other functions because they tend to focus on functional efficiency and preserving their power vis-à-vis other functions, and often lack mechanisms or incentives to coordinate or communicate effectively with them. In particular, functional team members are typically evaluated on their performance in the function, not on how well they interface with other functions.[90]

Furthermore, the lack of a common knowledge base across different functions leads to high levels of miscommunication between functional teams.[91] This makes functional teams less likely than cross-functional teams to identify problems that are likely to occur downstream in manufacturing or marketing before they become costly errors.[92]

Second, cross-functional teams provide a better balance between customer needs and design constraints.[93] Team members from marketing provide information about customer needs and preferences, while team members from engineering provide information about design principles and specifications, and team members from manufacturing provide information about production costs and quality control.[94] When product development teams are composed solely of technical people, they tend to focus too little on customer needs, leading to new products that fail to find a place in the market. In contrast, when product development teams are composed solely of marketing people, they tend to focus too much on customer needs, resulting in missed opportunities to generate breakthrough innovations.[95]

Because functional and cross-functional teams both have advantages and disadvantages for product development, you need to understand when each of these types is better than the other. Research has shown that functional teams are better than cross-functional teams when the rate of technical change is high, the length of the project is long, and the interdependency among product components is low; but, when the opposite is true, cross-functional teams are better.[96] When the rate of change is low, technical specialization is less important to successful projects than other factors, reducing the skill-development benefits of functional teams. When interdependency among the product components is high, cross-functional teams are more important because they facilitate the integration of components. When the length of the project is short, cross-functional teams have an advantage because they can implement all aspects of the project.

Moreover, cross-functional teams are better than functional teams for radical innovations but are worse for incremental innovations. For incremental innovations, which demand less information from fewer sources, cross-functional teams often burden projects with too much information from too many sources, which slows development efforts.[97]

Matrix Structure and Multiple Project Management

As Figure 15.5 humorously points out, most organizations have a structure in which employees report up the hierarchy, with all people ultimately reporting to the CEO.

FIGURE 15.5
Reporting Relationships

Almost all organizations have an organizational chart that shows who reports to whom.
Source: http://www.cartoonstock.com.

The hierarchical reporting structure is useful for many organizational activities, but it forces organizations to choose either a functional or a project team structure.

The advantages that both functional and cross-functional teams provide, lead many organizations to try to achieve the benefits of both by adopting **matrix structures**. As Figure 15.6 shows, a matrix is an organizational arrangement under which a person reports to two bosses, in this case a functional boss and a project team boss. The matrix structure provides the specialization benefits of a functional team with the interdisciplinary benefits of a cross-functional team.

FIGURE 15.6
A Matrix Structure

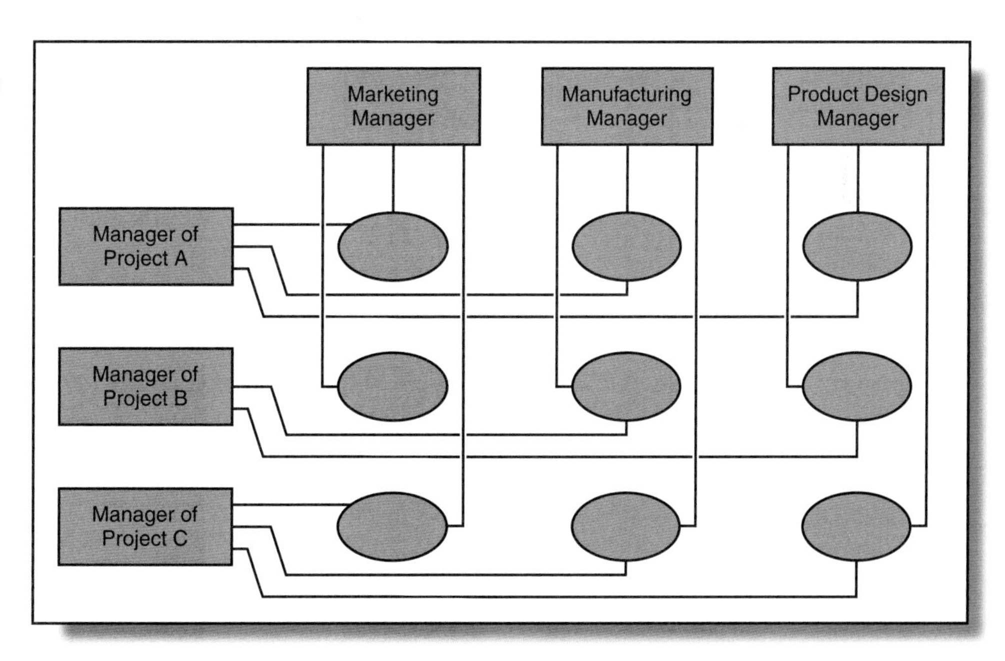

In a matrix organizational structure, employees report to two bosses; in this case the employees (ovals) report to a project manager and a functional manager.

However, using a matrix structure comes at a cost; employees need to manage two bosses.[98] Those of you who have had even one boss undoubtedly recognize how difficult it is to manage two bosses, each with his or her own preferences and agenda. Moreover, the managers in a matrix structure rarely have exactly equal authority because the employees who report to them have to be located somewhere, and the boss with whom they are colocated will tend to have more authority over them.[99]

If you add the limited authority of the managers (no one person has the authority to approve all of the employees' actions or spending) to the difficulty of managing two bosses of unequal authority, you will realize that it is often difficult for organizations with matrix structures to get much work done. Information slowdowns, conflicts between bosses, and lack of accountability plague these arrangements.[100]

Companies also use multiple project management to obtain the benefits of both functional and cross-functional teams. With multiple project management, employees sometimes work in functional areas and sometimes serve on cross-functional teams, giving them the experience of both.

Unfortunately, multiple project management is difficult to implement. It requires frequent coordination of technical personnel, project managers, functional department managers, and the executives that supervise them. Moreover, it also demands organizational and technical processes to coordinate people and capture and transfer knowledge in the organization.[101]

Lightweight, Heavyweight, and Autonomous Teams

If you use cross-functional teams for product development, you will also need to decide what type of team to use: lightweight teams, heavyweight teams, or autonomous teams (see Figure 15.7). **Lightweight teams** are composed of people who reside in their functional areas but have a liaison relationship with a project management committee. These teams are usually led by junior managers, who have limited authority, and who rarely control the resources that the project team needs to do its work.

Heavyweight teams are colocated teams composed of people from each of the functional areas of the organization. These teams are generally led by senior managers, who have the authority to make decisions about the project, and who control the resources that the project team needs to do its work. Because the team members typically report to, and are evaluated by, the team leader, as well as their functional department head, heavyweight team leaders have a great deal of power over team members.[102] Heavyweight team leaders usually are responsible for the integration of all work on the project across the different functional areas of the organization. Therefore, to be effective, they have to understand the different business functions, champion projects, and resolve conflicts.[103]

Autonomous teams are composed of people from different functional areas who are assigned to the team. Team members are colocated, and report only to the project team leader.[104] As with heavyweight team leaders, autonomous team leaders are responsible for the integration of work across functional areas of the organization.

Heavyweight and autonomous teams typically develop a **project charter**—a document that identifies the mission and goals for the project.[105] Project charters outline the time commitment that different employees will make to the project team, the project's budget, and its time line. These teams also typically have a **contract book**, which sets out the plan to achieve the goals set out in the charter, including the team's deliverables, resources, scheduling, and milestones.[106]

So how should you decide whether to set up lightweight, heavyweight, or autonomous cross-functional teams? You should consider the type of project that your

FIGURE 15.7 Autonomous, Heavyweight, Lightweight, and Functional Teams

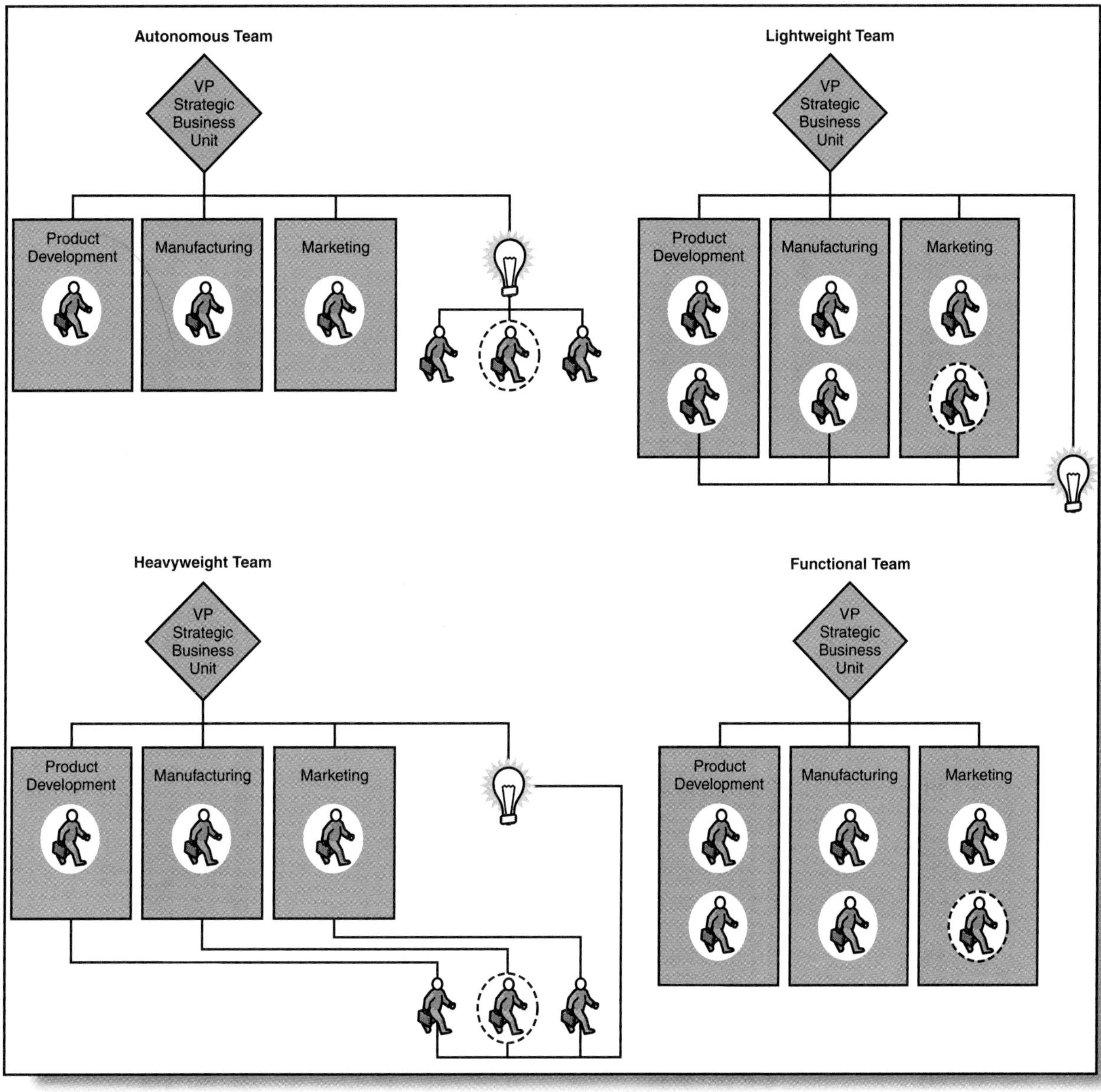

Autonomous Team
VP Strategic Business Unit
Product Development | Manufacturing | Marketing

Lightweight Team
VP Strategic Business Unit
Product Development | Manufacturing | Marketing

Heavyweight Team
VP Strategic Business Unit
Product Development | Manufacturing | Marketing

Functional Team
VP Strategic Business Unit
Product Development | Manufacturing | Marketing

Companies use a variety of different types of product development teams, each of which have advantages and disadvantages.

company is pursuing, the nature of the innovation that it is undertaking, and its capabilities. Heavyweight teams are needed for platform projects (projects that create the base for a new line of products) because these teams are very effective at integrating customer needs with design constraints. Moreover, platform projects often require leaders with a great deal of authority to ensure that adequate resources are provided to the team. However, heavyweight teams are problematic for derivative projects because they cannot be managed as efficiently as lightweight teams, are less subject to senior management control, and do not conform well to a functional organizational structure.[107]

Autonomous teams are typically needed for the development of breakthrough projects (projects to develop radical innovations) because these projects require focused attention.[108] However, autonomous teams are problematic for derivative projects because they operate outside of the organization's routines, hindering reintegration of the team and the products into the rest of the organization.[109]

Lightweight teams are very good at projects that exploit existing organizational capabilities, whereas heavyweight and autonomous teams are needed to create new capabilities.[110] Therefore, if your company does not currently have capabilities in the area in which you are conducting your product development project, then you will probably need to use heavyweight or autonomous cross-functional teams to develop the product.

Heavyweight and autonomous teams are more effective than lightweight teams when the project team is developing a radical innovation. Why? In addition to the point made earlier that autonomous teams can focus their attention on the innovation, heavyweight and autonomous teams have better communication, greater allegiance of team members to the project, and more varied knowledge to bring to bear on problems, than lightweight teams.[111] All of these things are very important when companies are developing innovations that are a break from the type of products that they have developed before.

In contrast, when innovation is incremental, lightweight teams are a better choice.[112] Lightweight teams are very effective for derivative products (products that are extensions of existing products) because these products build on ones that companies have developed before. Therefore, they do not require a great deal of championing or political power to implement. Moreover, they generate less conflict between the team leaders and senior management, which matters when the superior communication, team identity, and cross-functional problem-solving skills of autonomous and heavyweight teams matter less to the innovation process.[113]

Virtual Teams

With the rise of information technology, another type of team has emerged—the **virtual team**, or a team that is not colocated, but collaborates through electronic means, using Intranets and e-mail. Virtual teams reduce the cost of product development significantly, in some cases by as much as half the cost of in-person product development efforts. They also speed product development because, with team members in different geographic locations, virtual teams can work 24 hours a day on product development. Furthermore, virtual teams can incorporate more diverse knowledge of market needs and skills because they can be drawn from a wider range of locations than in-person teams.

However, virtual teams face important challenges. Team members have to rely on less rich methods of communication than face-to-face interaction, which means that subtle bits of information often fail to get transferred between them. Moreover, virtual team members engage in little informal interaction, such as having drinks or lunch together, which hinders trust building and increases conflict. As a result, virtual teams have more difficulty than face-to-face teams in exchanging tacit knowledge.[114]

So how should you manage a virtual team? You should bring team members together at an initial meeting to allow the team to gel, and then bring the team together periodically to make sure the cohesion holds. You should also identify clear goals for the team and make sure that all the team members identify with the team and its goals. That way, team members realize what they are trying to accomplish even if their coworkers are not there to remind them.

Key Points

- Product development teams are most effective when they are of moderate size and intermediate tenure.
- Functional teams have clearer supervision, greater specialization, and lower coordination costs than cross-functional teams; however, cross-functional teams have better integrative problem-solving capabilities, and can better link market needs to technical requirements.
- Because functional and cross-functional teams have advantages and disadvantages, companies sometimes use matrix structures or multiple product management to obtain the benefits of both types of teams; however, both of these arrangements entail high organizing costs.
- Cross-functional teams can take the form of lightweight teams, heavyweight teams, or autonomous teams; the three types of teams differ as to the authority of the team lead and the reporting relationships of the team members.
- The most appropriate type of cross-functional team depends on the type of innovation and the organization's capabilities.
- Virtual teams are teams that are not colocated, but collaborate through electronic means.

Discussion Questions

1. Some people argue that entrepreneurs are more innovative than the employees of large firms. Is this true? If so, why? If not, why might people in large organizations be more innovative and creative than those in start-ups?
2. What are the main idiosyncratic characteristics associated with scientists and engineers? What are the implications of these characteristics for management of the innovation process?
3. How do internal communication patterns affect innovation? Is there anything a company can do to make its communication patterns more conducive to innovation? Why or why not?
4. What incentives encourage innovation? Are there drawbacks to those incentives? Are new firms better or worse than established firms at providing these incentives? Why or why not?
5. Is it possible to identify potential employees who are more creative than others? If not, why not? If so, what attributes should you look for?
6. Is it possible to make the employees of your company more creative? If not, why not? If so, how can you do it?
7. What are the pros and cons of having cross-functional teams conduct product development?
8. Should you use autonomous, heavyweight, or lightweight product development teams? What factors should influence your choice?

Key Terms

Autonomous Teams: Teams composed of people from different functional areas who are assigned to the team, and who report only to the project team lead.

Contract Book: A document that defines the plan to achieve the goals set out in a project charter, including the team's deliverables, resources, scheduling, and milestones.

Creativity: The ability to come up with novel solutions to problems.

Cross-Functional Teams: Teams composed of people from more than one business discipline.

Functional Teams: Teams composed of people from a single business discipline.

Heavyweight Teams: Teams composed of a person from each functional area of the organization, led by a senior manager who has the authority to make all decisions about a project.

Intrinsic Motivation: Motivation that comes internally rather than from external rewards, like salary or bonuses.

Lightweight Teams: Teams that reside in functional areas, but have liaisons to a project management committee.

Matrix Structure: An organizational arrangement under which a person reports to two bosses.

Organization Culture: The set of beliefs, norms, values, and attitudes common to the members of an organization.

Project Charter: A document that identifies the mission and goals for a project.

Virtual Teams: Teams that are not colocated but collaborate through electronic means.

Putting Ideas into Practice

1. **An Exercise in Creativity** The purpose of this exercise is to use the SCAMPER technique to think of changes that you can make to an existing product. SCAMPER stands for substituting (components or materials), combining (mixing and integrating things), adapting (altering the function or use of something), modifying (changing scale, shape, or attributes), putting to another use, eliminating (removing elements and simplifying), and reversing (turning something inside out or upside down).[115]

   Identify an existing product. Then think of aspects of the product that you can substitute, combine, adapt, modify, put to another use, eliminate, or reverse to create a new product. Finally, make a list of all of the new product ideas that you thought of.
2. **Innovation at Google** Google approaches product development in the following way. The company accepts ideas from everyone in the organization, through a variety of sources, such as meetings, focus group sessions, and, of course, the Web. The company also has brainstorming sessions every week to come up with new product ideas.

   Once it has collected these ideas, the company then seeks to prioritize them on the basis of usefulness to users, difficulty of implementation, and effect on revenues.

   It then creates small teams of engineers to work on the ideas that are identified as a priority. If a project is large, it is subdivided so that all the work is done in teams of three. The team is colocated and works together for about three months. One person, the lead, is made responsible for achieving technical excellence.

   The product development efforts require minimal documentation. Every Monday, each employee sends a "snippet"—an e-mail message about what they are working on—to others in their work group.[116] The rest of the time, the team is free to work on the project without documenting its actions.

   Evaluate this approach to innovation. What do you think is good about it? What do you think are the drawbacks? Explain why you make this evaluation.
3. **Designing a Project Team** Suppose you were responsible for designing the project team to develop a new detergent with bleach for Procter & Gamble. Specify the type of project team that you would create to develop this product. State whether you would set up a functional or cross-functional team and explain the logic behind your recommendation. If you recommend the creation of a cross-functional team, state whether you believe the team should be a lightweight, heavyweight, or autonomous team. Explain the logic behind this recommendation as well.

Notes

1. Adapted from Bartlett, C. 2001. Microsoft: Competing on talent (A). *Harvard Business School Case*, Number 9-300-001.
2. Bartlett, C. 2001. Microsoft: Competing on talent (A) and (B). *Harvard Business School Teaching Note*, Number 5-302-010.
3. Bartlett, Microsoft: Competing on talent (A).
4. Bartlett, Microsoft: Competing on talent (A) and (B).
5. Bartlett, Microsoft: Competing on talent (A).
6. Ibid.
7. Ibid.
8. Ibid.
9. Ibid.
10. Bartlett, Microsoft: Competing on talent (A) and (B).
11. Bartlett, Microsoft: Competing on talent (A).
12. Ibid.
13. Bartlett, C., and M. Glinska. 2001. Microsoft: Competing on talent (B). *Harvard Business School Case*, Number 9-301-135.
14. Bartlett, Microsoft: Competing on talent (A).
15. Ibid.
16. Martins, E., and F. Terblanche. 2003. Building organizational culture that stimulates creativity and

innovation. *European Journal of Innovation Management*, 6(1): 64–75.
17. Dougherty, D., and C. Hardy. 1996. Sustained product innovation in large, mature organizations: Overcoming innovation-to-organization problems. *Academy of Management Journal*, 39(5): 1120–1153.
18. Narayanan, V. 2001. *Managing Technology and Innovation for Competitive Advantage*. Upper Saddle River, NJ: Prentice Hall.
19. Martins and Terblanche, Building organizational culture that stimulates creativity and innovation.
20. Day, G., and P. Schoemaker. 2000. Avoiding the pitfalls of emerging technologies. In G. Day and P. Schoemaker (eds.), *Wharton on Managing Emerging Technologies*. New York: John Wiley.
21. Sathe, V. 2003. *Corporate Entrepreneurship*. Cambridge, UK: Cambridge University Press.
22. Burton, T. 2004. By learning from failures, Lilly keeps drug pipeline full. *Wall Street Journal*, April 12: A1, A12.
23. Ibid.
24. Clarke, T. 2002. Why do we still not apply what we know about managing R&D personnel? *Research Technology Management*, 45(2): 9–11.
25. Ibid.
26. James, W. 2002. Best HR practices for today's innovation management. *Research Technology Management*, 45(1): 57–60.
27. Ibid.
28. Lublin, J. 2006. Nurturing innovation. *Wall Street Journal*, March 20: B1, B3.
29. James, Best HR practices for today's innovation management.
30. Cardinal, L. 2001. Technological innovation in the pharmaceutical industry: The use of organizational control in managing research and development. *Organization Science*, 12(1): 19–36.
31. Stern, S. 2004. Do scientists pay to be scientists? *Management Science*, 50: 835–853.
32. Florida, R., and J. Goodnight. 2005. Managing for creativity. *Harvard Business Review*, July–August: 1–7.
33. Rothwell, R. 1994. Industrial innovation: Success, strategies, trends. In M. Dodgson and R. Rothwell (eds.), *The Handbook of Industrial Innovation*. Aldershot, UK: Edward Elgar, 33–53.
34. Ancona, D., and D. Caldwell. 1992. Demography and design: Predictors of new product team performance. *Organization Science*, 3: 321–341.
35. Dougherty, D. 1990. Understanding new markets for new products. *Strategic Management Journal*, 11: 59–78.
36. Whalen, J. 2006. Bureaucracy buster? Glaxo lets scientists choose its new drugs. *Wall Street Journal*, March 27: B1, B3.
37. Afuah, A. 2003. *Innovation Management*. New York: Oxford University Press.
38. Poppendieck, M. 2005. Tactics and benefits of concurrent software development. *Concurrency*, 14(2): 1–119.
39. Maidique, M., and R. Hayes. 1984. The art of high technology management. *Sloan Management Review*, 25: 18–31.
40. Takeuchi, H., and I. Nonaka. 1986. The new product development game. *Harvard Business Review*, January–February: 2–11.
41. Ettlie, J. 2000. *Managing Technological Innovation*. New York: John Wiley.
42. Florida and Goodnight, Managing for creativity.
43. Rothwell, Industrial innovation: Success, strategies, trends.
44. Kirzner, I. 1997. Entrepreneurial discovery and the competitive market process: An Austrian approach. *Journal of Economic Literature* 35: 60–85.
45. Ancona, D., and D. Caldwell. 1992. Bridging the boundary: External process and performance in organizational teams. *Administrative Science Quarterly*, 37: 634–665.
46. Rizova, P. 2006. Are you networked for successful innovation? *Sloan Management Review*, 47(3): 49–55.
47. Ahuja, G. 2000. Collaboration networks, structural holes, and innovation: A longitudinal study. *Administrative Science Quarterly*, 45(3): 425–455.
48. Powell, W., K. Koput, and L. Smith-Doerr. 1996. Interorganizational collaboration and the locus of innovation: Networks of learning in biotechnology. *Administrative Science Quarterly*, 41: 116–145.
49. Holmstrom, B. 1989. Agency costs and innovation. *Journal of Economic Behavior and Organization*, 12: 305–327.
50. Whalen, Bureaucracy buster?
51. Graves, S., and N. Langowitz. 1993. Innovative productivity and the returns to scale in the pharmaceutical industry. *Strategic Management Journal*, 14(8): 593–605.
52. Sathe, *Corporate Entrepreneurship*.
53. Cyran, R., H. Dixon, and E. Chancellor. 2005. Should Microsoft break up on its own? *Wall Street Journal*, November 26–27: B16.
54. Hauser, J. 1998. Research, development, and engineering metrics. *Management Science*, 44(12): 1670–1685.
55. Fleming, L., and O. Sorenson. 2003. Navigating the technology landscape of innovation. *Sloan Management Review*, Winter: 15–23.
56. Woodman, R., J. Sawyer, and R. Griffin. 1993. Toward a theory of organizational creativity. *Academy of Management Review*, 18(2): 293–321.
57. Ibid.
58. Allen, K. 2003. *Bringing New Technology to Market*. Upper Saddle River, NJ: Prentice Hall.

59. Amabile, T. 1998. How to kill creativity. *Harvard Business Review*, September–October: 77–87.
60. Ibid.
61. Amabile, T. 1988. A model of creativity in organizations. In B. Staw and L. Cummings (eds.), *Research in Organizational Behavior* (10th edition). Greenwich, CT: JAI Press, 123–167.
62. Root-Bernstein, R. 1989. Who discovers and invents. *Research Technology Management*, 32(1): 43–52.
63. Allen, *Bringing New Technology to Market*.
64. Shapero, A. 1997. Managing creative professionals. In R. Katz (ed.) *The Human Side of Managing Technological Innovation*. New York: Oxford University Press, 39–46.
65. Damanpour, F. 1991. Organizational innovation: A meta analysis of effects of determinants and moderators. *Academy of Management Journal*, 34: 555–590.
66. Amabile, How to kill creativity.
67. Woodman, Sawyer, and Griffin, Toward a theory of organizational creativity.
68. Pawlak, A. 2000. Fostering creativity in the new millennium. *Research Technology Management*, November–December: 32–35.
69. Los, M. 2000. Creativity and technology innovation in the United States. *Research Technology Management*, November–December: 25–26.
70. Adapted from A Century of Innovation: The 3M Story http://multimedia.mmm.com/mws/mediawebserver.dyn?6666660Zjcf6lVs6EVs666IMhCOrrrQ-.
71. Schilling, M. 2005. *Strategic Management of Technological Innovation*. New York: McGraw-Hill.
72. Eisenmann, T., and K. Herman. 2006. Google Inc., *Harvard Business School Case*, Number 9-806-105.
73. Katz, R. 1997. Managing creative performance in R&D teams. In R. Katz (ed.), *The Human Side of Managing Technological Innovation,* New York: Oxford University Press: 177–186.
74. Iansiti, M. 1993. Real world R&D: Jumping the product generation gap. *Harvard Business Review*, May–June: 138–147.
75. Piskorski, M., and C. Knoop. 2006. Friendster (A). *Harvard Business School Case*, Number 9-707-409.
76. Katz, Managing creative performance in R&D teams.
77. Katz, R. 1982. The effects of group longevity on project communication and performance. *Administrative Science Quarterly*, 27: 81–104.
78. Dougherty, D. Forthcoming. Managing the "unmanageables" for sustained product innovation. In S. Shane (ed.), *Blackwell Handbook on Technology and Innovation Management*. Oxford: Blackwell.
79. Cusumano, M., and K. Nabeoka. 1998. *Thinking Beyond Lean*. New York: Free Press.
80. Clark, K., and S. Wheelwright. 1992. Organizing and leading "heavyweight" development teams. *California Management Review*, 34(3): 9–28.
81. Wheelwright, S., and K. Clark. 1992. *Revolutionizing Product Development*. New York: Free Press.
82. Iansiti, Real world R&D.
83. Afuah, *Innovation Management*.
84. Clark and Wheelwright, Organizing and leading "heavyweight" development teams.
85. Schilling, *Strategic Management of Technological Innovation*.
86. Clark, K., W. Chew, and T. Fujimoto. 1987. Product development in the world auto industry. *Brookings Papers on Economic Activity*, 3: 729–771.
87. Dougherty, D. 2001. Reimagining the differentiation and integration of work for sustained product innovation. *Organization Science*, 12(5): 612–631.
88. Sathe, *Corporate Entrepreneurship*.
89. Takeuchi and Nonaka, The new product development game.
90. Fojt, M. 1995. Focusing on customers. *Journal of Services Marketing*, 9(3): 29–31.
91. Schilling, M., and C. Hill. 1998. Managing the new product development process: Strategic imperatives. *Academy of Management Executive*, August: 67–81.
92. Brown, S., and K. Eisenhardt. 1995. Product development: Past research, present findings, and future directions. *Academy of Management Review*, 20(2): 343–378.
93. Ibid.
94. Schilling and Hill, Managing the new product development process.
95. Ettlie, *Managing Technological Innovation*.
96. Cusumano and Nabeoka, *Thinking Beyond Lean*.
97. Cardinal, Technological innovation in the pharmaceutical industry.
98. Afuah, *Innovation Management*.
99. Cusumano and Nabeoka, *Thinking Beyond Lean*.
100. Rizova, Are you networked for successful innovation?
101. Cusumano and Nabeoka, *Thinking Beyond Lean*.
102. Wheelwright and Clark, *Revolutionizing Product Development*.
103. Schilling, *Strategic Management of Technological Innovation*.
104. Ibid.
105. Clark and Wheelwright, Organizing and leading "heavyweight" development teams.
106. Schilling, *Strategic Management of Technological Innovation*.
107. Clark and Wheelwright, Organizing and leading "heavyweight" development teams.
108. Wheelwright and Clark, *Revolutionizing Product Development*.
109. Clark and Wheelwright, Organizing and leading "heavyweight" development teams.
110. Christiansen, C. 1999. *Innovation and the General Manager*. New York: McGraw-Hill.

111. Clark, Chew, and Fujimoto, Product development in the world auto industry.
112. Christiansen, C. 1997. *The Innovator's Dilemma.* Boston: Harvard Business School Press.
113. Wheelwright and Clark, *Revolutionizing Product Development.*
114. Schilling, *Strategic Management of Technological Innovation.*
115. http://www.mycoted.com/SCAMPER
116. Evelyn Rodriguez's notes from a presentation made to the Silicon Valley Product Management Association by Google product manager Marissa Mayer on January 8, 2003, http://evelynrodriguez.typepad.com/crossroads_dispatches/files/GoogleProductDevProcess.pdf.

Chapter 16

Organization Structure for Technology Strategy

Learning Objectives

After reading this chapter, you should be able to:

1. Explain the pros and cons of centralization for innovation.
2. Understand when organic and mechanistic organization structures are best for innovation.
3. Explain the role of organizational slack in the innovation process.
4. Understand the advantages and disadvantages that small and large firms have at innovation.
5. Identify the industry conditions that make small firms more innovative than large firms (and vice versa), and explain why they have this effect.
6. Understand why spin-off companies are founded, and how to incorporate spin-offs into an effective technology strategy.

7. Explain how new high-growth technology companies finance their operations.
8. Define *corporate venturing,* and explain the advantages and disadvantages of undertaking it.
9. Identify the different types of corporate venturing activities that firms undertake, and explain why firms might take one approach rather than another.
10. Explain the pros and cons of using acquisitions as a way to grow technology companies.

Corporate Venture Capital: A Vignette[1]

In 1991, the senior management at Intel decided that the company's strategic interests would be enhanced by supporting start-up companies that developed products and services that were complementary to its semiconductor operations. As a result, Intel unveiled a new strategy. It would make equity investments in start-up companies that were developing new products based on computing and communications platforms, through a venture capital fund, Intel Capital, which it set up as a subsidiary.

Since its inception in 1991, Intel Capital has made more than $4 billion worth of investments in more than 1,000 companies in over 30 countries. Among its investments have been companies that use Intel's Itanium chip to provide training and consulting, security software, and supply chain management;[2] companies that provide products built on the WiMAX standard for wireless data transmission (which, as Figure 16.1 shows, increase demand for laptop computers with Intel chips);[3] and companies that are advancing the development of computer memory.[4]

Intel has been quite successful with its venture capital fund. Approximately 160 of the companies in which it has invested have been acquired, and 150 have gone public. Moreover, many of these investments have been made as a lead, colead, or sole investor, leaving no question in the minds of observers about Intel's prowess in venture capital.

Currently, Intel Capital has six funds (with a capitalization of $1.25 billion) that make equity investments in start-ups developing computer hardware, software, networking, communications, and services around the world. Because Intel's venture capital fund is designed to support Intel's strategic interests, all of these funds invest only in companies that develop products that are complementary to both Intel's products and its technology architectures.

Intel Capital offers many benefits to its portfolio companies, some that independent venture capital firms cannot offer. It gives portfolio companies access to its laboratories and senior managers for assistance on matters ranging from the technical to the strategic. The company uses its connections with other financiers to help portfolio companies obtain follow-on financing, and its connections with customers and suppliers to help portfolio companies develop and sell their new products. The company sponsors events, such as its CEO Summit and Capital Technology Days, which help the founders of portfolio companies to network with customers and suppliers, as well as to learn from other entrepreneurs and industry leaders.

FIGURE 16.1 Intel and WiMAX

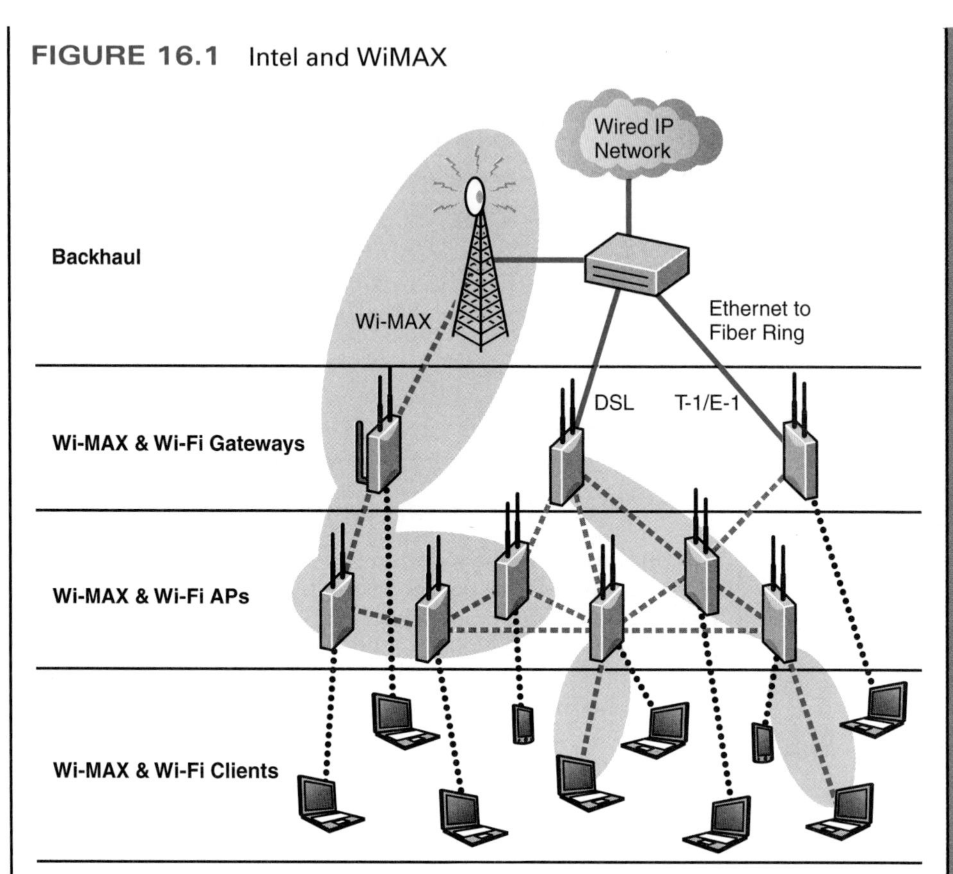

As part of its broadband portfolio, Intel Capital invests in companies that support the IEEE 802.16 standard for broadband wireless data transmission, which is the technology basis for WiMAX. Increasing the amount of broadband data available to PC users is a strong driver of demand for more powerful laptop and desktop computers, thus increasing the demand for more powerful Intel microprocessors.

Source: http://www.thealarmclock.com/euro/images/wimaxvis.png.

INTRODUCTION

To innovate successfully, companies need to be designed right. They need to be the right size and strike the right balance between centralization and decentralization, organic and mechanistic structures, slack and efficiency, and developing innovations in-house and creating new companies. Get this balance wrong and a company will fail to innovate or, worse yet, create new competitors that exploit the very technologies that it failed to take advantage of.

This chapter discusses the relationship between organizational structure and innovation. The first section explains how to design organizations to make them more innovative, focusing on centralization versus decentralization, organic versus mechanistic structures, and slack versus efficiency. The second section explains why the innovation process is different in large and small firms, and discusses the advantages and disadvantages that each type of firm has at this activity under different conditions. The third section explains why people start companies to exploit innovations that they identified during their prior employment, focusing on the effects of incentives and technology to explain the formation of spin-offs. The fourth section describes the sources of equity capital for high-growth new companies. The fifth section discusses corporate venturing, describing the advantages and disadvantages of this activity, and comparing different ways in which companies engage in it. The final section evaluates acquisition as a strategy for growing technology companies.

Basic Organization Structure

The structure of your organization affects its ability to innovate. Researchers have identified three dimensions that are particularly important in this regard: centralization, organicity, and slack. As a technology manager or entrepreneur, you need to understand how these three dimensions of organization structure affect innovation so that you can design an appropriate structure for your company.

Centralization Versus Decentralization

If you are interested in fostering innovation, you might ask: Should I develop a centralized or decentralized organizational structure? (**Centralization** is the extent to which decision making in an organization is controlled at the top.) The answer to this question is less clear than you might think. Whether centralization or decentralization is better for innovation is a matter of debate among researchers, especially since the growth of computing technology has changed the nature of communication in organizations.[5] (See Table 16.1 for a summary of the two points of view.)

Benefits of Decentralization

Some observers have argued that decentralized organizations are more innovative than centralized ones, and have offered several reasons why. First, decentralization increases the amount of information that flows to the company from external stakeholders because it provides more points of contact between the organization and its external environment. Access to information from outside the organization is important to innovation because it provides the company with feedback from external stakeholders, which is useful for developing products valued by the marketplace.

Second, decentralization increases the number of people in the organization who can engage in innovative activity. Increasing the number of potential innovators is helpful because people are limited in their cognitive capacities. Their ability to think creatively is affected by the information that they have, and by their past experiences. Because no two people's backgrounds or information sets are the same, different people

TABLE 16.1
The Advantages and Disadvantages of Decentralization
You need to balance the benefits and costs of decentralizing innovation when deciding what organization structure to adopt.

Advantages	Disadvantages
Improves access to information	Makes coordination more difficult
Enhances idea generation	Leads to avoidance of radical innovation
Increases market orientation	Causes conflict with ongoing operations
Reduces financial risk to the overall organization	Hinders strategic management of innovation
Makes it easier to get government aid for innovation	Results in underinvestment in large-scale innovation
Facilitates innovation in noncore areas	

will come up with different ideas. This means that the headquarters staff of a large corporation cannot think of all new ideas that might benefit a company. By involving more people in the innovation process, decentralization increases the diversity of information and experience of the innovators, thus increasing the number of new ideas suggested.

Moreover, decentralization facilitates the evaluation of new ideas. The cognitive capacity of headquarters personnel limits an organization's ability to evaluate new ideas. If too many ideas get sent up to headquarters, they stack up waiting for the headquarters staff to evaluate them. Allowing evaluation to occur directly in the units that identify the ideas, rather than requiring the ideas to be evaluated at headquarters, removes this obstacle. For example, as Microsoft grew large, its centralized structure made it very slow to take advantage of innovative opportunities. Therefore, when Steve Ballmer became CEO, he divided the company into seven different divisions and gave each division head profit and loss responsibility, with the goal of speeding up decision making.[6]

Third, decentralization makes R&D activities more market oriented, which increases the likelihood that innovations will be valued by customers.[7] In centralized organizations, R&D personnel tend to receive their assignments from headquarters staff and are often isolated from the marketplace, with little or no contact with customers or their needs. Consequently, they tend to focus on developing new technology or advancing science, rather than on market-oriented customer problem solving.[8]

At the same time, employees in operating units, who are close to customers and have information about their needs and preferences, often find it difficult to obtain assistance from R&D laboratories. To obtain assistance, the operating units need to make their requests through headquarters, which leads to garbled messages to R&D personnel, delay in evaluation, and a rejection of many requests for assistance, all of which hinder innovation.

Fourth, decentralization makes the organization better able to bear the financial risk of undertaking innovative activity. When innovation is decentralized, each part of the company engages in its own efforts. As a result, the financial downside of a failed effort by one unit does not affect the performance of another unit, thus providing the benefits of diversification.

Fifth, decentralization makes it easier to obtain government aid for R&D. Many nations (and state governments within the United States) will financially support innovation only if that innovation occurs within its geographic boundaries. Thus, by decentralizing innovation efforts, organizations can increase the amount of financial support that they can obtain from government entities, and reduce the portion of the cost that they have to bear themselves.

Sixth, decentralization facilitates innovation in areas that are different from the core operations of the company. When innovation occurs in new areas—for instance, a chemical company innovating in pharmaceuticals—autonomy from the core activities of the company is useful because the routines and procedures that the company undertakes in its core activities may not be appropriate for the new business. Decentralization helps companies to develop new routines and operating procedures because it permits the physical separation between new and core operations that makes it easier for the new operation to be run differently.

Benefits of Centralization

Researchers have noted several benefits to innovation that centralization provides. First, centralization reduces the cost of coordinating innovation efforts across different parts of the organization.[9] Communication and resource sharing between parts of decentralized organizations is often limited by the independence of different units. Centralization increases the likelihood that different parts of the organization will communicate and share talented personnel because headquarters can compel different units to work together.[10] In addition, managers in centralized organizations are more likely to work on innovations that demand competencies not directly under their control because managers in different units know that they can rely on headquarters to compel other units to work with them.[11] Furthermore, centralization facilitates the acquisition of specialized equipment and facilities, which may not make sense to operate below a certain size, because the coordination by headquarters ensures that multiple units will make use of those resources.

Second, centralization improves performance at innovation by giving technical personnel freedom from operational activities, which often conflict with the activities necessary for innovation. When employees undertaking innovation also have operational responsibilities, they have to limit the time and attention that they devote to innovation. Moreover, operational responsibilities usually get first priority on their time because these activities are more time sensitive.[12] Because performance at innovation is at least partially a function of the time and attention people devote to it, limiting these resources hinders performance at innovation.

Third, centralization encourages companies to invest in radical innovation. Managers in decentralized organizations have an incentive to focus on making incremental improvements to existing product lines at the expense of more radical innovation because incremental innovation tends to generate unit-specific benefits, whereas radical innovation tends to produce firm-wide gains. Moreover, the managers of decentralized units are rarely evaluated on, or rewarded for, undertaking firm-wide innovation.

Fourth, centralization encourages companies to undertake large-scale innovation. Decentralized units are often too small for managers to have the budgetary authority to undertake large-scale innovation or the diversification to bear its risk.[13] Moreover, centralization creates the incentive to invest in innovations that benefit more than one unit in the organization, and the scale necessary to amortize investments of greater magnitude.[14]

Fifth, centralization facilitates the development of a corporate strategy toward innovation.[15] Centralization allows you to use command and control to exploit synergies between different parts of your organization. This is important because companies that leverage these synergies perform better at innovation than ones that do not. Centralization also allows companies to take advantage of existing technological and market capabilities when they innovate, which, as Chapter 11 explained, also enhances performance.

3M's recent centralization of R&D activity illustrates the value of centralization to the development of a corporate strategy for innovation. By changing its structure so that the managers of all units undertaking R&D report to the same senior manager, 3M now has someone who can look for strategic synergies across all R&D efforts, thus increasing the likelihood that such synergies will be found and exploited.[16]

Mechanistic Versus Organic Organization Structures

Organization structures can be organic or mechanistic. Companies with **organic** structures are loosely controlled, with nonhierarchical communication patterns and unspecialized jobs. They have few rules and informal and unstandardized procedures. Mobilization of resources occurs largely through employee commitment and enthusiasm.

GETTING DOWN TO BUSINESS
Decentralizing R&D at GlaxoSmithKline[17]

When Tadataka Yamada became CEO of the pharmaceutical firm GlaxoSmithKline, he implemented a major reorganization of R&D activities at that company. He changed the process from a centralized one in which a committee of R&D directors decided what drug prospects to fund, to a decentralized one in which the company's scientists were divided into six centers of excellence, each led by a senior vice president who had responsibility for the group's budget and project selection.[18]

The restructuring also changed the role of the company's scientists in product development. Under the old process, discoveries were sent to a preclinical unit once a compound had been identified. As a result, the scientists had essentially no contact with product development. However, under the new system, the scientists remained involved with the development of the drugs through the end of Phase IIa clinical trials, requiring them to pay more attention to product development issues throughout the drug discovery process.[19]

To support this new structure, GlaxoSmithKline changed how it rewarded its scientists. Under the new structure, the scientists were given stock options if their research efforts were instrumental in discovering a high-quality drug target. This was a significant change from the old approach in which the scientists received no financial reward for discoveries that lead to major drugs.[20]

In contrast, companies with **mechanistic** structures are tightly controlled, with hierarchical communication patterns and specialized jobs. They have many written rules and standardized procedures. Mobilization of resources occurs largely by authority.

Unlike the discussion of centralization and decentralization in the previous section, the choice of whether to use organic or mechanistic structures to encourage innovation is quite clear. Researchers have found that organic structures are better for innovation than mechanistic ones.[21]

Organic structures permit greater organizational flexibility and adaptability than mechanistic ones. This is important because innovation is uncertain, and success often comes from having the flexibility to change strategy when circumstances demand it.[22]

Cross-functional activity, which facilitates innovative problem solving, is also easier to undertake in organizations with organic structures because people are linked horizontally, rather than just to supervisors and subordinates, as they are in hierarchical organizations (like the one shown in Figure 16.2). Because innovation requires the acquisition of information and resources from other parts of the organization, the greater degree of horizontal communication makes organic structures better suited for innovation than mechanistic ones.[23]

Innovation is also easier to accomplish in organic organizations because they are bound less to rules than mechanistic ones. Strict adherence to rules and policies, and close monitoring of behavior, creates an incentive for employees to follow procedures, and limits their willingness to experiment, thus hindering creativity.[24] It also leads people to concentrate on minor improvements with short-term outcomes for which the link between actions and outcomes is easier to gauge, and which are more likely to be successful.[25] Finally, strict rules and policies demotivate and drive away creative people, who often break rules, defy authority, and favor freedom.[26]

However, you should not rush out and set up an organic organization structure just yet. You need to understand the costs of using this type of arrangement. While organic structures are better for innovation, they are also less efficient than mechanistic ones.

FIGURE 16.2 Hierarchical Organization

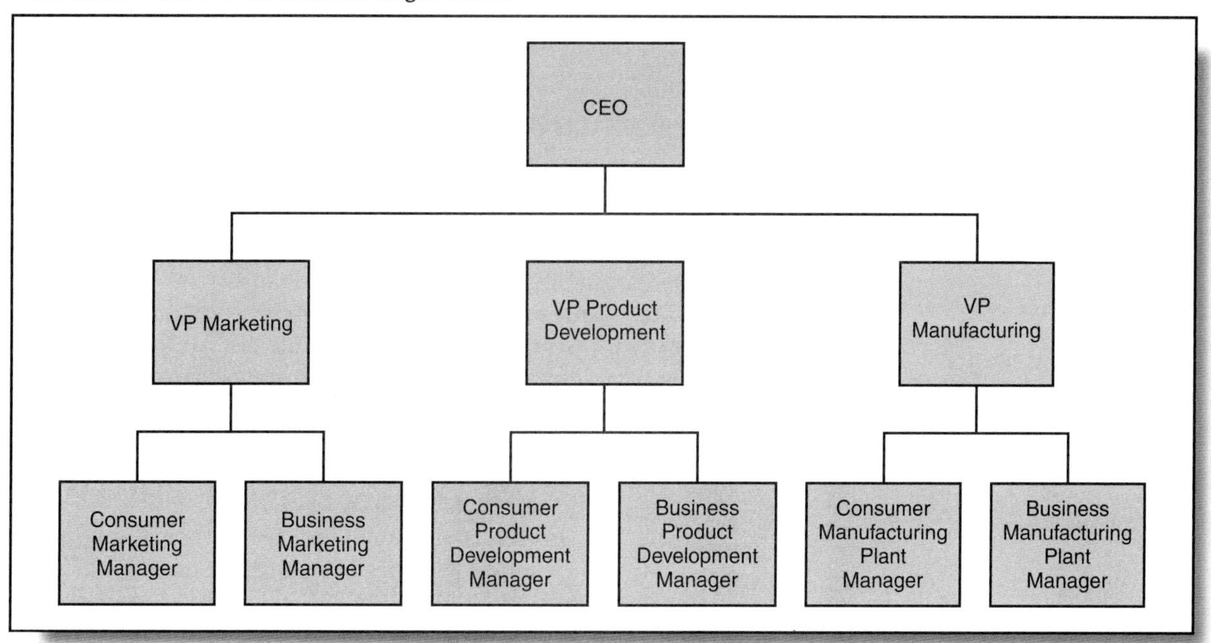

Cross-functional problem solving is difficult to accomplish in a hierarchical organization, such as this one, because the only way for the consumer and business product development managers to exchange information with their counterparts in marketing and manufacturing is for the information to travel through the organization hierarchy.

Mechanistic structures make work more uniform, thus allowing greater monitoring and evaluation of employees, which facilitates organizational efficiency.

Moreover, the advantages of organic structures for innovation are much greater for radical innovation than for incremental innovation. Remember the Abernathy-Utterback model that was discussed in Chapter 2? It explained that firms tend to have more organic structures during the era of ferment when a radical technological innovation is being introduced into an industry and product innovation is very high, and tend to have more mechanistic structures during the era of incremental change, when a dominant design is set, organizational efficiency is important, and the focus of innovation is on process change. Research has shown that the lack of formalization, better cross-functional communication, greater flexibility, and more participative decision making present with organic structures are more important to the development of radical innovations than to the creation of incremental ones.[27] In contrast, having efficient decision-making processes, authority structures, and communication patterns outweighs these benefits for more predictable, incremental, innovation.[28]

Efficiency and Slack Resources

To innovate, organizations need to maintain slack resources.[29] Innovation often requires employees to spend time and money on activities that are not directly relevant to the firm's current products and services, and to interact with people inside and outside the organization who are not part of the current production process. Therefore, successful high-technology firms build in the slack necessary for innovation. They offer employees the freedom to pursue innovative projects,[30] provide seed grants so that they can pursue their ideas, and hire people whose skills are not necessary for current operations, but that might be useful for future products and services.[31] For example, at Google, employees are allowed to spend up to 20 percent of their time on projects of their own choosing,[32] a policy that has led to the creation of Google News and Orkut, the company's social networking Web site.[33]

Efforts to make organizations more efficient usually involve the elimination of these kinds of slack resources because they create a drag on efficiency. Investment in innovation is inefficient because change is uncertain, creating dead-ends in which costs that are incurred are never recouped, and because successful innovation often results in products and processes that cannibalize existing activities. Moreover, the gains, if any, that come from investment in innovation occur in the future, not in the present when the costs of innovating are incurred. Therefore, companies facing financial problems often cut R&D spending to generate short-term efficiency gains.

However, highly innovative technology companies resist the urge to cut R&D expenditures to enhance organizational efficiency because such cuts reduce their ability to develop new products and services in the long run. For example, despite the widespread financial pressures on General Motors and its shedding of manufacturing jobs, the company remains one of the largest investors in R&D of any company in the world. The company has sought to preserve long-term research projects in such areas as electric and fuel cell vehicles despite the need to cut costs around the organization.

Key Points

- Three dimensions of organization structure affect innovation: its level of centralization, whether it is organic or mechanistic, and its degree of resource slack.
- Decentralization enhances innovation by providing access to information, increasing the number of people involved in the process, making R&D more market-oriented, diversifying investment in innovation, facilitating access to government aid for it, and encouraging innovation in noncore areas.
- Centralization enhances innovation by reducing the cost of coordination across the organization, providing freedom from operating responsibility, encouraging investment in radical innovation, motivating large-scale innovation, and facilitating a strategic approach to it.
- Organic structures are better for innovation than mechanistic structures because they permit greater organizational flexibility, encourage cross-functional activity, and are bound less to rules; however, these benefits come at the expense of lesser organizational efficiency.
- To innovate, organizations need to maintain slack resources; cutting R&D expenditures leads to an improvement in short-term efficiency but at the expense of a reduced ability to conduct long-term innovation.

Does Size Matter—For Innovation?

Large and small firms have different advantages at innovation. Small firms are less bureaucratic,[34] have simpler and faster communication, and make decisions more rapidly. Furthermore, their smaller scale allows them to colocate marketing and engineering, and to promote fewer talented researchers to management positions, leaving a larger number of them to develop new products and services.[35] As a result, small firms produce their innovations at a lower cost of R&D, and at a higher rate per employee, than large firms. Their innovations also tend to be more technically important, on average, than large firm innovations.

On the other hand, large firms can hire a more talented and more diverse group of employees than small firms, which improves the quality of their human capital. They are also more likely to have the resources to invest in R&D projects, and to generate the slack necessary to innovate.[36] Their size permits them to exploit economies of scale and learning curves in R&D, to diversify their innovation efforts across a wider range of activities,[37] and to reduce their cost of capital.[38] Finally, large firms can use market power to appropriate the returns to innovation, and vertical integration to eliminate problems with reliance on contractual modes of business.[39]

Because firm size provides both advantages and disadvantages to innovation, many large firms take strategic action to preserve the benefits of being large, while reducing the drawbacks. The most common of these actions is to divide the company into smaller units.[40] Dividing up the company makes communication processes more effective, speeds the pace of decision making, gives employees a sense of ownership and responsibility,[41] and attracts stronger employees. At the same time, it allows companies to preserve the option to exploit the benefits of size in areas, like manufacturing and marketing, where being big might offer significant advantages.

Many small firms also take strategic actions to obtain the benefits of being large, while retaining the advantages of being small. The most common of these actions is to join up with other companies in loosely coupled networks. By joining a network, companies can remain small for the parts of the innovation process for which small size is beneficial, while becoming large for the parts for which large size is an advantage.[42]

As Table 16.2 shows, there are also important industry differences in whether small or large firms are more innovative. Small firms tend to be more innovative than large firms in industries that are younger, less R&D intensive, less concentrated, less capital intensive, less advertising intensive, and more reliant on skilled labor.[43] They are also more innovative in industries, like computer software, in which products are simple and composed of a single system that can be developed by small teams, than in industries, like automobiles, in which products are complex and composed of multiple systems that need to be created by large teams.[44]

Key Points

- Small and large firms have different advantages at innovation.
- Many large firms seek to obtain the advantages of small firm innovation by dividing themselves into smaller units, while many small firms seek to obtain the advantages of large firm innovation by joining networks.
- Research has shown that the relative advantages of large and small firms at innovation depends on the industry in which the firms operate; small firms are more innovative in younger industries; industries that are less R&D intensive, less capital intensive, less advertising intensive, and more reliant on skilled labor; and industries that make simple products that can be developed by small teams.

TABLE 16.2
Small and Large Firm Innovation
Whether small or large firms are more innovative depends on the industry; the ratio of large to small firm innovation ranges from 31.000 in the case of aircraft to 0.067 in the case of measuring and controlling devices.

Selected Industries	Ratio of Large to Small Firm Innovations
Aircraft	31.000
Pharmaceutical preparations	9.231
Photographic equipment	8.778
Office machinery	6.710
Surgical appliances and supplies	4.154
Industrial inorganic chemicals	4.000
Semiconductors	3.318
Environmental controls	2.200
Special industry machinery	2.048
Radio and TV equipment	1.153
Surgical and medical instruments	0.833
Electronic components	0.740
Fabricated metal products	0.706
Electronic computing	0.696
Industrial trucks and tractors	0.650
Valves and pipefitting	0.606
Instruments to measure electricity	0.596
Optical instruments and lenses	0.571
Scientific instruments	0.518
Plastics products	0.268
Measuring and controlling devices	0.067

Source: Based on information contained in Acs, Z., and D. Audretsch. 1988. Innovation in large and small firms: An empirical analysis. *American Economic Review*, 78(4): 678–690.

Spin-off Companies

As Figure 16.3 shows, new firms are sometimes founded when an existing company creates a new organization separate from the parent company. However, not all spin-off companies are created with the blessing of the parent organization. Sometimes people create spin-offs by discovering ideas for new businesses while working for companies and then quitting those companies to exploit the ideas on their own. In fact, research shows that more than 80 percent of all firms are founded in the same industry as the one in which their founders were previously employed,[45] and more than 60 percent of firm founders start companies to serve the same or similar customers as their prior employers.[46] Therefore, employees of existing companies often choose to start their own companies even though they could pursue the same innovations on behalf of their employers. As a technology entrepreneur or manager, you need to understand why people do this.

Sharing Rewards

Many companies do not share the rewards from innovation with their employees. In fact, some even establish policies that prohibit employees from sharing the profits from innovation because they are concerned that giving employees a share of

FIGURE 16.3
Spin-Off Companies

"Our parent company has opted to spin us off."

Sometimes, employees create new companies by quitting their jobs in established companies to start new ones; other times existing companies create them.
Source: United Features Syndicate Inc., March 12, 1996.

the profits will give them too much of an incentive to focus on innovation, and that sharing profits will create inequity in compensation (hard to believe in an era of phenomenal CEO compensation!). As a result, many employees see little reason to exploit the innovations that they have discovered on behalf of their employers, and instead quit to start their own businesses. For instance, Xerox's policy against giving innovators a share of the profits from innovation led one of its employees, Robert Metcalfe, to quit working at Xerox PARC and found 3Com to exploit a technology—the Ethernet—that he came up with while at Xerox.

The lack of shared rewards accounts for the most basic explanation for why people create spin-off companies. In the absence of receiving a share of rewards from innovating, the employee's motivation is quite obvious. If the employee quits work to start a company to exploit the innovation, he or she will earn the profits generated by that innovation. But if the employee stays with the company, he or she will earn nothing by exploiting it. Thus, even if starting the company is risky, the employee has a strong incentive to do it, as long as the expected value of the innovation exceeds his or her salary.

Bargaining over Profits

Another reason why employees form spin-off companies is that they disagree with their employers about how to divide the profit from innovation between them. It is easy to see why an employer and employee would disagree about this division. Because the innovation is uncertain, its valuation is subjective. The employee's valuation will be more positive than the employer's because, as the source of the innovation, the employee has private information about its value that he or she cannot—or will not—disclose to the employer. Of course, if the employee's estimate is higher than the employer's, then the employee will view starting a company as a better choice than exploiting the innovation on behalf of the employer.[47]

Take, for example, the story of Dr. Alice Huxley, a biochemist who had been working for Novartis on a drug to fight an enzyme that raises blood pressure. In the late

1990s, Novartis decided to abandon the development of the drug because it didn't think the drug could be manufactured profitably. Huxley disagreed with the company's assessment, quit, licensed the drug, and set up a biotech firm.[48]

Source of Value

The explanations for spin-off company formation so far assume that the employee is willing to exploit the innovation on behalf of his or her employer if the two sides can come to an agreement about the innovation's value. But that assumption might not be true. If employees are unwilling to exploit innovations on behalf of their employers, then, as Figure 16.4 shows, a key explanation for spin-off company formation lies in the ability of employers to preclude their employees from exploiting innovations that were identified while in their employ.

This ability depends on whether the key ingredient that makes the innovation possible lies in human or physical capital. Employers can more easily preclude their employees from taking physical assets that are valuable to the employers and using them to start new businesses than they can preclude employees from taking valuable human capital. The reason is that the U.S. legal system—and the legal systems of most countries—views physical assets as the property of the employer and human assets as the property of the employee.

Take, for example, a new business making ski bindings. The company can prevent its employees from quitting and starting companies to make ski bindings using the employer's equipment. However, the employer is not going to be able to stop the employees from using what is in their heads—their expertise in fitting ski bindings to boots—to start similar companies.

The strength of the employer's intellectual property rights also affects the ability to preclude employees from exploiting innovations that were identified while in their employ. Companies can much more easily maintain control over innovative ideas if they can obtain patents to protect them. If the technologies are patented, then the employer can sue employees for infringement if they use the technologies to start companies.

However, if the innovations are not patented, then the situation is more difficult. The employer can try to use the noncompete and nondisclosure agreements discussed

FIGURE 16.4
Who Owns Ideas?

Most companies use employment agreements to maintain control over the intellectual property developed by their employees.

Source: United Features Syndicate, August 17, 2002.

in Chapter 9 to prevent employees from using the innovations to start companies. For example, Tandy Corporation made it difficult for Jeff Hawkins, who had developed the C language enhancements for his personal digital assistant product when working for GRiD Systems, to found Palm Computing because the company owned the rights to all the products developed by employees of GRiD Systems, which Tandy had acquired.[49]

But, as Chapter 9 explained, the effectiveness of these agreements is quite limited, and they are not equally enforceable in all states. (They are particularly difficult to enforce in California.) Moreover, if the innovations are based on general know-how, then the employees will be able to use them when they start new companies because employers cannot preclude former employees from making use of general knowledge that they gained during their employment.

Key Points

- New firms often form when people, who develop innovations while working for other companies, start their own companies to exploit them.
- The most basic explanation for why people create spin-off companies not sanctioned by the parent organization is that their employers will not share the rewards of innovation with them.
- Many companies will not share the profits from the innovations identified by their employees because they do not want to give their employees strong incentives to innovate, and because they do not want to create inequity in compensation.
- Spin-off companies also form because innovations are uncertain and employers and employees disagree over their expected value.
- Innovations that are based in human capital, and that are not protected by patents, are more likely to lead to spin-offs because firms can more readily preclude their employees from using the company's physical assets and intellectual property to start new businesses than they can preclude them from using their own human capital.

Venture Capitalists, Business Angels, and Corporate Investors

New technology companies often need to raise money from external equity investors. The sheer cost of developing and launching new products—several hundred million dollars in biotechnology—means that individual entrepreneurs often cannot afford to finance their own businesses. Moreover, the need to incur costs for several years until revenues are earned means that debt, which requires payments on a fixed schedule, often cannot be serviced. Furthermore, the market, technology, and competitive risks faced by start-ups demand that investors earn a very high rate of return, sometimes as much as 100 percent per year. That level of return usually means taking equity because usury laws preclude charging interest rates that high.

External equity investment in technology start-ups typically comes from three sources: corporations, business angels, and venture capitalists. Many high-tech corporations make strategic investments in new companies to gain access to their technologies or products, or to evaluate them as acquisition targets. These corporations also make investments to marry the new business's technologies or products with their marketing or manufacturing assets to generate advantage over their competitors.

Business angels, private individuals who invest their own money in new companies, also supply capital to high-tech start-ups. These investors, some of whom are wealthy former technology entrepreneurs, often invest in new businesses to remain involved in the entrepreneurial process and to help the communities in which they live and work. They sometimes provide hands-on assistance to the entrepreneurs that they finance.

Venture capital firms, organizations—generally limited partnerships—that raise money from university endowments, pension funds, and wealthy individuals and invest that money in high-technology start-ups, are a third source of capital. Venture capital firms are run by general partners who make the investment decisions and manage the investments on behalf of the limited partners who provided the money. In addition to providing capital, venture capitalists offer assistance to start-ups, helping to identify key managers, initial customers, and valuable suppliers, and developing the businesses for sale to larger firms or to go public.

The companies that raise money from these three sources can be very high-potential businesses. They may operate in a high-growth industry, have a strong competitive advantage, offer a product desired by the marketplace, have an experienced management team, and have a strategy to exit through sale to a larger company or through an initial public offering.

Investors sometimes demand a large portion of the start-up's equity in return for providing capital. Depending on the stage of development of the company, they might seek an annual rate of return of between 20 percent and 100 percent (see Figure 16.5). To generate this rate of return, the investor has to take a large portion of the company.

To protect themselves against the risks of investing in new companies, investors sometimes impose restrictions on the companies in which they invest. They might purchase convertible preferred stock so that they have a liquidation preference if the business doesn't do well, but can own common stock at the time of exit. They might require entrepreneurs to agree to covenants that keep the latter from buying assets, issuing stock, hiring key managers, or engaging in other activities without the investors' permission. Furthermore, they might provide money in stages so that they have a right to abandon the investment if it turns out not to be a very good one.

FIGURE 16.5
Rates of Return and Financing Rounds

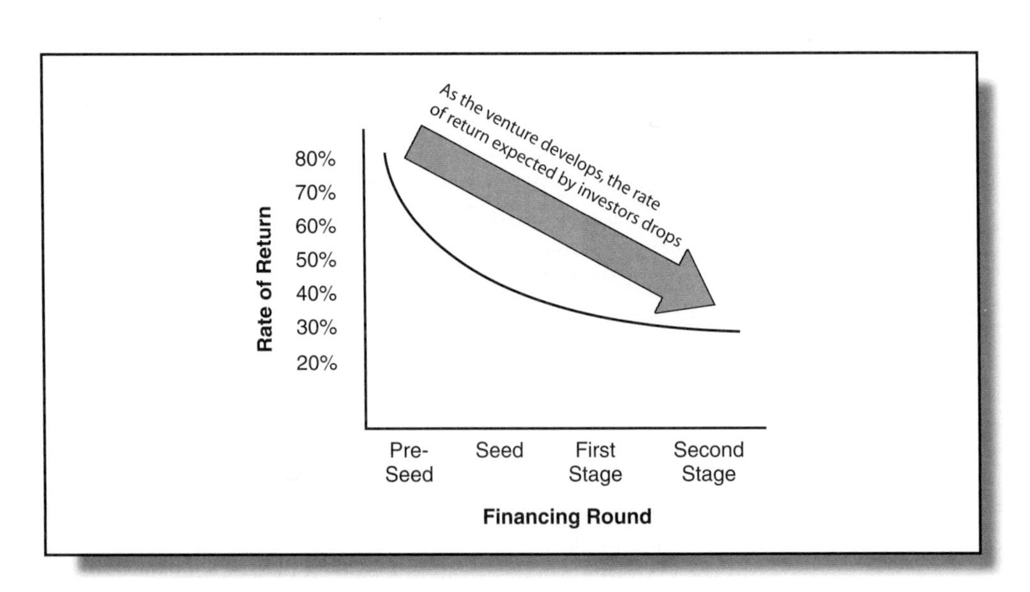

The rate of return that investors demand decreases as the venture becomes more developed.

Key Points

- High-technology start-ups sometimes need to raise external equity from corporations, business angels, or venture capitalists.
- Only a small portion of start-up companies obtain external equity—those in high-growth industries that have a product that customers want, a strong competitive advantage, and an experienced management team are more likely to do so.
- Equity investors in new ventures sometimes demand high rates of return and impose significant restrictions on the actions of entrepreneurs to compensate for the high risk of investing in those businesses.

Corporate Venturing

Corporate venturing, the creation of new businesses by existing organizations, is a common business activity among large companies in technology-intensive industries. In 2003, for example, more than 500 of the largest companies in the United States engaged in some type of formal corporate venturing activity.[50] Typically, corporate venturing takes the form of corporate venture capital funds or new venture groups.

Corporate venture capital funds are entities that invest a business's money in start-up companies in return for an equity interest in them. These companies sometimes invest in start-ups that spin off from the corporate parent, and other times they invest in new businesses founded by independent entrepreneurs.

In 2003, approximately 233 U.S. companies operated corporate venture capital units;[51] and corporate venture capital accounted for almost 17 percent of the $20 billion in venture capital invested in the United States.[52] However, as Figure 16.6 shows, the number of companies in which these funds invested, and the amount provided, is currently far below its peak in 2000.

When companies establish corporate venture capital units, they can do so on their own, or in conjunction with independent venture capitalists. For example, Intel set up its own venture capital fund, while Texas Instruments developed a corporate venture capital fund in partnership with Granite Ventures, and Philips Research created its fund in partnership with New Venture Associates.

New venture groups are corporate units that create new companies under the umbrella of the parent firm. Unlike corporate venture capital funds, new venture groups only invest in new businesses that exploit opportunities or technologies developed in the parent firm, rather than those created by independent companies.

Corporate venture groups are an important part of the technology strategy of many large companies. As Figure 16.7 shows, while the number of corporate venture groups peaked in 2000, it still remains higher today than in the 1990s.

The corporate new ventures group model of corporate venturing differs from the corporate venture capital fund model in several ways. Corporate venture capital funds use a venture capital approach to corporate venturing, making larger investments, coinvesting with other financial institutions, staging their investments, and seeking a profitable exit through IPO or acquisition. In contrast, corporate venture groups follow a corporate approach, making smaller investments alone, through corporate budgets, and seek to maintain ownership of the new companies that they finance.[53] Corporate venture capital funds also usually have the goal of generating a

FIGURE 16.6
Corporate Venture Capital

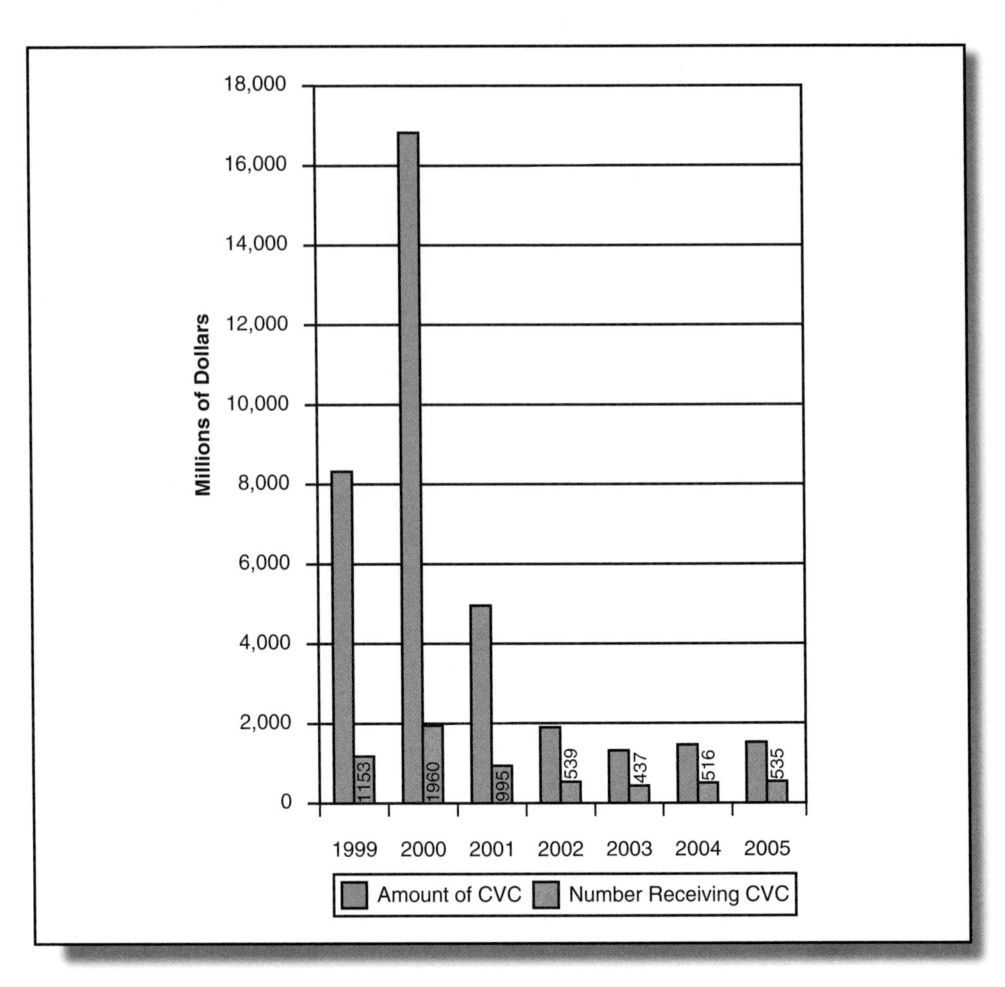

The number of companies receiving corporate venture capital and the amount of corporate venture capital investment have declined dramatically since the end of the Internet boom.

Source: Adapted from data contained in http://www.kedrosky.com/images/corp-vc.gif.

financial return for the parent company, and so make broader scope investments, seek more diversification, and care less about strategic alignment than corporate venture groups, which typically have the goal of putting the parent company's ideas, unused patents, and technologies to good use,[54] which leads them to make a smaller number of lower risk investments.[55]

The two forms of corporate venturing are also managed differently. The human resource practices, governance, and compensation arrangements of corporate venture capital funds are modeled on venture capital firms, while those of internal corporate venture groups are modeled on corporations.[56] For instance, venture groups are usually managed by a company hierarchy; whereas corporate venture capital funds are typically run by an independent board of directors.[57] With internal venture groups, the compensation of entrepreneurs is similar to that of corporate managers; with corporate venture capital funds, it is similar to that of independent entrepreneurs.[58] And to measure success, the venture capital funds typically use capitalized return on investment, while the venture groups use incremental increases in profit.[59]

FIGURE 16.7
Number of Firms with Corporate Venturing Units

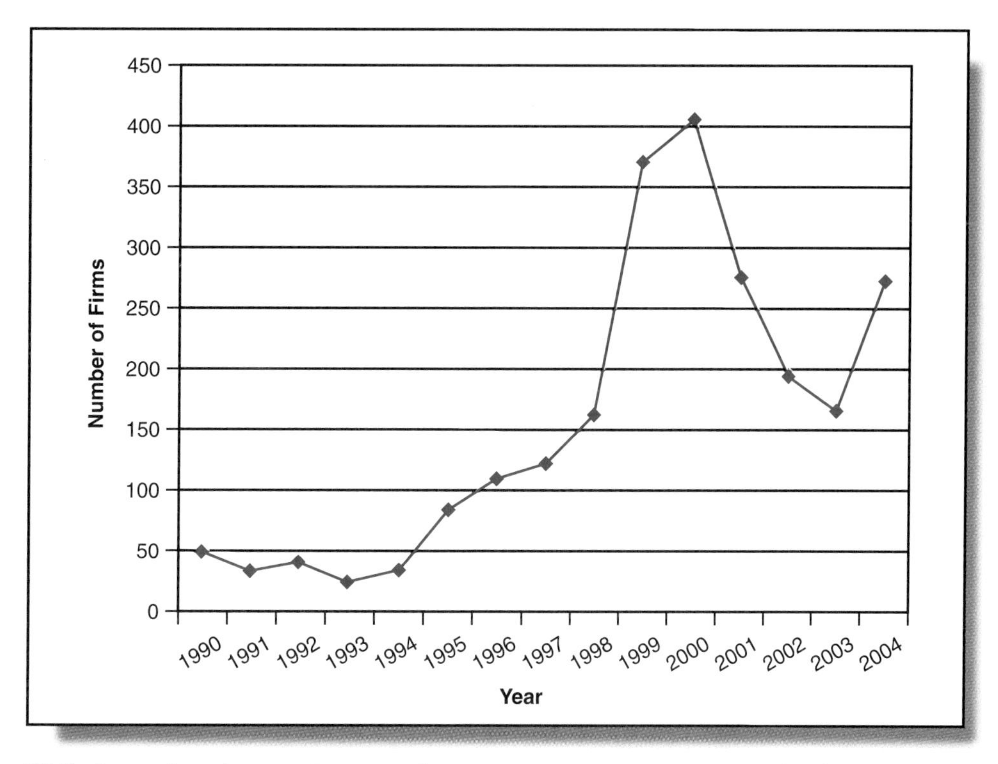

While the number of companies engaged in corporate venturing rose to great heights in 2000 and 2001 and then fell, it remains much higher than it was in the early 1990s.

Benefits of Corporate Venturing

Researchers have identified five benefits that corporate venturing brings to large, established, technology companies. First, it provides a mechanism to capitalize on radical innovations developed in R&D laboratories.[60] When the R&D arms of large companies develop radical new technologies or technologies that need to be applied in different markets from those in which the companies are found, their operating units are usually unwilling to replace their core technologies with these new technologies, or to serve markets that they have not served before. Corporate venturing provides a way for established companies to benefit from these new technologies, by facilitating the development of new companies that exploit them.[61]

Second, corporate venturing provides a way for companies to use equity as an incentive to motivate key employees responsible for innovation, which reduces the likelihood that those employees will quit to form new companies. Equity incentives are necessary to motivate employees to bear the risk of developing radical innovations. However, large, established companies find it difficult to give equity to key innovators because of the norm of providing similar compensation to managers in similar positions in operating units.

Third, corporate venturing offers a positive way to maintain control over technology developed by employees. While companies can rely on litigation to deter their employees from using technology developed while in their employ, litigation is not always the best solution to this problem. As Chapter 9 explained, many U.S. states limit the terms of employment contracts, hindering the ability of companies to sue to collect damages. Moreover, a litigation-oriented strategy hampers a company's ability to hire the best scientists and engineers, and creates conflict with noncorporate research partners, such as universities and research institutes, because these groups do not want to work with companies that might engage in legal action against them. Finally, litigation is a "negative" strategy. While it allows a business to stop its employees from using its intellectual property to start companies, it does nothing to make productive use of the technology itself.[62]

However, if established companies invest in the new companies founded by their former employees, and establish the right of first refusal to acquire the products or companies once they have been developed, they can use corporate venturing to exploit their technology in a positive way. For instance, in the case of Alice Huxley's biotechnology spin-off from Novartis mentioned earlier in this chapter, her former employer made a $3 million venture capital investment in the new business in return for a portion of the equity, and reserved the right to repurchase the company and the product in the future.[63] That way if the new technology proved to be a success, Novartis could take advantage of it.

Fourth, corporate venturing provides a way for companies to develop innovations that do not appeal to their existing customers, or are outside their core strategies, and to exit businesses or shed technologies that are not central to their current plans.[64] Because corporate venturing can be used to create new organizations to exploit innovations, it provides companies with entities that are independent of their operating companies, which need to satisfy their mainstream customers. As a result, corporate ventures help companies to exploit underserved new markets by providing them with a mechanism to go after small market niches with products that do not appeal to their existing customers.

For the same reason, corporate venturing allows firms to invest in businesses that might have value in the future, but which, at present, represent markets that are too small to meet the growth demands of large firms.[65] For example, Cisco recently created an emerging products group charged with developing non-networking products that employees think make sense to develop. The group collects ideas, evaluates them, and figures out how to bring them to market.[66] By developing these ideas as new businesses rather than trying to exploit them through Cisco, the group can take advantage of products whose customer base is initially too small to get the attention of Cisco's operating managers.

Fifth, corporate venturing allows companies to develop new technologies that are outside of their core capabilities but, nevertheless, support mainstream operations. By using corporate ventures as a way to support independent companies that develop new process technologies, test new equipment, develop new tools, exploit new materials, and develop complementary technologies, large established companies can enhance demand for their mainstream products and services.[67]

Drawbacks of Corporate Venturing

While corporate venturing offers the benefits described above, it also has several important drawbacks, of which you need to be aware. First, as Figure 16.8 shows, when companies engage in corporate venturing, they usually need to take the strategy of the parent firm into consideration and can't maximize financial returns, like independent venture capitalists.[68] For example, Mitsubishi Corp.'s venture capital unit, Captech Corp., only invests in start-ups in electronics materials and high-tech metal products, such as solar batteries and catalysts, and British Gas's corporate venturing unit only invests in technologies for natural gas, because these investments further the goals of the parent companies.[69]

FIGURE 16.8
The Importance of Financial Goals in Corporate Venturing

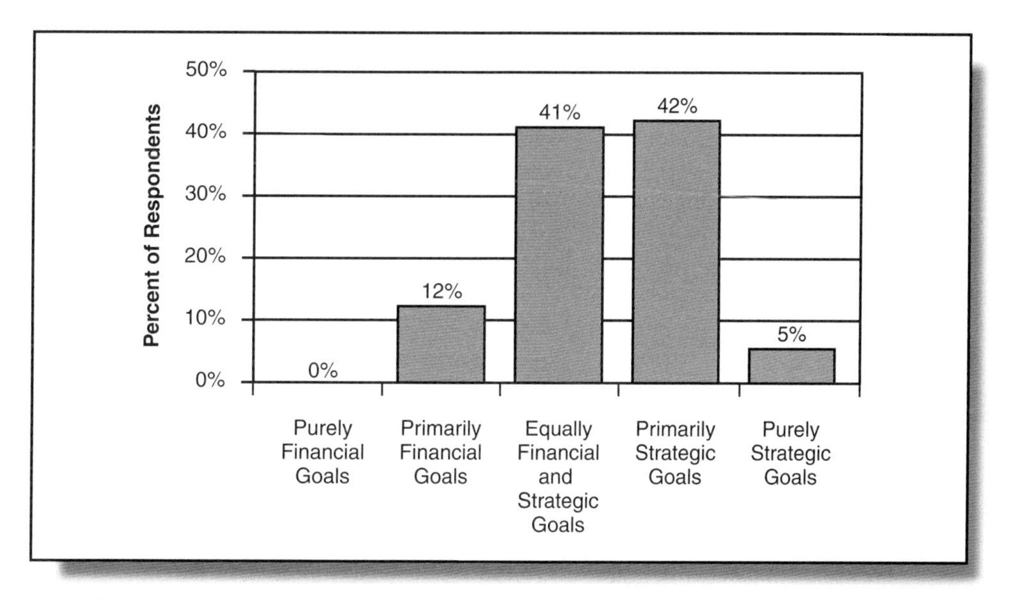

According to data collected by the consulting firm Arthur D. Little, only 12 percent of firms have purely or primarily financial objectives in mind when conducting corporate venturing, while 47 percent have purely or primarily strategic goals.

Source: Adapted from Arthur D. Little. 2002. *Venturing for Innovation*. Thalwil, Switzerland, December, http://www.adlittle.com/insights/studies/pdf/corporate_venturing_study_report.pdf.

In fact, companies are often dissatisfied with corporate ventures that have generated strong financial returns but have not met their strategic goals. For example, Xerox shut down its internal venture capital unit, Xerox Technology Ventures, after it generated an internal rate of return of 56 percent on an investment in the spin-off Documentum because that company had no strategic ties to Xerox.[70]

The focus of corporations on achieving strategic goals through corporate venturing means that they can't coinvest easily with independent venture capitalists. A focus on strategic goals leads corporations to avoid investments in new companies that exploit technologies and markets that are not a strategic fit. Because these technologies and markets sometimes have greater potential than those which are a strategic fit, this approach constrains corporations from maximizing the financial returns to their investments, which is the goal of independent venture capital firms.[71]

Moreover, because new ventures are uncertain, they often evolve in different directions from those that the founders had intended. This evolution creates tension between corporate investors and independent venture capitalists. While independent venture capitalists see as positive any evolution that benefits new ventures, corporations often resist changes that move new ventures in directions that are not aligned with their strategies. For example, Texas Instruments's corporate venture capital partnership with Granite Ventures was adversely affected when several portfolio companies shifted into areas that were not of strategic interest to Texas Instruments because Granite Ventures's partners saw those changes as beneficial, while Texas Instruments's senior management did not.

Furthermore, independent venture capitalists believe that focusing investment on a single company's technology limits the breadth of their options. As a result, they tend to avoid coinvesting with corporations that make such a focus a requirement for their corporate venturing activities.

Finally, independent venture capitalists do not like to give corporations the right to acquire portfolio companies at fair market value plus a set percentage, which many corporations demand as a condition for their investment.[72] Independent venture capitalists see this arrangement as "leaving money on the table," and prefer to sell portfolio companies to the highest bidder.

Second, corporate venturing imposes several major risks on companies that engage in it. Companies risk their reputations if their corporate ventures do something unethical or illegal, as might occur if a chemical company's corporate venture polluted the environment. Moreover, they risk intellectual property lawsuits if they give their technical employees too much access to potential portfolio companies because such involvement invites charges of intellectual property theft.[73] Consequently, companies typically need to erect barriers between their corporate venturing units and the rest of their operations, using only people in the corporate venturing units to evaluate businesses and technologies that come from outside.[74]

Third, corporate venturing imposes additional legal, financial, and accounting burdens on public companies. Investors in start-up companies need to revalue their investments as the companies obtain additional rounds of funding. Down rounds (additional rounds of investment in start-up companies at lower valuations than the previous rounds) and shutdowns of start-up companies create volatility in the investors' financial performance. This volatility is problematic for public companies, which have to provide the results of their investments in their quarterly reports.[75]

Moreover, corporate investors cannot easily take board seats at start-up companies because such action might create fiduciary conflicts with their parent companies.[76] Board members of start-ups need to act in the best interest of their shareholders, which could easily conflict with the responsibilities of the corporate investors to act in the best interest of the parent company's shareholders, particularly when corporate venture capitalists make strategic investments on behalf of their parent companies.

Fourth, corporate venturing often interferes with the ongoing operations of parent companies. Not only do corporate ventures compete with operating units for internal resources, they sometimes compete with their parent firms in the marketplace. In fact, established companies often face a tension between allowing their corporate ventures to compete with their operating companies, enhancing the former's financial returns, and limiting the actions of the ventures, reducing their financial returns, but protecting the performance of the operating companies.[77]

Moreover, corporate venturing often forces large companies to give up many of their advantages at innovation. Because corporate ventures often need to operate independently from their parent companies to be successful, large companies need to give up the benefits of scale economies when they engage in corporate venturing. They also find it more difficult to make use of assets, such as intellectual property, in both the existing and new operations because these assets need to be assigned either to the parent companies or the corporate ventures.[78] And they find it harder to learn from experience because corporate ventures have no mechanism for transferring knowledge to their parent companies or to other corporate ventures, the way that units of operating companies do.[79]

How Much Independence?

You will need to decide how independent to make your corporate ventures. Performance is often hindered if corporate ventures are not set up as separate companies because the new businesses need to establish different compensation

TABLE 16.3

The Pros and Cons of Corporate Venturing

Like many aspects of technology strategy, corporate venturing has advantages and disadvantages that you need to consider when deciding whether or not to engage in it.

PROS	CONS
Provides a mechanism to capitalize on radical innovation	Imposes financial, legal, and accounting burdens
Offers a way to use equity as an incentive to motivate employees	Is at odds with maximizing financial returns on investments
Provides a positive approach to control knowledge spillovers	Imposes additional risks on companies
Helps companies to develop technologies that support their mainstream operations	Interferes with ongoing operations
Frees companies from dependence on mainstream customers	

schemes, rules, operating procedures, structures, performance measures, and cultures from the parent companies to be successful.[80] Insufficient independence can create conflict between the parent companies and new ventures over management, marketing, or human resource efforts,[81] especially when the new ventures need to hire people who have novel abilities and skills.[82] For example, Teradyne experienced problems with its semiconductor testing corporate venture because it did not make the new venture's human resource policies independent of the parent company's policies. When Teradyne had to lay off workers because of a slowdown in its business, the corporate venture faced a lot of pressure not to hire needed employees to conform to the policies enacted in the rest of the organization.[83]

It is particularly important to set up a separate company for a new venture that exploits radical or competence-destroying technology. These technologies tend to draw on a different set of organizational competencies and require hiring different types of employees than those present in the parent company. They also tend to cannibalize the markets for the parent company's products. As a result, new ventures that exploit competence-destroying technologies tend to cause too much conflict to work effectively with the rest of the organization. For example, Swagelok, a manufacturer of valves and fittings, had to create a new company to exploit a new heat-treating process to make high-performance steel for automotive, oil and gas, aerospace, and defense industries. The new technology overturned received wisdom about stainless steel, allowing surface hardness to be increased threefold and corrosion resistance to be enhanced 10,000-fold.[84] Therefore, it was hard to fit the business within the confines of Swagelok's main operations.

However, making corporate ventures independent is not without cost. Using the marketing, manufacturing, and technology assets of parent companies helps corporate ventures to perform better than independent start-ups. Moreover, by using corporate procurement, new ventures can obtain products and services at lower prices than independent start-ups have to pay.[85] Perhaps most important of all, affiliation with parent companies gives new ventures access to customers and the credibility to sell to them. For example, Documentum, a Xerox Technology Ventures portfolio company, was successful, in part, because of the credibility and customer access that came from its association with Xerox.[86]

So how independent should you make your corporate ventures? As Figure 16.9 shows, the answer depends on the relationship between the corporate ventures and the core operations of your company.[87] If the corporate ventures are strategically important and operationally related to your core capabilities, then the new ventures should be closely tied to your core businesses through shared

FIGURE 16.9
How Independent Should a Corporate Venture Be?

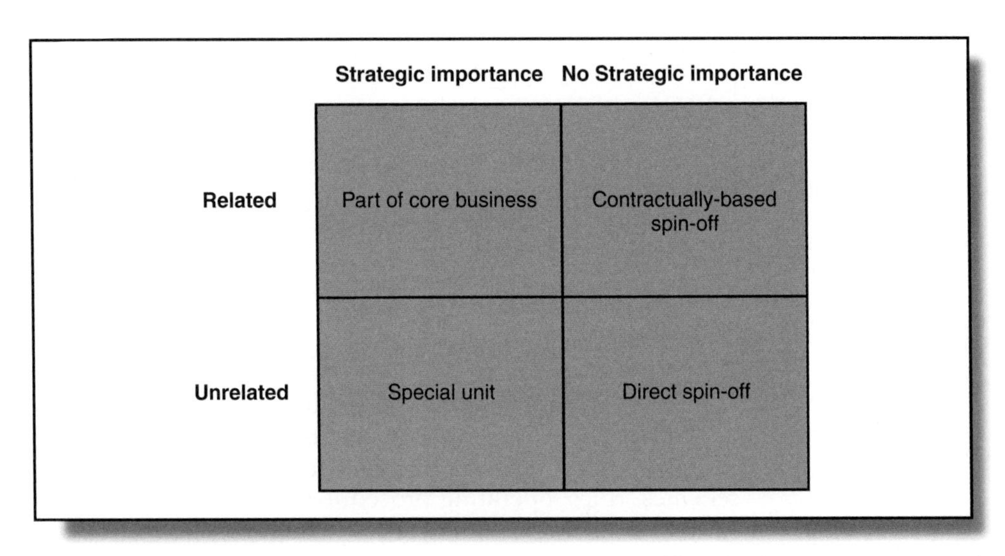

You should consider the strategic importance of a new venture and how it relates to your company's core capabilities when deciding how much independence to give it.

Source: Adapted from Burgelman, R., C. Christiansen, and S. Wheelwright. 2004. *Strategic Management of Technology and Innovation.* New York: McGraw-Hill Irwin.

technology or product use. If the corporate ventures are strategically important, but unrelated to your core businesses, then establishing independent organizational units makes the most sense. If the corporate ventures are strategically unimportant and are unrelated to your core businesses, then you should spin off the companies, but hold equity in them or reserve a right to buy them back. Finally, if the corporate ventures are strongly related to your core businesses, but are strategically unimportant, then contractual-based spin-offs make the most sense.[88]

Finally, you will need to design the right incentives for the employees of your new ventures. You will need to compensate them in a way that imitates the equity incentives of start-ups, through phantom stock or other form of compensation that is paid only if the venture succeeds. Otherwise, your employees will be unwilling to take basic research ideas and turn them into applied products and services.[89]

For example, the absence of proper incentives at Xerox has hindered that company's performance at corporate venturing. Its long-time policy of restricting employees in R&D laboratories from undertaking further development of their inventions, let alone giving them equity in ventures that develop those inventions, has led entrepreneurial Xerox researchers to start companies to exploit technologies developed at Xerox, and has kept the company from profiting from their efforts.[90]

Key Points

- The two main forms of corporate venturing are corporate venture capital funds and new venture groups.
- The corporate venture capital fund model differs from the new venture group model on several dimensions, including financial goals, types of investments, human resource and governance practices, and performance measures.

- Corporate venturing benefits large, established technology companies by allowing them to exploit radical technologies created in their R&D labs; permitting them to use equity as an incentive to motivate employees; allowing them to maintain control over their technology without resorting to litigation; permitting them to develop innovations that do not appeal to their existing customers; and allowing them to develop businesses that are outside of, but support, their mainstream operations.
- Corporate venturing suffers from several drawbacks, including the inability to maximize financial returns; the imposition of additional risks on companies; the creation of financial, legal and accounting burdens; and interference with the ongoing operations of the parent company.
- To successfully manage a corporate venture, you must balance the benefits of independence against the benefits of ties to the parent company and develop the right compensation scheme for the staff of the venture.

ACQUISITIONS

As an alternative to corporate venturing, you can use **acquisitions**—the purchase of one company by another—to expand and take advantage of technologies developed by other companies.

There are several advantages to expanding through acquisition:

1. Acquisitions allow you to gain control over proven new technical capabilities and intellectual property. For example, Novartis acquired biotechnology pioneer Chiron to gain access to the latter's vaccine development capabilities and to take advantage of the increasing size of the vaccine market that has resulted from fears of Avian flu.[91]
2. Acquisitions allow you to gain knowledge about how to serve a market that you would like to enter. For instance, Cisco acquired Scientific Atlanta in part to learn how to serve markets other than the network equipment market.[92]
3. Acquisitions allow you to put your capabilities to work in new markets, increasing the potential return on your investment in their development. For example, Cisco's acquisition of Scientific Atlanta was designed, in part, to use Cisco's international sales force to sell Scientific Atlanta's television set boxes, which allowed Cisco to exploit its international business capabilities across a wider range of products.[93]
4. Acquisitions allow you to enhance efficiency by combining operations and eliminating duplicative assets. For example, SBC Communications's acquisition of AT&T allowed the combined company to reduce R&D activities, headquarters staff, and other operations that were duplicated across the two firms.[94]
5. Acquisitions allow you to achieve synergies across different products. For instance, Cisco's acquisition of Linksys allowed it to achieve synergies between its activities and those of Linksys, providing the necessary equipment for home telecommunications networks more efficiently.[95]
6. Acquisitions allow you to manage the risk of investing in the development of new technologies. By making investments in other companies that develop new technologies, firms can create the option to acquire those companies and obtain access to successful new technologies that cannot be transferred

through market mechanisms. This approach particularly benefits large firms, which find it easier to purchase and integrate small new firms into their organizations than to develop new technology from scratch inside the organization.

On the other hand, acquisitions have several drawbacks. Perhaps the most important of these is that acquisitions almost always seem better in theory than they are in practice. Almost all companies that undertake acquisitions believe that their performance will be enhanced by the acquisition. However, the returns to shareholders of acquiring companies tend to be rather poor, especially in the years immediately following an acquisition. While we don't know exactly why these returns are so low, they likely reflect the inability of companies to estimate accurately the costs and benefits of merging organizations. Thus, from a purely financial point of view, acquisitions are rarely a good idea for technology companies.

Acquisitions are also not a very good approach to obtaining access to another company's human capital. While you don't have to worry about the loss of physical and financial assets of an acquired company immediately after the acquisition, you do have to be concerned about the loss of human capital. Research shows that many managerial employees of acquired companies leave in the first year after an acquisition.[96] Thus, acquisitions are much more effective in obtaining access to capabilities that are embedded in physical or financial assets, like pieces of intellectual property, than they are in obtaining access to capabilities that are embedded in human capital, like the expertise of sales personnel.

Horizontal mergers—mergers between companies at the same stage in the value chain—tend to be more successful, on average, than vertical mergers—mergers between companies at different stages in the value chain. The reason for this appears to be that companies have much more knowledge about how to manage additional operations at the same stage of the value chain than at upstream or downstream stages, and because they can more easily achieve efficiencies when operations at the same stage of the value chain are combined.[97]

Research has also shown that acquisitions of related companies are more successful than acquisitions of unrelated companies.[98] Again the reason has to do with capabilities and efficiencies. Firms that acquire unrelated companies often don't understand the target companies' businesses or how to manage them, and they can rarely achieve any efficiencies from combining the companies.

Key Points

- Acquisitions—the purchase of one company by another—allow firms to gain access to technical capabilities, intellectual property, and market knowledge; extend the reach of existing capabilities to new markets; eliminate duplicative assets; establish synergies between the operations of different organizations; and reduce the risk of exploring new technologies.
- Acquisitions are difficult to implement in practice, leading to depressed financial performance of acquirers in the period immediately following acquisitions.
- Acquisitions are more effective for obtaining financial and physical assets than for obtaining human capital because many employees of target companies leave the joint company in the year following the acquisition.
- Horizontal acquisitions are more effective than vertical acquisitions, which are more effective than acquisitions of unrelated companies.

Discussion Questions

1. Is there a best way to structure a company for innovation? If so, how? If not, why not? What are the advantages and disadvantages to innovation of the different types of organizational structure?
2. Are large or small firms more innovative? Why? Are there conditions that affect which group is more innovative? If so, what are they?
3. What can large firms do to be more innovative? What can small firms do to be more innovative? Why do these things make each group more innovative?
4. Why do spin-offs occur? What reasons do you think are the most important, and which are the least important? Why?
5. Is engaging in corporate venturing a good idea for large, established companies? Is a company better off growing through acquisition? Explain your answer.
6. What is the difference between corporate venture capital and other forms of corporate venturing? Why might a company choose corporate venture capital over other models of corporate venturing?
7. How closely tied should corporate ventures be to their parent companies? Why?

Key Terms

Acquisition: The purchase of one company by another.

Business Angels: The private individuals who invest their own money in new companies.

Centralization: The extent to which decision making in an organization is controlled at the top.

Corporate Venture Capital Fund: An entity through which established companies invest in start-up companies in return for an equity interest in them.

Corporate Venturing: The creation of new businesses by an existing company.

Mechanistic: An organization structure that is tightly controlled.

New Venture Group: A corporate unit that creates new companies under the umbrella of the parent firm.

Organic: An organization structure that is loosely controlled.

Venture Capital Firm: An organization that raises money from university endowments, pension funds, and wealthy individuals and invests that money in high-technology start-ups.

Putting Ideas into Practice

1. **Spin-off Companies** The purpose of this exercise is to evaluate company policies toward spin-off companies. Assume that you are in charge of setting the human resource policies at Dow Chemical Company. Should you encourage or discourage the creation of nonsponsored spin-off companies by your employees? Assuming that you would like to discourage these spin-offs, what policies would you suggest implementing? Why do you recommend these policies? Now assume that you would like to encourage spin-offs but capture the benefits generated by them. What policies would you recommend implementing to achieve that goal? Why do you recommend these policies?
2. **Organizational Structure** Assume that you have been tasked with developing a plan to improve Texas Instruments's performance at innovation. (If you don't know anything about the company, take a look at their Web site: www.ti.com.) What recommendations would you make for the company's organizational structure? (Remember to address whether the company should be centralized or decentralized, adopt a mechanistic or organic structure, and generate slack or focus on efficiency. Also provide any recommendations that you have to avoid the disadvantages of large size, while preserving its benefits). Why do you make these recommendations?
3. **Corporate Venture Feasibility** Identify a new technology developed in an R&D lab of a major corporation. Write the executive summary of a business plan for a corporate venture intended to commercialize the new technology. Remember to describe the technology and explain why it is worth commercializing through corporate venturing. Also identify at least three key issues that need to be resolved in order to commercialize the technology through a corporate venture, and formulate a plan for resolving those issues. Finally, specify the structure of the corporate venturing arrangement that your firm will adopt, and explain why you have chosen that structure.

Notes

1. Adapted from http://www.intel.com/capital.
2. Intel. Intel 63 Fund, http://www.intelportfolio.com/cps/portlist_fund.asp?fund=64.
3. Shieber, J. 2006. Intel increases its foreign investments. *Wall Street Journal*, February 2: B3.
4. http://www.intel.com/pressroom/archive/releases/20030924corp.htm
5. George, J. and J. King. 1991. Examining the computing and centralization debate. *Communications of the ACM*, 34(7): 62–72.
6. Yoffie, D., D. Mehta, and R. Seseri. 2006. Microsoft in 2005. *Harvard Business School Case*, Number 9-705-505.
7. Argyres, N., and B. Silverman. 2004. R&D, organization structure, and the development of corporate technological knowledge. *Strategic Management Journal*, 25: 929–958.
8. Burgelman, R., C. Christiansen, and S. Wheelwright. 2004. *Strategic Management of Technology and Innovation*. New York: McGraw-Hill/Irwin.
9. Argyres and Silverman, R&D, organization structure, and the development of corporate technological knowledge.
10. Rothwell, R. 1994. Industrial innovation: Success, strategies, trends. In M. Dodgson and R. Rothwell (eds.), *The Handbook of Industrial Innovation*. Aldershot, UK: Edward Elgar, 33–53.
11. Burgelman, Christiansen, and Wheelwright, *Strategic Management of Technology and Innovation*.
12. Sathe, V. 2003. *Corporate Entrepreneurship*. Cambridge, UK: Cambridge University Press.
13. Ibid.
14. Argyres and Silverman, R&D, organization structure, and the development of corporate technological knowledge.
15. Schilling, N. 2005. *Strategic Management of Technological Innovation*. New York: McGraw-Hill.
16. Stevens, T. 2004. 3M reinvents its innovation process. *Research Technology Management*, March–April: 3–5.
17. Adapted from Huckman, R., and E. Strick. 2005. GlaxoSmithKline: Reorganizing Drug Discovery (A). *Harvard Business School Case*, Number 9-605-074.
18. Whalen, J. 2006. Bureaucracy buster? Glaxo lets scientists choose its new drugs. *Wall Street Journal*, March 27: B1, B3.
19. Huckman and Strick, GlaxoSmithKline: Reorganizing Drug Discovery (A).
20. Ibid.
21. Peters, T. 1994. *The Tom Peters Seminar: Crazy Times Call for Crazy Organizations*. New York: Vintage Books.
22. Christiansen, C. 1997. *The Innovator's Dilemma*. Boston: Harvard Business School Press.
23. Rizova, P. 2006. Are you networked for successful innovation? *Sloan Management Review*, 47(3): 49–55.
24. Kanter, R. 1988. When a thousand flowers bloom: Structural, collective, and social conditions for innovations in organization. *Research in Organizational Behavior*, 10: 169–11.
25. Cardinal, L. 2001. Technological innovation in the pharmaceutical industry: The use of organizational control in managing research and development. *Organization Science*, 12(1): 19–36.
26. Schilling, *Strategic Management of Technological Innovation*.
27. Ettlie, J., W. Bridges, and R. O'Keefe. 1984. Organization strategy and structural differences for radical versus incremental innovation. *Management Science*, 30(6): 682–695.
28. Olson, E., O. Walker, and R. Ruekert. 1995. Organizing for effective product development: The moderating role of product innovativeness. *Journal of Marketing*, 59(1): 48–62.
29. Sathe, *Corporate Entrepreneurship*.
30. Maidique, M., and R. Hayes. 1984. The art of high technology management. *Sloan Management Review*, 25: 18–31.
31. Gupta, A., and A. Singhal. 1993. Managing human resources for innovation and creativity. *Research Technology Management*, 36(3): 41–48.
32. Lublin, J. 2006. Nurturing innovation. *Wall Street Journal*, March 20: B1, B3.
33. Eisenmann, T., and K. Herman. 2006. Google Inc., *Harvard Business School Case*, Number 9-806-105.
34. Griliches, Z. 1990. Patent statistics as economic indicators: A survey. *Journal of Economic Literature*, 28: 1661–1707.
35. Graves, S., and N. Langowitz. Innovative productivity and the returns to scale in the pharmaceutical industry. *Strategic Management Journal*, 14(8): 593–605.
36. Kotabe, M., and K. Swann. 1995. The role of strategic alliances in high technology new product development. *Strategic Management Journal*, 16: 621–636.
37. Ibid.
38. Acs, Z., and D. Audretsch. 1988. Innovation in large and small firms: An empirical analysis. *American Economic Review*, 78(4): 678–609.
39. Ibid.
40. Maidique and Hayes, The art of high technology management.

41. Zenger, T., and W. Hesterly. 1997. The disaggregation of corporations: Selective intervention, high-powered incentives, and molecular units. *Organization Science*, 8: 209–223.
42. Schilling, *Strategic Management of Technological Innovation*.
43. Acs and Audretsch, Innovation in large and small firms.
44. National Academy of Engineering. 1995. *Risk and Innovation*. Washington, DC: National Academies Press.
45. Young, R., and J. Francis. 1991. Entrepreneurship and innovation in small manufacturing firms. *Social Science Quarterly*, 72(1): 149–162.
46. Cooper, A., and W. Dunkelberg. 1987. Entrepreneurship research: Old questions, new answers and methodological issues. *American Journal of Small Business*, 11(3): 11–23.
47. Audretsch, D. Forthcoming. Knowledge spillover entrepreneurship and innovation in large and small firms. In S. Shane (ed.), *Blackwell Handbook of Technology and Innovation Management*. Cambridge, UK: Blackwell.
48. Whalen, J. 2007. Small firms seek to profit from drug giants' castoffs. *Wall Street Journal*, February 7: A1, A13.
49. Hart, M. 1996. Palm Computing, Inc. (A). *Harvard Business School Case*, Number 9-396-245.
50. Sheahan, M. 2003. Corporate spin can't mask the VC units' blunders. *Venture Capital Journal*, March 1.
51. Ibid.
52. http://www.1000ventures.com/business_guide/corporate_vinvesting_external.html
53. Chesbrough, H., and S. Socolof. 2000. Creating new ventures from Bell Labs technologies. *Research Technology Management*, March–April: 13–17.
54. Cambell, A., J. Birkinshaw, A. Morrison, and R. Batenburg. 2003. The future of corporate venturing. *Sloan Management Review*, 45(1): 30–37.
55. Chesbrough and Socolof, Creating new ventures from Bell Labs technologies.
56. Ibid.
57. Ibid.
58. Ibid.
59. Ibid.
60. Cherbourg, H. 2002. Graceful exists and missed opportunities: Xerox's management of its technology spin-off organizations. *Business History Review*, 76(4): 803–837.
61. Cherbourg and Socolof, Creating new ventures from Bell Labs technologies.
62. Lane, D. 2000. Intel Capital: The Berkeley networks investment. *Harvard Business School Case*, Number 9-600-069.
63. Whalen, Small firms seek to profit from drug giants' castoffs.
64. Cherbourg, Graceful exists and missed opportunities.
65. Chesbrough, H. 2001. Intel Capital: The Berkeley networks investment. *Harvard Business School Teaching Note*, Number 5-601-121.
66. White, B. 2007. Cisco's homegrown experiment. *Wall Street Journal*, January 23: A14.
67. Chesbrough, Intel Capital.
68. Chesbrough and Socolof, Creating new ventures from Bell Labs technologies.
69. http://www.1000ventures.com/business_guide/corporate_investing_external.html
70. Chesbrough, Graceful exists and missed opportunities.
71. Ibid.
72. Markham, S., S. Gentry, D. Hume, R. Ramachandran, and A. Kingon. 2005. Strategies and tactics for external corporate venturing. *Research Technology Management*, 48(2): 49–59.
73. Chesbrough, Intel Capital.
74. Markham et al. Strategies and tactics for external corporate venturing.
75. Yoffie, D. 2005. Intel Capital (A). *Harvard Business School Case*, Number 9-705-408.
76. Lane, Intel Capital.
77. Chesbrough, H. 2000. Designing corporate ventures in the shadow of private venture capital, *California Management Review*, 42(3): 31–49.
78. Cambell et al. The future of corporate venturing.
79. Chesbrough, Graceful exists and missed opportunities.
80. Day, G., and P. Schoemaker. 2000. A different game. In G. Day and P. Schoemaker (eds.), *Wharton on Managing Emerging Technologies*. New York: John Wiley.
81. Chesbrough and Socolof, Creating new ventures from Bell Labs technologies.
82. Markham et al. Strategies and tactics for external corporate venturing.
83. Bower, J. 2005. Teradyne: The Arora project. *Harvard Business School Case*, Number 9–397–114.
84. Funk, J. 2007. Swagelok to market super stainless steel. *The Plain Dealer*, January 16: C1.
85. Bower, Teradyne: The Arora project.
86. Chesbrough, Graceful exists and missed opportunities.
87. Chesbrough, H. 2004. Making sense of corporate venture capital. In R. Burgelman, C. Christiansen, and S. Wheelwright (eds.), *Strategic Management of Technology and Innovation*. New York: McGraw-Hill/Irwin.
88. Burgelman, Christiansen, and Wheelwright, *Strategic Management of Technology and Innovation*.

89. Chesbrough and Socolof, Creating new ventures from Bell Labs technologies.
90. Chesbrough, Graceful exists and missed opportunities.
91. Hamilton, D. 2005. Novartis agrees to acquire the rest of Chiron for $5.1 billion. *Wall Street Journal*, November 1: A6.
92. Richtel, M., and K. Belson. 2005. Cisco announces Scientific-Atlanta deal. *New York Times*, November 19: B3.
93. Ibid.
94. Berman, D., and J. Singer. 2005. Big mergers are making a comeback as companies, investors seek growth. *Wall Street Journal*, November 5: A1, A2.
95. Richtel and Belson, Cisco announces Scientific-Atlanta deal.
96. Wheelwright, S. 2000. Cisco Systems Inc: Acquisition Integration for Manufacturing (A) and (B), *Harvard Business School Teaching Note*, Number 5-600-134.
97. Berman, D., and J. Singer. 2006. Blizzard of deals heralds new era of megamergers. *Wall Street Journal*, June 27: A1, A16.
98. Ibid.

Name Index

E

F

G

H

I

U

V

W

X

Y

Z

Subject Index

D

E

F

G

H

I

J

S

T

U

V

W

X